Quantitative Analysis of Ecological Networks

Network thinking and network analysis are rapidly expanding features of ecological research. Network analysis of ecological systems includes representations and modelling of the interactions in an ecosystem, in which species or factors are joined by pairwise connections. This book provides an overview of ecological network analysis including generating processes, the relationship between structure and dynamic function, and statistics and models for these networks. Starting with a general introduction to the composition of networks and their characteristics, it includes details on such topics as measures of network complexity, applications of spectral graph theory, how best to include indirect species interactions, and multilayer, multiplex and multilevel networks. Graduate students and researchers who want to develop and understand ecological networks in their research will find this volume inspiring and helpful. Detailed guidance to those already working in network ecology but looking for advice is also included.

Mark R. T. Dale is Professor in Environmental Science at the University of Northern British Columbia. His research interests include the spatial structure of plant communities and the development and evaluation of numerical methods to answer ecological questions, including graph theory and network complexity. His graduate students have worked in a diverse set of systems from prairie to alpine and at a range of spatial scales from plant neighbour competition to landscape disturbance patterns. He wrote *Spatial Pattern Analysis in Plant Ecology* (Cambridge University Press, 1999) and *Applying Graph Theory in Ecological Research* (Cambridge University Press, 2017) and was co-author, with Marie-Josée Fortin, of *Spatial Analysis: A Guide for Ecologists* (Cambridge University Press, 2005, 2nd ed. 2014).

Marie-Josée Fortin is a University Professor in Ecology and Evolutionary Biology at the University of Toronto. She is a Fellow of the Royal Society of Canada and holds a Tier 1 Canada Research Chair in Spatial Ecology. Her research endeavours focus on conservation biology issues. She investigates how ecological processes and environmental factors affect species persistence, species dispersal, and species range dynamics. Her research is at the interface of several disciplines (spatial ecology, conservation, forest ecology, disturbance ecology, community ecology, landscape genetics, spatial epidemiology, spatial statistics, spatially explicit modelling, and network theory) where the most important challenging problems lie.

Quantitative Analysis of Ecological Networks

MARK R. T. DALE
University of Northern British Columbia

MARIE-JOSÉE FORTIN
University of Toronto

CAMBRIDGE
UNIVERSITY PRESS

CAMBRIDGE
UNIVERSITY PRESS

University Printing House, Cambridge CB2 8BS, United Kingdom

One Liberty Plaza, 20th Floor, New York, NY 10006, USA

477 Williamstown Road, Port Melbourne, VIC 3207, Australia

314–321, 3rd Floor, Plot 3, Splendor Forum, Jasola District Centre, New Delhi – 110025, India

79 Anson Road, #06–04/06, Singapore 079906

Cambridge University Press is part of the University of Cambridge.

It furthers the University's mission by disseminating knowledge in the pursuit of education, learning, and research at the highest international levels of excellence.

www.cambridge.org
Information on this title: www.cambridge.org/9781108491846
DOI: 10.1017/9781108649018

First published 2021

Printed in the United Kingdom by TJ Books Limited, Padstow Cornwall

A catalogue record for this publication is available from the British Library.

Library of Congress Cataloging-in-Publication Data
Names: Dale, Mark R. T. (Mark Randall Thomas), 1951- author. | Fortin, Marie-Josée, 1958– author.
Title: Quantitative analysis of ecological networks / Mark R.T. Dale, University of Northern British Columbia, Marie-Josée Fortin, University of Toronto.
Description: Cambridge, UK ; New York, NY : Cambridge University Press, 2021. | Includes bibliographical references and index.
Identifiers: LCCN 2020040227 (print) | LCCN 2020040228 (ebook) | ISBN 9781108491846 (hardback) | ISBN 9781108740715 (paperback) | ISBN 9781108649018 (epub)
Subjects: LCSH: Ecology–Statistical methods. | System analysis. | Quantitative research.
Classification: LCC QH541.15.S72 D35 2021 (print) | LCC QH541.15.S72 (ebook) | DDC 577.072/7–dc23
LC record available at https://lccn.loc.gov/2020040227
LC ebook record available at https://lccn.loc.gov/2020040228

ISBN 978-1-108-49184-6 Hardback
ISBN 978-1-108-74071-5 Paperback

Contents

Colour plates can be found between pages 118 and 119.

Preface

Ecology is about understanding how organisms interact with other organisms and the environment they inhabit (i.e. fundamental and realised niches). It is easy to imagine an individual organism of any kind as a dot with all sorts of arrows impinging upon it; an arrow can represent abiotic factors (temperature, light, etc.), along with many arrows for all the other organisms (biotic factors, intra- and inter-specific interactions) that affect it. Ecology aims therefore to determine the magnitude and rate associated with some of the arrows and identify which are the most important and why. Each organism also has its own effects on the same list of factors, even if the effects are small, so we can also imagine arrows going out from the same dot, one to each of the same factors (they can be dots too). Again, a challenge is to determine the associated weights and importance of the arrows, some of which are directed towards other organisms. As soon as we consider more than a single organism, even just a few, we immediately have a complex structure of dots and arrows: an ecological network!

It is an obvious step to consider ecological systems as ecological networks, and as such to assess how network theory (concepts and methods) might be applied to them. Network theory and the mathematics of graph theory that underlie network analysis provide simple concepts that can applied to systems that are complex both in structure and dynamics. It is those concepts that allow us to provide a sorted set of methods for the quantitative analysis of ecological networks, together with thoughts and advice on how best to proceed. Through the years, the need to take a network analysis framework to study complex systems has arisen in many fields (physics, computer science, communication science [transportation, electricity, social], and bio- and eco-informatics), and there is a challenging diversity of approaches, methods and measures that should be understood, or at least sorted, before applying them to our own data. The overarching goal of this book is to help ecologists in selecting the appropriate network methods to represent, analyse and model their ecological systems using network theory.

As the title suggests, this book is for ecologists with an interest in ecological networks. Perhaps they have a project in mind or even some potential network data "in hand" but are not sure how to proceed or on which network characteristics to concentrate. This book provides both background and some advice on how to answer interesting ecological questions in the context of network analysis. It does cover some general network material (that is needed as background), but to the breadth of books like Newman (2010) or Estrada (2012). It was partly inspired by Kolaczyk (2009), but it is less technical in tone and detail and more inclusive of recent developments. All three of the books just cited, together with Kolaczyk and Csárdi (2014) which supplies R code, are excellent resources for more information, as needed. The other inspiration for our current book is Dale (2017), which focused on applying graph theory in ecology, and mentioned, but did not cover, network analysis. The current project moves forward into network analysis from that work but does not assume its material as background.

As is often the case, the book evolved as it was being written. Initially, there was no separate chapter on multilayer and multilevel networks, but it became obvious that it was both necessary and crucial – hence Chapter 4. Similarly, we had not explicitly included causal networks in the original outline, although it now seems central to the presentation. Hence, now a central theme of the book is to stress the interplay between the structure and the dynamics (or between form and function) of these complex networks from ecological systems. Last, the ending of the book started out as only the summary diagram "It all ties together," but when we got to the end, there was more that needed to be said – hence Chapter 5 (we did keep it short). One factor in driving the evolution of this book is the speed at which the field of the study of ecological networks, let us call it "network ecology," is growing, developing and advancing. It is very challenging just to keep pace with developments, but it is also very satisfying to see the subdiscipline getting the attention and application we think it should.

There are many people and organisations to thank for supporting the project, starting with Natural Sciences and Engineering Research Council (NSERC), NSERC Canada Research Chair, the University of British Columbia, and the University of Toronto for financial support. We thank Alex Aravind, Charlotte Brown, Cheryl Smyth, Conan Veitch, and J. C. Cahill for their reading of chapters and their thoughtful advice for improvements. We especially thank J. B. Grace for permission to use material from an unpublished manuscript on statistical models and causal analysis as the basis for Table 3.4.

Writing this book has been an exciting and rewarding journey of discovery through the rapidly changing landscape of network ecology. No book can be completely up to date, but the hope is that the course it has followed is in the right direction for continuation into the future. It is also our hope that the book can serve an as exciting and helpful guide for others on a similar journey of discovery.

1 Ecological Processes and Network Systems

1.1 Introduction

This book is about ecological networks, and networks in ecology include almost anything that can be represented by a set of points joined in pairs by lines or "edges" (Newman 2010). By choosing the right network characteristics to investigate, network analysis allows us to develop concepts and to test hypotheses about ecological systems and processes that we could not express or investigate otherwise.

We will begin by suggesting how ecological questions and the quantitative analysis of the ecological networks are tied together. We cannot expect a simple one-to-one relationship between network structure or topological properties and network function and the related ecological processes, but we can provide some of the best matches. To illustrate the close connection between ecological questions and network analysis, Table 1.1 gives several examples of such questions and the network characteristics that might be used to help answer them.

We could cite many more examples of ecological questions for ecological systems for which the network approach seems like the best approach for finding answers, but the preceding list provides a good sample. It is obvious, too, that several network properties and measures appear frequently in the list, for example: degree, centrality, modularity (and clusters), and connectivity. These concepts and their measures will be described in detail in the chapters that follow. More characteristics and more in-depth analysis will be required to answer these queries for ecological networks, but the intention is to indicate where our analysis should start. Greater detail on the recommended quantitative analysis will be provided in the chapters that follow.

1.2 Network Analysis in Ecology

The phenomena and systems studied by ecologists span a huge range of spatial scales from molecules to ecosystems, but any objects we study have relationships with other objects in their system, thus creating the system's structure. This structure can be treated as a network based on those relationships and then represented by a graph with the objects as the nodes (the points) and the relationships between them as the edges (the lines), as in Figure 1.9. Two important features of ecological networks are that (1) characteristics of the nodes and edges can be described qualitatively and quantitatively

Table 1.1 Ecological questions for network analysis and key network properties to analyse them

The network terms may be unfamiliar, but they will be explained in later sections of the book (Chapters 2–5).

1. Can dispersal (field observation, genetic analysis) be predicted from the spatial structure (size, shape, interpatch distances) of habitat patches (vegetation type)?
   (Fall et al. 2007; Minor & Urban 2007; Urban et al. 2009)
   Node weights represent patch area, quality, or population density.
   Edge density and length or cost represent the difficulty of dispersal.

   What are the effects of dispersal and/or route topology on metapopulation persistence?
   (Shtileman & Stone 2015)
   Indegree (number of directed edges into the node) and outdegree (number of directed edges out from the node) indicate rates or ease of immigration and emigration.
   ➔ Figure 1.1
2. How can patterns of interindividual proximity or contact (frequency, order, relative timing) predict the spread of disease (transmission mode)?
   (Hamede et al. 2009; Marsh et al. 2011; Silk et al. 2017)
   Determine the temporal paths and distances to diseased or susceptible individuals.
   ➔ Figure 1.2
3. How do patterns of species interactions (positive or negative; weak or strong) change along environmental gradients?
   Are there consistent differences among positive versus negative interactions?
   (Pellissier et al. 2018)

   What are the association patterns among ecosystems?
   (Williams et al. 2014)
   Measure the similarities of the network structures between sites.
   Examine the node degree distribution and betweenness centrality as elements of association patterns.
   ➔ Figure 1.3
4. How can we best understand how interspecific interactions themselves interact (e.g. predation rate affects herbivore competition) and include these effects in the network?
   (Golubski & Abrams 2011)

   How important are multitaxa interactions in communities?
   (Connor et al. 2017; Layeghifard 2017)
   Modify network structure to allow edges to be secondary nodes.
   Create a correlation network for the community and determine the pattern of the strongest edges.
   ➔ Figure 1.4
5. What characteristics of trophic networks (compartmentalisation, number of levels, specialists, omnivores) are associated with system robustness and resilience?
   (Dunne et al. 2002)
   Measure connectivity, longest paths, and look for (low?) modularity.
   Are the trophic levels ambiguous or strict in a multilayer network structure?
   ➔ Figure 1.5
6. How does a multilayer temporal data structure affect the interactions of individuals and the analysis of those interactions?
   (Boccaletti et al. 2014; Kivelä et al. 2014)

   What factors can predict the spread of disease through a mobile population?
   (Craft et al. 2010; Godfrey 2013; Stella et al. 2018)
   Use distinctive edges to connect instances of the same individual in different times.

Table 1.1 *(cont.)*

Identify paths through spatial layers as time slices and the nodes as moving individuals; proximity (or contact) allows the formation of path elements within a layer.
➔ Figure 1.6

7. How are the different "layers" of interactions (plant-pollinator, pollinator-antagonist, plant-herbivore, etc.) integrated?
How to determine the key species or key interactions (frequencies, intensities) in such a complex multi-interaction network?
(De Domenico et al. 2015)
Study multilayer paths and subgraph structures. Investigate centrality measures of nodes as indicators of keystoneness
➔ Figure 1.7

8. How do phenological marker events (date of first leaf, first flower, . . .) observed at many sites (altitude, latitude, distances) over many years, flow through space and time, and how is the flow affected by changing climate?
(Schröder et al. 2014)
What processes mimic the spread of disease through a sessile population?
(Peyrard et al. 2006)
These require the study of network dynamics and flow, and the comparison of changing network characteristics.
Percolation is a network process determining the spreading of a state through adjacent nodes or through edges that share a node.
Percolation is a model of processes like disease or fire.
➔ Figure 1.8

9. How does metacommunity diversity depend on the arrangement of habitat patches?
(Chisholm et al. 2010)
Connectivity is a general measure of how difficult it is to break a network into pieces by removing nodes or edges; the more removals required indicates better connectivity.

10. How can we determine spatiotemporal hot-spots of disease in a population?
(Craft et al. 2010)
Multilayer cluster analysis examines networks of several spatially explicit layers, each representing a time slice, for clumps of disease that are close in space and in time.

11. How are species functional roles related to their pollination interactions?
(Coux et al. 2016)
Examine the nodes' normalised degree (corrected for network size). Can nestedness indicate interaction asymmetry?

12. What are the similarities and key differences among the "versions" of an ecological network represented in a multilayer network?
(De Domenico et al. 2013; Kao & Porter 2017)
Examine multilayer clusters and multilayer centrality and compare the results with the same characteristics measured within layers.

13. How does the multilayer structure of a network affect robustness and resilience?
(Pilosof et al. 2017)
Focus on the interactions between layers as interlayer edges and compare with aggregated within layer characteristics.

14. What are the key species or key interactions in a multi-interaction (multilayer) network?
(De Domenico et al. 2015)
Measure multilayer centrality of nodes, also called versatility.
Compare with the within layer characteristics.

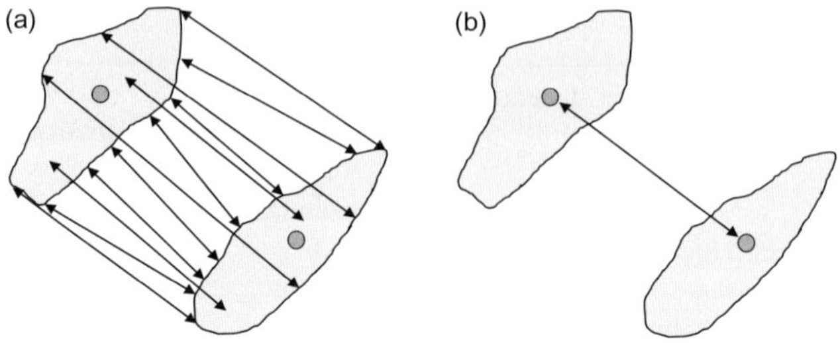

Figure 1.1 Dispersal between patches may follow (a) any one of several paths, not just the shortest or easiest ("least cost"), but (b) all of these can be summarised by a single path from centroid to centroid, a sum of all paths.

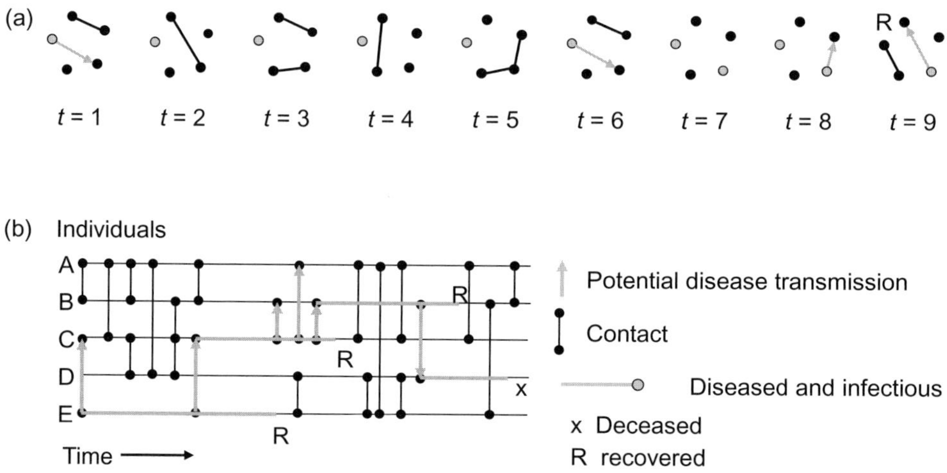

Figure 1.2 Temporal network of inter-individual contacts and disease transmission. Pairwise contacts recorded at time intervals: (a) explicit network format and (b) time-only graph format.

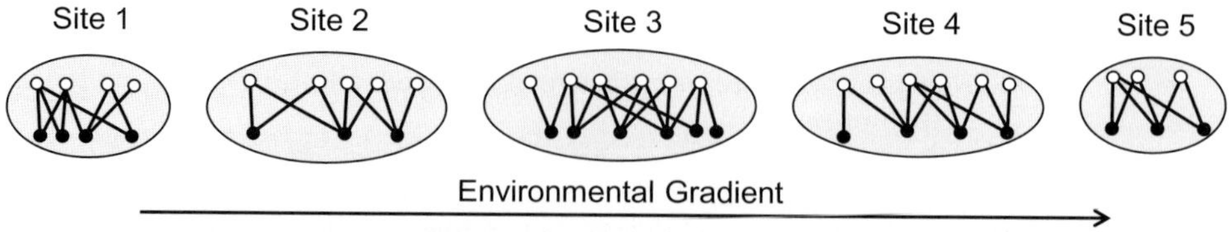

Figure 1.3 Locations as layers in a multilayer network. Pollination mutualisms in patches on an environmental gradient: bipartite graphs of plants (black dots) and pollinators (white dots) at five sites. Examine the structure for differences in degree (specialist vs. generalist), degree distributions, neighbour degree correlation, nestedness, and so on.

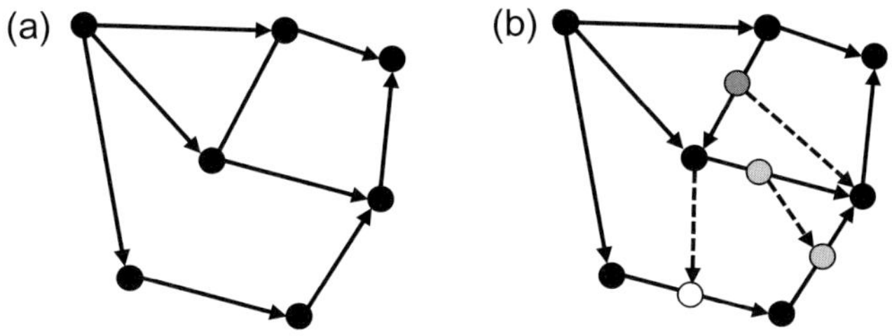

Figure 1.4 Interactions among interactions. Starting with 9 two-species directed interactions (edges) among 7 species (nodes) in (a), modifications can occur in which edges now act secondarily as nodes. The secondary interactions are shown by coloured edges. One modification is for a third node to affect another two-species interaction (black species node to blue interaction node). Another variant is for a two-species interaction to affect a third individual species (red interaction node to black species node). Last, one two-species interaction can affect another two-species interaction (green interaction node to green interaction node). (A black and white version of this figure will appear in some formats. For the colour version, please refer to the plate section.)

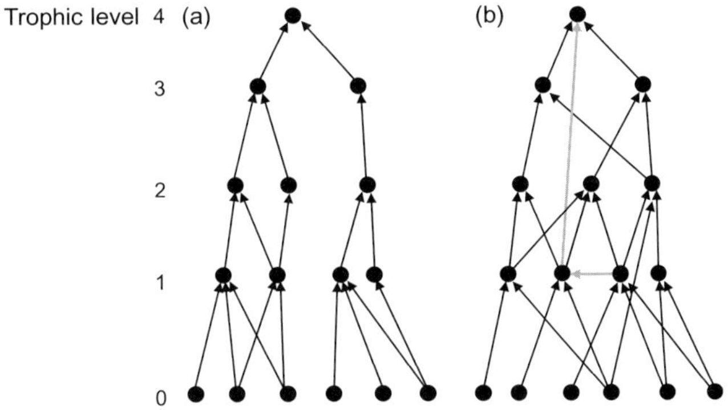

Figure 1.5 What trophic network characteristics are associated with robustness/resilience? (a) The first food web has a strict hierarchy: orderly, fewer links, modular. (b) The second is less orderly: exceptional links in grey, more links overall, not modular.

(Delmas et al. 2019) and (2) they and their characteristics tend to exhibit spatiotemporal dynamics. Therefore, many ecological questions or hypotheses related to network systems can be evaluated only through quantitative analysis (Dale & Fortin 2010; Dale 2017). Furthermore, the dynamics studied in network analysis can be divided between the dynamics ***of*** the network, changes in the structure or topology of the network itself, such as adding or deleting servers in the internet, and the dynamics ***on*** the network, involving changes in network characteristics within a stable structure, such as passing packets of information through the connections of the internet (Figure 1.10; Peterson et al. 2013).

Graphs (such as Figure 1.9) can be used to represent ecological networks. Graph theory treats networks formally as mathematical structures and an extensive body of

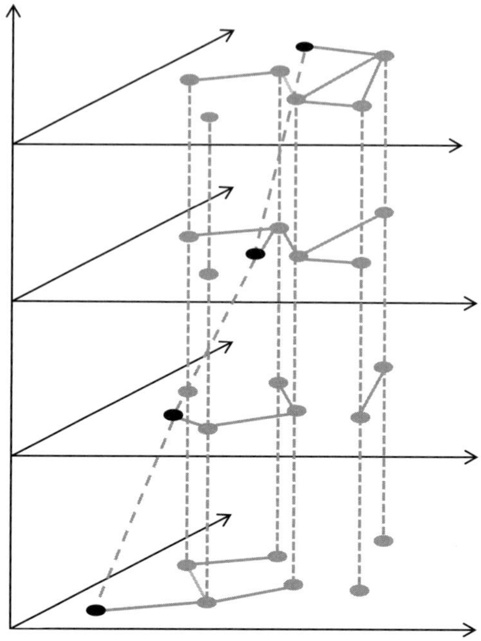

Figure 1.6 Multilayer network (possibly space and time). Layers show shared nodes and changing edges. Identity edges (dashed lines) join instances of the same node, even when it moves with respect to the others.

knowledge underlies the mathematics of network analysis (Harary 1969; Dehmer & Emmert-Strieb 2015). A broad range of applications of graph theory already exist in ecological research, as described in Dale (2017), but although this book on network analysis follows from that work, we will not assume the material as background knowledge. We will provide whatever graph theory is needed in context so that the presentation of the material on networks is complete. Like the graphs that represent them, networks can occur fully abstracted from the dimensional world of the original phenomena. On the other hand, graphs and networks may be embedded in one or several dimensions, such as time only, the spatial plane, or the four dimensions of spatiotemporal data. In addition to the useful aspects of graph theory, representing a network as a graph enhances our ability to visualise important aspects of the data, although that too can be challenging for very large networks (see Newman's 2010 for an example of the internet network of 2003). The future inevitably will see very large ecological networks, especially from spatiotemporal data sets, and we will have to follow the best approaches available to understand them, and effective visualisation should be a key element.

If graph theory is the main theoretical backbone of network analysis, statistical analysis is the analytical framework to quantify relationships between objects. The starting point for network analysis is complex and structured data, where complexity includes dependence, internal structuring, and heterogeneity over a range of scales (Barrat et al. 2008). Even with such a level of complexity, there is insight to be gained

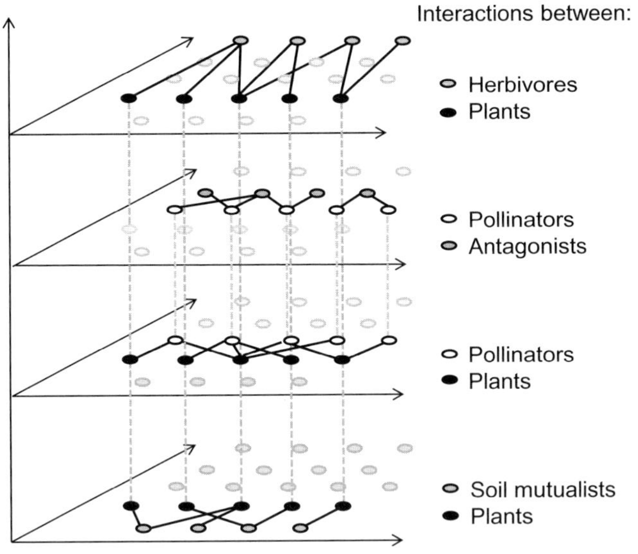

Figure 1.7 Multi-interaction network as layers. Layers show species nodes and edges for particular interactions: herbivory, pollinators and their antagonists, pollination mutualism, and soil-based mutualisms (e.g. mycorrhizae). Identity edges (dashed lines) join instances of the same node. (A black and white version of this figure will appear in some formats. For the colour version, please refer to the plate section.)

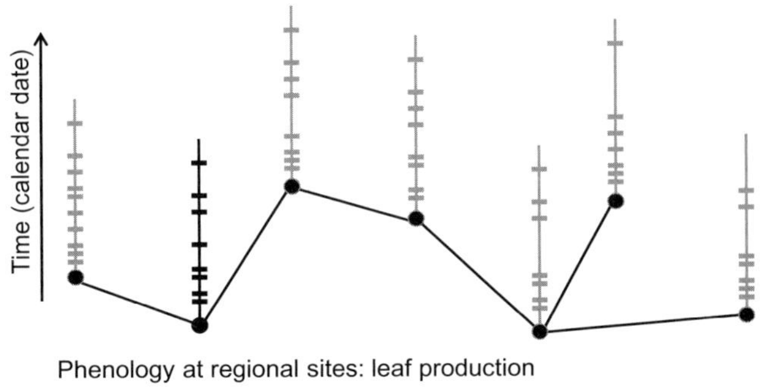

Figure 1.8 Phenology with permanent nodes and continuous time. Topological definition of permanent edges, minimum spanning tree (MST). One kind of event, leaf production indicated by bars (a single site indicated by black): examine for burstiness and synchrony of events, perhaps as function of absolute or relative locations (can be aspatial).

in evaluating network characteristics and their dynamics while comparing to those generated based on random dynamic graphs (Durrett 2007). Nowadays, given the availability of more complex data sets, novel approaches to statistical analysis and interpretation of networks are needed (Crane 2018). Such analytical developments can help to assess networks' emergent properties, their degree of self-organisation, and their potential to evolve through time given their resilience or redundancy.

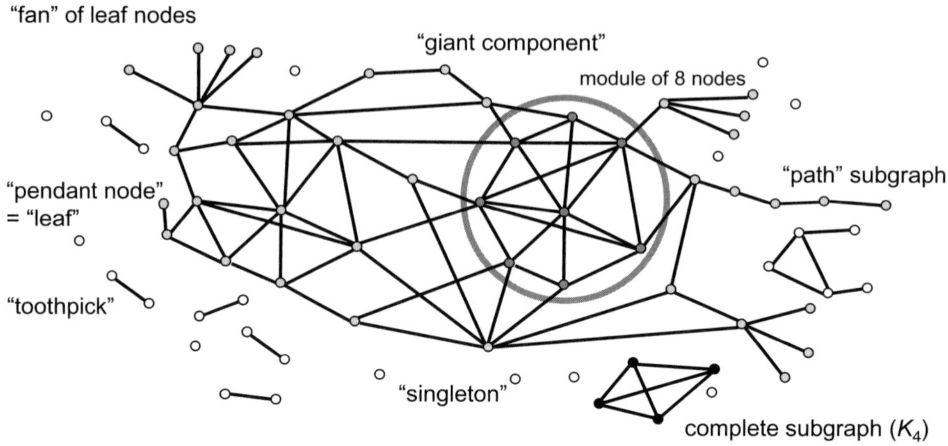

Figure 1.9 A graph of a network. Dots represent the basic units of the ecological system and are the nodes of the network graph. Lines represent relationships between units in the network and are the edges of the network graph. The various structural elements of the network graph are labelled.

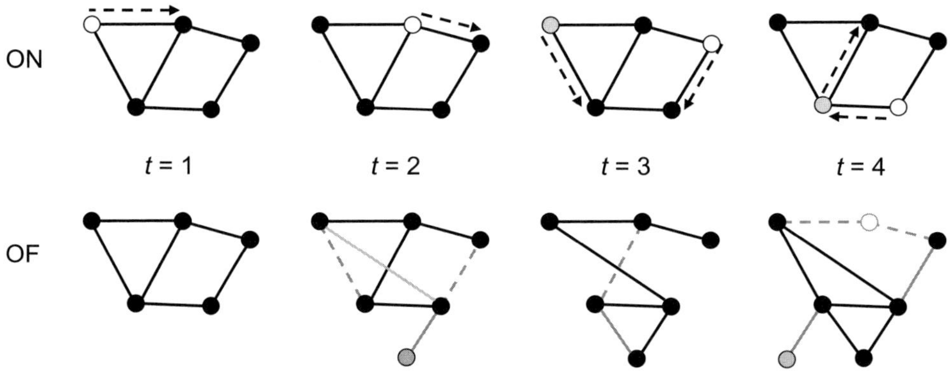

Figure 1.10 Dynamics ***on*** a network: Weights or labels may change but nodes and edges neither lost nor gained, e.g., node state passed on to neighbour (or not) as indicated. The topological structure of the network is maintained. Dynamics ***of*** a network: the configuration changes as nodes and edges are lost (white nodes and dashed grey edges) or gained (grey) through time.

The current popular view of network analysis is based mostly on social and information networks which tend to be sparse in their edge density and which may focus on hubs and centrality and power laws and small worlds (but see Newman 2010; Estrada 2012). Yet, most ecological networks (trophic networks, correlation networks, and phenology networks, among many) are different in their topological structure, edge density, and function. By choosing the right characteristics to investigate, network analysis allows us to develop concepts and to test hypotheses that we could not express otherwise. In this book, we highlight how many of the most important ecological questions can be answered only in the context of networks and network analysis. The questions, and the kinds of network analysis that apply, can be organised in several categories based on the dimensionality, what the edges represent, the interaction between structure and function, and the overall structuring of the network itself. After presenting a few examples, we will discuss several categories into which our examples may fall.

Current popular ideas that are based on social and information networks are correct about one feature shared by those examples and most ecological networks: they are both *complex*! The title of a well-cited book is *The Structure of Complex Networks*, (Estrada 2012). Ecology studies natural systems that are both complex and dynamic, consisting of organisms and their environment. The organisms interact with each other and there are reciprocal effects of the organisms on the environment and the environment on the organisms. The environment changes through space and time, and the organisms change and evolve, responding to the environment and the interactions they experience, giving rise to a complex adaptive system (Levin 1998). The spatiotemporal structure of an ecological network system and the key roles of the many interactions within it, together mean that many ecological questions cannot be answered or even formulated (!) without reference to the networks in which they all occur.

Later in the book (Chapter 2, Section 2.4), we will provide a discussion of the measurement of the complexity of the network itself, but the first complexity faced in network ecology is collecting and compiling the data on which the network is to be based. We will not provide a detailed guide to this part of the endeavour; it will vary greatly with purpose and system, but we need to acknowledge the importance and difficulties faced at this stage of a study. As just one example, consider the efforts of Timóteo et al. (2018) to investigate seed dispersal interactions in four habitats in Gorongosa National Park, Mozambique. They collected 1399 fecal samples from 98 animal species and found 12,159 undamaged seeds of 94 plant species from 29 dispersers. Each observed plant–disperser combination creates an edge between the two nodes representing them in the network, and edges occur only between the two groups, not within. That data provided 508 network edges; other observations gave 85 edges and camera traps provided 15 more. The network was treated as a multilayer structure, with each habitat being treated as a layer. The full network consists of 608 edges between 32 animal species and 101 plant species; the proportion of edge positions that are realised or the *connectance* is 608/3232, or about 20%. Each node and each edge have a quantitative value which is a numerical count or an interaction frequency of the seeds of a plant species associated with the activities of a particular seed disperser. The authors' own evaluation of the effectiveness of sampling was that the effort found 77% of the disperser species and 44% of the plant species. Primates were the most frequent dispersers and the study indicated the importance of the most versatile dispersers, those that were effective across habitats, the network layers. The network characteristic that was the most important in the study was found to be modularity: the existence of subsets of species that interact in and across different layers.

Incomplete data are inevitable in field studies, and there are some useful methods available to help understand the effects of the incompleteness (de Aguiar et al. 2019).

For example, Timóteo et al. (2018) used a non-parametric method for mark-and-recapture data to estimate the numbers of species missed (see Chao 1987). We will have more detailed comments on sampling and missing data in Chapter 5, Section 5.2 on constructing and reconstructing networks.

In addition to your own data, many data sets for network ecology are now archived and available online. For these, while the initial data collection has been accomplished, comparative or summary research requires compiling the data from more than one source or and ensuring that the comparisons are based on the same "currency" and have appropriate spatial and temporal scales. The archived data sets will also be subject to missing data, and the preceding comments about dealing with missing observations applies equally here.

1.3 Classification of Network Questions

There are many ways in which ecological networks and the questions that pertain to them can be classified. An obvious characteristic is how the network is embedded in dimensionality: time only, space only (two dimensions or three), space-and-time, not at all. Many ecological networks are about interactions between organisms, and so the interactions included can be the basis for categories of networks, as well as the relationship between network structure and the functioning of the system it represents. Similar to the criterion of dimensions, networks can also be classified by the criterion of their own structural framework, such as monolayer versus multilayer. These suggested criteria for classifying ecological networks and their associated questions are not exclusive, and so a single example may fall into several categories.

1.3.1 Criterion 1: Structure-and-Function

For many ecological networks, the truly important questions are those concerning the relationship between the structure and the function of the network (clearly these may also be about interactions or about relationships in the dimensions of space and time). How does network topology affect the dynamics of its system? How does the dynamics of and on a network affect its structure? These are the basic "structure–function" questions that can be expressed in many different ways and applied to many different systems: "pattern" versus "process," "topology" versus "dynamics," and so on. Another version of the same questions concerns what we can learn or deduce about network function from its structure, and what additional information is required. This is the inspiration for the use of various kinds of subgraphs (e.g. motifs and graphlets) to investigate the relationship between structure and function (e.g. Pržulj 2007; Sarajlić et al. 2016).

Familiar structure–function questions include: What structural characteristics indicate a keystone species in a food web (Lupatini et al. 2014)? What characteristics make established communities more resistant to invasion (Lurgi et al. 2014)? What trophic network characteristics contribute to robustness or resilience (Dunne et al.

2002)? What network metrics are associated with metapopulation persistence (Zamborain-Mason et al. 2017)? Such structure-function questions lead naturally to the consideration of causal relationships in an ecological system: why is species X the keystone species in this system? (Or, under what conditions is it a keystone species?) Is the direction of control bottom-up or top-down, and why? (Or, under what conditions is. . .?) What causes populations of species X to have multiyear cycles in this region? Why do other populations follow or not follow such cycles? Many questions include a "why?" and it is for good reason that a recent book on the scientific analysis of cause and effect is called *The Book of Why* (Pearl & MacKenzie 2018). Those considerations and questions then lead to the analysis of causal networks, which we describe in Chapter 3, where they fit with correlation networks, the structure–function relationships, and network dynamics.

1.3.2 Criterion 2: Interactions

Another basis for network classification is the kind(s) of interactions depicted by the edges. Are these trophic interactions? Mutualistic? Antagonistic? Do the edges have signs? Do the edges have rates or other weights? Are the relationships symmetric or asymmetric? If asymmetric is the network graph a digraph in which edges have directions ($A \rightarrow B \neq B \rightarrow A$)? Transitive or intransitive? One kind of interaction or several? If several, does it result in a multilayer network, with one layer for each kind of interaction? How important are indirect interactions in the system of interest? Many important ecological questions can be placed in the category of questions about interactions.

1.3.3 Criterion 3: Dimensions of Space and Time

A network is made up of nodes that are joined together by edges (connections, interactions, links, are the essence of networks), and an obvious example network is a set of locations in space, such as landscape patches, with edges representing connections (whether physical or functional) between them. A clear example of a spatial network question is how patch diversity in a metacommunity depends on the arrangement of habitat patches (Chisholm et al. 2010); because this question can be investigated at a single point in time, this is a space-only case. Time-only examples are not common in ecology (permanent sample plots often have a spatial component), for instance seasonal diversity patterns in marine systems (Gilbert et al. 2012) or the analysis of contact or proximity networks related to the spread of disease (Hamede et al. 2009; Craft et al. 2010; Marsh et al. 2011). They also arise naturally in graphical or simulation models of iterative change of state for a particular system. These models may be formalised using a *configuration graph*: the nodes represent the possible states of the system and the edges are one-iteration transitions between these states that are allowed or observed. Originally from the study of games and puzzles (e.g. tic-tac-toe or Rubik's cube), such graphs are used in graphical and simulation models of changing ecological networks with transitions of landscape elements among states or stages such as forest, grassland, and agriculture (e.g. Daniel et al. 2018;

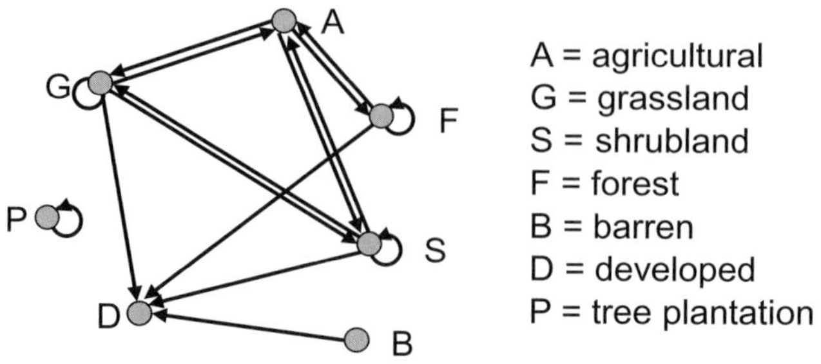

Figure 1.11 Configuration graph of state transitions for landscape patches, with loops, based on a case study on Hawai'i. A = agricultural; G = grassland; S = shrubland; F = forest; B = barren; D = developed; P = tree plantation. [Based on Daniel, C. J., Sleeter, B. M., Frid, L., & Fortin, M.-J. (2018). Integrating continuous stocks and flows into state-and-transition simulation models of landscape change. *Methods in Ecology and Evolution*, **9**, 1133–1143 under Creative Commons Attribution 4.0 International license.]

Figure 1.11). In that particular example, the network has loops indicating the potential for state persistence from one iteration to the next. For obvious reasons, such models are also called "state-and-transition" models, with different kinds of rules being possible for determining the transitions. Any configuration graph provides a clear indication of the relationship between the graph's structure and the system's dynamics because the graph codifies which transitions can occur, and often their rates, probabilities, or effective influences. In some cases, the transitions between states can be determined by using Markov chains in which a transition probability depends only on the current state, not on the whole history of past states and transitions, as seen in studies of successional sequences (Horn 1975, 1976).

In other cases, states can be spatially explicit such that the transitions depend on the spatial locations and the neighbouring locations (Daniel et al. 2016). The spatial neighbouring could be based on grid-data or spatial network forming network-of-networks (here, spatial connections and state transitions).

There are many spatiotemporal questions for network analysis. For example, spatially explicit epidemiological studies of the spread of disease, parasites or invasive species through populations, metapopulations, or metacommunities (Craft & Caillaud 2010; Godfrey et al. 2010; Danon et al. 2011; Stella et al. 2018). This parallels the spread of disturbance in a forested landscape (Fall et al. 2004; Daniel et al. 2018). Models to describe or predict spread or movement include "gravity" models, in which the strength of the interaction between two nodes varies with the product of their "mass" (e.g. patch area) divided by their "distance" squared, analogous to Newton's law of gravitational attraction. For network flow, the product of two measures of "mass" divided by distance squared may predict the flow between the nodes, for example, the flow of individuals between two subpopulations in a landscape based on of the patches' densities and their geographic separation.

Related to spread or flow in a network, another question would be "How can we determine spatiotemporal hot-spots of disease in a population network?" This question can be seen as the network equivalent of using scan statistics to find hot spots in spatiotemporal data (Glaz et al. 2001) or as a multilayer version (the layers are different times) of patch detection in spatial analysis (Dale & Fortin 2014).

Many ecological networks are dynamic so that change (i.e. processes) should be measured, parameterised, and modelled using temporal data of time series such as time-since-disturbance and phenological networks. Phenological networks record the sequential appearance of seasonal marker events in plants and animals. Large-scale patterns and changes in climatic conditions are studied by examining the phenology of chosen species in geographic networks of sites over many years. This research is critical to understanding climate change and its effects on ecological systems. Phenological networks also provide the opportunity for investigating a range of different kinds of ecological and evolutionary questions. What are the benefits and risks for greater or less synchrony of equivalent events (e.g. flowering or fruiting) in different sympatric species? How close or loose does that synchrony have to be for the benefit or risk to be effective? At what spatial distances do the forces for or against synchrony act (see Johansson et al. 2015; Chiune & Régnière 2017)?

Network approaches can be also be used to compare the effects of disturbances using just "before" and "after" network data. For example, what are the differences in the trophic network before and after the functional loss (or gain) of a species (e.g. severe decline in cod biomass in the Baltic Sea; Tomczak et al. 2013)? How do pollinator networks change post-disturbance?

1.3.4 Criterion 4: Multilayer

Another category in classifying networks is the overall structure or framework, as determined by the nature of the data. Data that divide naturally into two or more locations, two or more time periods, two or more kinds of interactions among species, two or more systems in a common space, and so forth, give rise to multilayer networks, in which each layer has its own network. Multilayer networks and related structures, such as multiplex and multilevel, will be discussed in detail in Chapter 4.

As an example of multilayer, the study by Timóteo et al. (2018) of seed dispersal interactions in four habitats, each habitat provides a layer of the overall network, and each has its own bipartite subnetwork of plants and their seed dispersers. This framework gives rise to obvious questions about the differences between and variability among the subnetworks in the different layers. It is also appropriate for more general questions such as: Does increasing the number of layers increase or decrease stability (Radicchi & Bianconi 2017)? How does the plant-pollinator network change through time (year or season) or among locations (Burkle et al. 2013; Gray et al. 2018)?

1.4 Concluding Comments: Questions and Answers

Many of the ecological systems we study are natural networks and abstracted networks and network graphs provide the best format for portraying and understanding

Table 1.2 Organisation of network analysis material, by (I) categories and (II) instances

I. Categories	Structural: topological[a]	Structural: composition and configuration[b]	Functional	Dynamics: ON network	Dynamics: OF network
Concepts	Basic elements (unweighted): nodes, edges, subgraphs (motifs, graphlets); graph theory	Weighted or labelled elements: nodes, edges, subgraphs (motifs, graphlets, digraph); network theory	Weighted (signed, directional) nodes, edges, subgraphs (motifs, graphlets); edges that can affect nodes' labels or weights; network theory	Node's labels or weights as well as edge's weights can change through time but there is no loss of nodes or edges; flow network; dispersal network (meta-pop, -comm, -ecosystem); network theory	Node and edge weights can change through time where both nodes and edges can be lost or gained; network flow; network theory
Analysis	Degree of the node distribution; connectance; modularity; clusters; spectral properties (incidence matrix, adjacent matrix)	Weight degree distribution; neighbouring weighted edge distribution; frequency of subgraphs; spectral graph (Laplacian matrix); (dis-) Assortativity for node and edge properties	Hypergraph; correlation network; spectral graph (Jacobian matrix); frequency of subgraphs' effects on network's function	Hypergraph; correlation network; spectral graph (Jacobian matrix); frequency of subgraphs' effects on network's function	Adaptive network

Models	Configuration; random; constrained randomisation; probabilistic	Random node or edge weight or identity (label) assignment	Path analysis; Structural Equation Model (SEM); gravity models; graphical and causal models	State-and-transition matrix	Flow
II. Instances					
Species interaction network	Interaction network; bi-, multi-partite	Weights and strengths or frequency	Indirect interactions; bi-, multipartite	Changing rates in the same structure	Changing nodes or edges
Spatial network	Landscape connectivity	Neighbour hierarchies of size and distance	Metapopulations; metacommunities	Metaecosystem; propagule and other agent movements	Metaecosystem; disturbance changes nodes and edges
Temporal network	Phenological trends	Synchrony of events; event clumping	Event orders changing	Sequential/temporal legacy	Sequential/temporal legacy
Spatiotemporal network	Landscape dynamics	Weights and strengths or frequency	Indirect interactions	Flow not changing network	Flow; percolation; spread of disease, pest, invasion; fire
Multilayer network	Multipartite	Layers as interaction types	Layers as interaction types or time slices		

[a] Purely topological: it is mainly graph theory and does not use any quantitative information fully.
[b] Beyond purely topological: it is where the categorical or quantitative information begins to be considered or included.

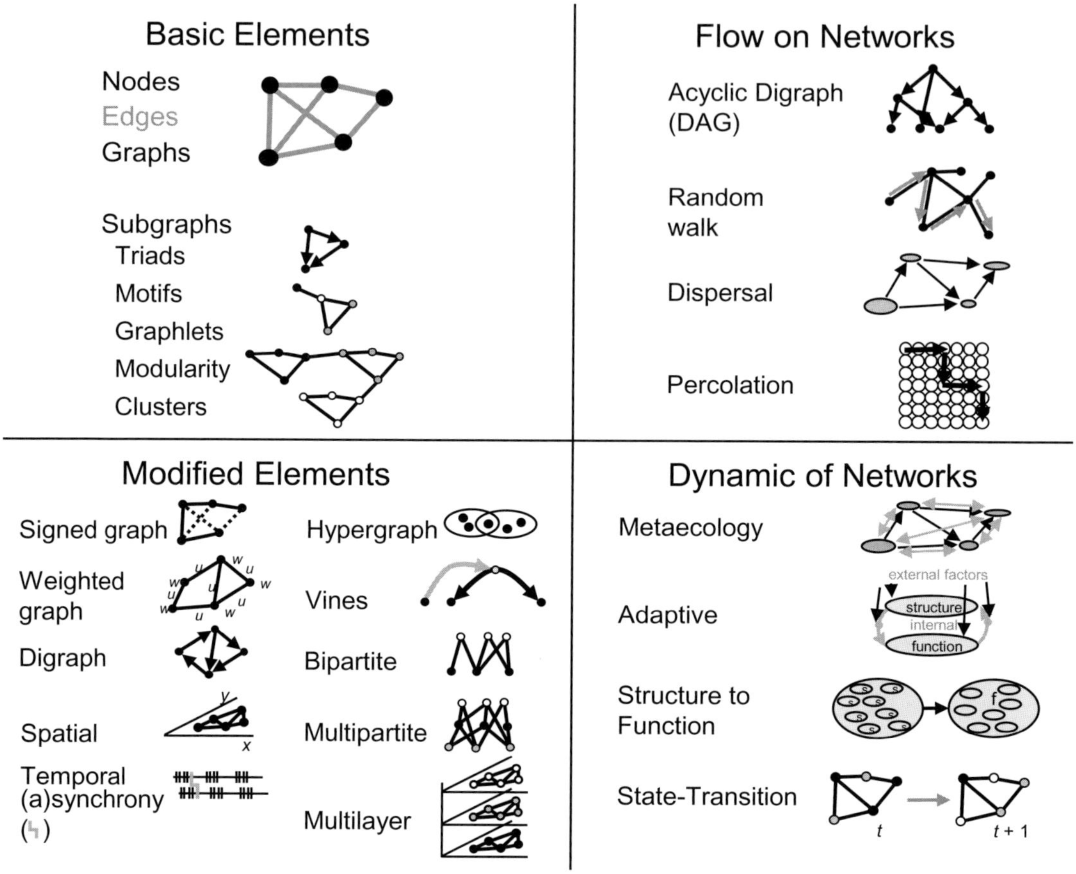

Figure 1.12 Elements and properties of networks: Basic elements: nodes, edges, subgraphs, etc. Modified elements: signed, directed, weighted graphs . . . Flow on networks: DAGs, random walk, dispersal, percolation. Dynamics of networks: metaecology, structure and function in adaptive networks.

them, and so the quantitative analysis of these networks is the most promising approach to answering many important ecological questions. How to proceed in any particular example may not be always obvious, but the place to start is with the basic structural properties of networks, laying the groundwork for statistics and other approaches to their quantitative analysis (Chapter 2). With that accomplished, we can then proceed to evaluating the dynamics of ecological networks and the relationships between structural properties and dynamic behaviour in simpler networks (Chapter 3). The last step is to consider the more complicated network structures created by various elaborations of the basic network form, the "multi-something" structures introduced in Section 1.3, for example, several layers, several levels, and so on (Chapter 4).

Network analysis and its application to ecological studies are both developing with great speed. That rapid development makes the whole cluster of topics extremely exciting as well as demanding, but it also makes it impossible for a book to keep up with and report on the most recent advances. The purpose of this book, however, is not to provide a complete synoptic summary of the field, but to provide a guide for researcher on approaches, applications, and trends to watch for further initiatives and advances. Table 1.2 offers two views of the material, with both categories and instances organised according to network structure as topology, structure as composition and configuration, functional aspects, and dynamics on and of the network. Many of the basic elements and properties of networks referred to in Table 1.2 are illustrated in Figure 1.12. This scheme may help organise the concepts presented throughout the book and draw them together, but it is incomplete. The best we will be able to offer by way of a whole summary is our "it all ties together" schematic figure at the conclusion of the book (Figure 5.3) which sketches out the relationships among the structures, measures, and approaches in the quantitative analysis of ecological networks. That summary should provide an "atlas" for analysis that is more inclusive and far-reaching than a simple *X*-to-*Y* "road map" of methods. It should help researchers find the best route of their own progress in that endeavour, but without trying to include absolutely all possible side branches and detours.

2 Structural Properties of Networks

2.1 Introduction

Characteristics and measures for the structural properties of networks are provided in great numbers and diversity (Costa et al. 2007; Dale & Fortin 2010; Rayfield et al. 2011; Delmas et al. 2019). For example, Costa et al. (2007) discussed 31 such measures, making it a daunting task to understand the choices, evaluate their properties, and determine the best for a given purpose. Just organising the material is a challenge. This chapter will start at the local scale, with the properties of nodes and edges, then moving to the global scale of characteristics of networks in their whole graphs, beginning with different approaches to analysis (e.g. spectral graph theory and information measures) ending with a discussion of the properties that are evaluated. The focus here is primarily structural properties; network dynamics and the relationship between structure and function will be covered in Chapters 3 and 4.

The most basic properties of a network graph G are the number of nodes, n, sometimes called the *order*, and the number of edges, m, sometimes called the *size*. A related global property is *connectance*, which is the proportion of places where edges occur relative to the total number of places at which edges can occur. Most structural properties of networks are best approached by expressing the network graph in matrix form. There are several network matrices commonly used and one is the adjacency matrix, **A**; it is an $n \times n$ matrix of elements a_{ij} which take the value 1 when there is an edge from node i to node j and 0 otherwise. For an undirected graph, this matrix is symmetric as in Figure 2.1a, upper panel, but for a directed graph it may be asymmetric as in the lower panel of Figure 2.1a. The matrix form is both convenient for a discussion of network properties and a prerequisite to quantitative analysis. We will not emphasise the topic, but an important consideration for subsequent analysis is how the required computing resources increase with the network's number of nodes and edges, or the size of its matrix, especially with the trend towards large and very large network data sets in ecological applications.

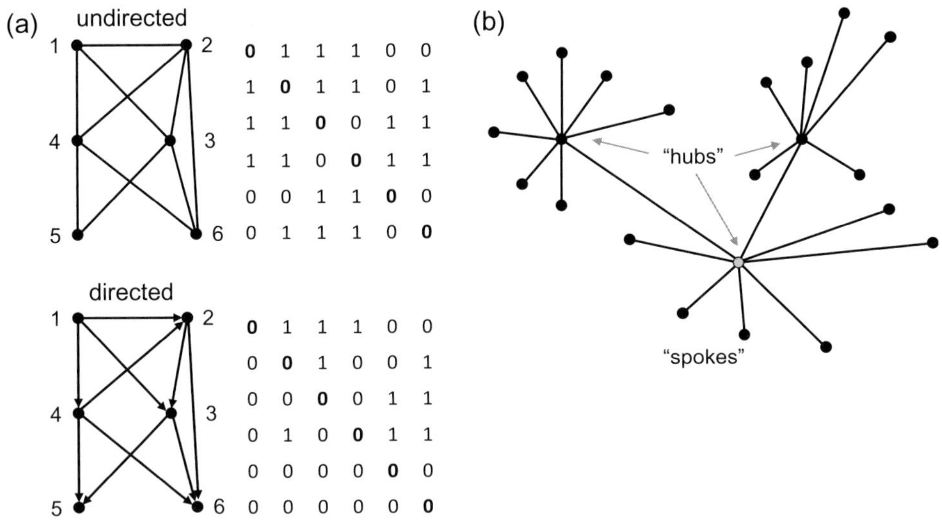

Figure 2.1 Network representations as matrices and graphs. (a) Network graphs and their adjacency matrices, undirected and directed. (b) Hub and spoke network graph.

2.2 Properties of Nodes and Edges

2.2.1 Degree and Degree Distribution

Because a network is made up of nodes joined together by edges (joining together is the essence of being a network), one of the fundamental properties of a network is the number of nodes to which each node is joined. That is a local characteristic called a node's *degree*, simply the number of edges attached to it; in an undirected graph, this is the row or column sum of the adjacency matrix. The node degrees are sometimes given their own matrix, **D**, with the degrees on the main diagonal and 0's elsewhere. The sum of the nodes' degrees is $2m$, and the connectance is $2m \div n(n - 1)$. For digraphs, graphs with directed edges, the total degree separates into *in-degree* (arrows towards the node, or node as sink; the column sum of the adjacency matrix) and *out-degree* (arrows away from the node, or node as source; the row sum of the adjacency matrix). Similarly, in a signed graph, the positive degree is usually calculated separately from the negative degree. A signed digraph therefore will have four classes of degree: positive and negative in-degree and positive and negative out-degree. In the graph of a weighted network, the concept of degree may be extended to being the sum of the weights of all attached edges at a node, sometimes called *node strength* (Kolaczyk & Csárdi 2014), rather than just the count of edges attached. For a weighted digraph, the weight degree of a node can be divided into out-weight (or out-strength) and in-weight (or in-strength) for obvious reasons.

Network graphs with no signs or directions on the edges can be partially characterised by the distribution of the degrees of all the nodes. For example, the degree distribution of a hub-and-spoke structure (Figure 2.1b) will have a small number of nodes with high degree (called the hubs) and many nodes of degree 1 (called the spokes).

Although the degree distribution is important as a description of characteristics of a network graph, its real importance lies in what it can (or cannot) tell about how the network will function or about the mechanism by which the network was generated. The various models of network development have different generating processes with differing amounts and forms of randomness and constraint in their formation. The degree distribution can disprove or confirm which models are compatible with observed network structure. The third way in which the degree distribution is important lies in its potential effects on the function of the network, for example, its second-order degree characteristics in particular: are hubs joined to hubs? These reasons explain the emphasis on this characteristic in the general network literature, and a few details will illustrate these comments.

Beginning with $p(k)$ as the proportion of nodes in a given network with degree $a_{ii} = k$, the fully random graph network model (known as Erdös–Rényi; Newman 2010; Dale 2017), in which edges are added to a set of nodes to join randomly chosen pairs, produces a degree distribution that is approximately Poisson:

$$p(k) \sim e^{-\kappa}\kappa^{k}/k!, \tag{2.1}$$

where κ is the mean or average degree. This distribution is unimodal with few nodes of very large or very small degree and the largest frequencies occur around the mean value, κ.

A second possible distribution for the degrees in a network is the exponential distribution (Newman 2010):

$$p(k) \sim e^{-k/\kappa}. \tag{2.2}$$

This distribution has the highest frequencies at the smallest degrees, with frequency falling off smoothly (and, yes, exponentially) with increasing degree.

In many real networks that have been studied, the distribution of degrees follows a power law (called Barabási–Albert; Newman 2010; Dale 2017), at least approximately, which is

$$p(k) = Ck^{-\theta}, \tag{2.3}$$

where C and θ are constants. This relationship suggests plotting $p(k)$ against k, both on logarithmic scales, because then

$$\log\,[p(k)] = \log\,[C] - \theta \log\,[k], \tag{2.4}$$

and θ can be estimated from the slope of the best fit straight line. Such networks are also referred to as "scale free" because the power law means they look the same at any scale. Many real networks follow a power law only approximately, often undershooting a "good" straight line fit at both the smallest and largest values (sketched in Figure 2.2; see also Shalizi 2006; Kolaczyk 2009; Newman 2010; Estrada 2012).

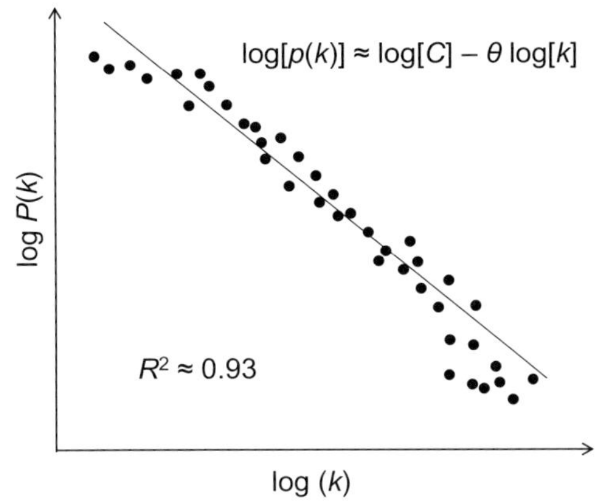

Figure 2.2 Possibly a "power law" distribution of node degrees as in a scale-free network. Log scale for both k and $P(k)$. $\log[p(k)] \approx \log[C] - \theta \log[k]$.

A Pearson correlation coefficient may quantify how well the line fits the data and is sometimes referred to as the "straightness" (Costa et al. 2007) of the distribution. Shalizi (2006) explained several statistical issues in determining whether a distribution really follows a power law, such as the reliance on R^2 to evaluate the fit of a straight line to the data. R^2 can be impressively high even if the distribution does *not* follow a power law (suggesting "*any* smooth curve looks like a straight line, if you confine your attention to a sufficiently small region . . ."; Shalizi 2006). The conclusion is that few analyses are able to confirm that a distribution (e.g. of node degrees) actually follows a power law as opposed to some other "heavy tailed" distribution (Shalizi 2006).

A fourth-degree distribution commonly referred to in network literature is the Watt–Strogatz model or "small world"; it is a compact distribution that is unimodal with a short and heavy tail. The path length (smallest number of edges to connect) between any two nodes in the network tends to increase with $\log(n)$, so that large networks have smaller diameters than random graphs of the same parameters, hence the "small world" allusion (Newman 2010; Dale 2017).

Related to the node degree distribution, another property that can both characterise a network and affect its function is the second-order degree distribution; this is the distribution of node-neighbour degrees. What is the distribution of the degrees of nodes that are attached to nodes of low degree compared to that for nodes of high degree? This characteristic can be assessed based on the conditional probability that a neighbour of a node with degree k has degree h: $P(h|k)$, usually summarised by the average neighbour degree for nodes of degree k (Costa et al. 2007):

$$\kappa_{NN}(k) = \sum_{h=1} hP(h|k). \tag{2.5}$$

When $\kappa_{NN}(k)$ increases with k, there is positive correlation of node and neighbour degrees, which is said to be assortative for degree. When $\kappa_{NN}(k)$ is a decreasing function of k, it is disassortative for degree. The standard Pearson correlation coefficient will evaluate the correlation of neighbouring nodes' degrees and then determine the goodness of fit with any reference network model (e.g. Erdös–Rényi or Barabási–Albert), or evaluate the consistency among networks of the same type, for example, ecological compared to social compared to technological (see Newman 2010). Kolaczyk and Csárdi (2014) have provided an example from yeast Protein–Protein Interaction (PPI) data; nodes of high degree tend to be adjacent to similar nodes (i.e. high degree), but nodes of low degree are adjacent to both low and high (Kolaczyk & Csárdi 2014).

The concept of node degree can be augmented for any of these applications in a straight-forward way when the edges have weights, by summing the weights of edges incident at each node to give the "weighted degree distribution" or the distribution of node strength (see Kolaczyk & Csárdi 2014).

2.2.2 Local Density

Extending the concept of node degree, the number of neighbours, and that of the distribution of the degrees of node neighbours, leads to the concept of the local density of edges, which may vary across a network. Local density is one characteristic related to a network's *cohesion*, that is, how well it sticks together rather than tending to fall apart (Kolaczyk 2009). Many networks are not uniform in either local density or in cohesion, often consisting of one large highly connected unit (the "giant component") together with many small connected components or unconnected nodes (Figure 1.9).

The greatest local density occurs in complete subgraphs, those with all possible edges for h nodes, designated K_h, and called *cliques* (see K_4 in Figure 1.9, a kite-shaped subgraph indicated by darker nodes). Large cliques tend to be rare in networks, generally, but the occurrence of 3-cliques, which are triangles, is often a characteristic of interest, especially in the study of social networks (see Kolaczyk 2009). Because cliques are rare, even for small values of h, they may not be useful indicators of local density and what are called "n-plexes" may be preferred as alternatives. A subgraph of h nodes is an n-plex ($h > n$), if no node has degree less than $h - n$. For example, the wheel-shaped subgraph labelled "module of 8 nodes" with darker nodes in Figure 1.9 is a 4-plex, because all the nodes have degree of at least 3 ($h = 7$; $n = 4$). This terminology seems a bit confusing because there is a similar concept of a subgraph called a "k-core": a subgraph is a k-core if all nodes have degree of at least k (Kolaczyk 2009); the node-counting is reversed. For example, the same wheel subgraph already indicated in Figure 1.9 as a 4-plex is both a 2-core and a 3-core. The tracking of k-cores has proven useful in a number of studies of biological networks, including the prediction of protein function (Altaf-Ul-Amin et al. 2003).

A very narrow version of local density is to look at the local *clustering coefficient* of a single node: it is the proportion of connected node triples with node i central in the triple that are complete triangles of edges (Costa et al. 2007):

$$C_i = N_\Delta(i) \div N_3(i), \tag{2.6}$$

where

$$N_\Delta(i) = \sum_{k>h} a_{ij}a_{ik}a_{jk} \text{ and } N_3(i) = \sum_{k>h} a_{ij}a_{jk}. \tag{2.7}$$

These can be adjusted to include weights if they are present in the network.

In a digraph, the triangles are either transitive ($A \rightarrow B$, $B \rightarrow C$, $A \rightarrow C$) or intransitive (cyclic: $A \rightarrow B$, $B \rightarrow C$, $C \rightarrow A$). A refinement is therefore to calculate the proportion of all digraph triangles that are transitive; call it P_t. For random directed networks, the expected value of P_t is 0.75 (six of the eight possible digraph triangles are transitive; Dale 2017), leading to the suggestion of an overall transitivity measure of $t_\Delta = 4(P_t - 0.75)$ (Shizuka & McDonald 2012).

Another local measure that is not strictly related to density is the *local cyclic coefficient* based on the sizes of the smallest cycles that include node i and its neighbours (Kim & Kim 2005). Define S_{ijk} as the size of the smallest cycle through those nodes; then the node's cyclic coefficient is

$$\vartheta_i = \frac{2}{k_i(k_i - 1)} \sum_{k>h} \frac{a_{ij}a_{ik}}{S_{ijk}}, \tag{2.8}$$

and the coefficient for the whole network is the average over all nodes:

$$\vartheta_\bullet = \frac{1}{n} \sum_i^n \vartheta_i. \tag{2.9}$$

For example, in the undirected graph in Figure 2.1, node 1 forms triangles with neighbours 2 and 3 and 2 and 4, but with 3 and 4 the smallest cycle is length 4; $\theta_1 = (3 + 3 + 4)/3 = 3.33$. The local cycle coefficients for the other nodes range between 3.0 and 4.0, with an average of $\theta_\bullet = 3.5$. On the other hand, a ring graph of six nodes gives $\theta_\bullet = 6$. For more open connected networks, this measure should be very useful in combination with other structural measures.

2.2.3 Connectivity

Connectivity is another aspect of network cohesion. It is related to the density of edges because it refers to how well connected the network is: the number of edges (or the number of nodes) that must be removed in order to disconnect the network graph into two or more components. (The network is connected if there is a path between any two nodes; a path is an alternating sequence of nodes and the edges joining them, with no element used more than once.) The higher the number of either nodes or edges that must be removed, the greater the connectivity. A network is *k-node connected* if the removal of any k nodes does not disconnect its graph. Similarly, the graph is *k-edge connected* if the removal of any k edges leaves the network connected.

For directed edges and a digraph‘s network, we distinguish between *strongly connected* when there is a directed path between any two nodes that follows the directions of the edges. If there is at least one path between any two nodes but only when the directions of the edges are ignored, it is *weakly connected.*

2.2.4 Centrality

There are many ways of determining how important any node is in the structure of a network. Leaf nodes are considered peripheral, for example, and the opposite of peripheral are nodes of high centrality: nodes that are essential in many of the longest paths through the network. Centrality measures have been applied frequently in ecology, for example, to evaluate the importance of landscape patches to connectivity and the local dispersal of organisms (see Estrada & Bodin 2008) and to evaluate how network structure affects individual parasite loads or diversity (Chen et al. 2008; Anderson & Sukhdeo 2011; Duboscq et al. 2015). There are many versions of this concept, including "closeness centrality," "betweenness centrality," and "eigenvalue centrality" (see Koschützki 2008; Kolaczyk 2009; Newman 2010; Estrada 2012; Dale 2017). Table 2.1 provides a summary of many of these measures.

These measures listed in Table 2.1 are for the betweenness of **nodes**, but we can develop an equivalent set of measures based on the importance of **edges** in the shortest paths through a graph. One such measure is *edge betweenness centrality*, defined as the number of the shortest paths in the network that include the edge (Girvan & Newman 2002). Following the node betweenness definition, given earlier, it compares the number of shortest paths between nodes u and w that include edge ε, $\sigma(u, w|\varepsilon)$, with the total number of such paths $\sigma(u, w)$. Scaled to run between 0 and 1, it is

$$C_b(\varepsilon) = \sum_{u \neq v}^{N} \sum_{w \neq u \neq v}^{N} \frac{2\sigma(u, w|\varepsilon)}{(N-1)(N-2)\sigma(u, w)}. \tag{2.10}$$

Similar adjustments may be considered to convert other centrality measures from node-based to edge-based, although not always successfully. An alternative is to convert the whole graph to its *line graph*, in which the edges become nodes, and then apply the standard node centrality measures to these edges-become-nodes (Kolaczyk & Csárdi 2014).

2.2.5 Partitions and Clusters

For any structure made up of units joined in pairs, one characteristic of interest is how the units can be divided or grouped into cohesive subsets. The partition should create clusters of nodes that share particular properties, often mutual connectedness, and with less similarity between nodes in different clusters than within. Partitioning can be hierarchical, with subsets nested within subsets, or it can be based on finding a predetermined number of clusters of a given size (Hastie et al. 2009). Finding partitions in a network is related to the familiar clustering approach for ecological data, and one important method for determining the partitions of a network is that same hierarchical clustering (see Legendre & Legendre [2012] for the ecological versions of these techniques). The second main approach to partitioning networks described by Hastie et al. (2009) is probably not familiar to ecologists. It is *spectral partitioning*, and it uses the relationship between characteristics of a network's connectivity and the

Table 2.1 Centrality measures for networks

- *Degree centrality*: A node's degree, the number of edges attached to it, is one measure of its importance, sometimes called "degree centrality." In digraphs, there is a distinction between in-degree and out-degree (the number of edges where node v is the receiver versus the number of edges where v is the source).
- *Betweenness centrality*: For node v, this is the proportion of the shortest paths (path length) between any pair of nodes that passes through node v. It compares the number of shortest paths between nodes u and w that include node v, $\sigma(u, w|v)$, with the total number of such paths $\sigma(u, w)$. Scaled to run between 0 and 1, it is

$$C_b(v) = \sum_{u \neq v}^{N} \sum_{w \neq u \neq v}^{N} \frac{2\sigma(u, w|v)}{(N-1)(N-2)\sigma(u, w)}. \tag{2.11}$$

  See Dale (2017, figure 3.8b). For a technical discussion, see Gago et al. (2015).
- *Closeness centrality*: For node v, closeness centrality measures the average path length, $\delta(u, v)$, from node v to all other nodes, u. Nodes with shorter distances have greater closeness centrality. For N nodes, closeness centrality, scaled to run from 0 to 1, is

$$C_c(v) = \frac{N-1}{\sum_{u \neq v}^{N-1} \delta(u, v)}. \tag{2.12}$$

  See Dale (2017, figure 3.8c) and see Kolaczyk (2009).
- *Eigenvector or spectral centrality*: Includes all possible paths, not just the shortest, accounting for the greater importance of shorter paths by a path weight based on length, γ: α^γ for some constant α. The actual calculation is from the adjacency matrix **A** and its eigenvalues and eigenvectors: $\lambda_1(\mathbf{A}) = \alpha^{-1}$ is the largest eigenvalue and $\mathbf{c}_e$ is the corresponding eigenvector; $\mathbf{c}_e(u)$ the coordinate of node u on $\mathbf{c}_e$. There are several versions of the measure possible, but the usual one is (Kolaczyk 2009):

$$C_e(v) = \alpha \sum_{(u,v) \in E}^{\mathbf{D}(v,v)} \mathbf{c}_e(u). \tag{2.13}$$

  See Dale (2017, figure 3.8d).
- *Spatial centrality*: For a spatial graph, path length centralities can be modified by replacing geodesic distance with physical or geographic distance: $d(u, v)$ instead of $\delta(u, v)$. A simple weighting of a node's geographic centrality in a spatial graph is the additive inverse of its distance from location of the centre of gravity of all N nodes, γ, giving an index constrained to run between 0 and 1:

$$C_g(v) = 1 - \frac{d(v, \gamma)}{\max_u [d(u, \gamma)]}. \tag{2.14}$$

  This weighting can modify existing measures of centrality into a spatially responsive form. It can also be the basis for weighting measures of edge centrality, as opposed to the measures of node centrality discussed so far (Dale 2017, figure 3.8e).

Adapted from Dale (2017).

eigenvalues of its adjacency or Laplacian matrix (Kolaczyk 2009). These methods for determining partitions will be described in detail in Section 2.3.2.

2.2.6 Assortativity

Assortativity refers to the tendency of like things to be found together. More specifically, the nodes display positive assortativity if nodes belonging to the same category or nodes with similar weights tend to be neighbours with edges between them, as with positive autocorrelation or like–like join counts (also called homophily). Negative assortativity means that nodes with different weights or different labels tend to be joined by edges, like negative autocorrelation (also called heterophily). As a specific example, already mentioned, networks can be assortative for or disassortative for node degree.

In general, we can define a "type mixing matrix" M where e_{st} is the count of edges joining a node of type s to a node of type t (or the total weight of those edges in a weighted network), and the elements of M are the normalised intertype counts, $e_{st} \div \Sigma\Sigma\, e_{st}$. Costa et al. (2007) suggested the following measure of assortativity:

$$Q = \frac{Tr(M) - \left\|M^2\right\|}{1 - \left\|M^2\right\|}. \tag{2.15}$$

More explicitly, Kolaczyk (2009) expressed this concept with f_{ij} as the fraction of edges between types of nodes i and j, defining the assortativity coefficient as

$$r_a = \frac{\sum_i f_{ii} - \sum_i f_{i\bullet} f_{\bullet i}}{1 - \sum_i f_{i\bullet} f_{\bullet i}}. \tag{2.16}$$

It is the same as Q, of course, just expressed differently! The value lies between -1 and +1, although -1 is not actually possible because even if there are no like–like edges, the minimum value of r_a (see Kolaczyk 2009) is

$$r_a = \frac{-\sum_i f_{i\bullet} f_{\bullet i}}{1 - \sum_i f_{i\bullet} f_{\bullet i}}. \tag{2.17}$$

A refinement of assortativity coefficients, which are usually applied as 'global' measures to whole networks, is to consider local assortativity at the level of small subgraphs or even at the level of individual nodes (Piraveenan et al. 2008, 2010). It is possible, then, to have a match or a mismatch between the local and global characteristics (Piraveenan et al. 2010), such as disassortative hubs in a disassortative network, for example, Crystal River food web (see Piraveenan et al. 2010) or disassortative hubs in an assortative network, for example, human PPI (see Piraveenan et al. 2010). A similar approach has been advocated by Park and Barabási (2007), who studied two-type networks by two characteristics: *dyadicity* $= D$ (the frequency of like–like joints for nodes of class 1) and *heterophilicity* $= H$ (the frequency of unlike joins). They found that different kinds of networks occupied different parts of a D–H phase diagram, showing the ability of these characteristics to differentiate among network types. These various local approaches to studying assortativity are clearly methods that should be explored further for other ecological networks, but they will seem familiar to ecologists because they are essentially join-count statistics.

2.2.7 Subgraphs, Motifs, and Graphlets

Network structure can be analysed using subgraphs in a number of different ways. The most common approaches are based on *motifs* and on *graphlets*. Motifs are small subgraphs of defined structure, either for simple graphs or for digraphs (see Figure 2.3 for digraphs). They may be partial subgraphs, possibly omitting edges between nodes that are included; for example, the three-node triangular motifs also include one or more 3-chains (Figure 2.3). The different motifs are counted in a whole network with the idea that frequently occurring motifs are important and can signal some important functional relationships in the network (Milo et al. 2002). The counts of motifs can be portrayed in a diagram called the *motif spectrum* and observed spectra can be compared with those generated by computer models, for example, comparing the motif spectra of empirical with simulated food webs (Paulau et al. 2015). This approach was helpful in identifying the hierarchical patterns in the observed trophic networks. In a further refinement of the motif approach, in addition to comparing the motifs of trophic networks (in pitcher plant communities) with random networks,

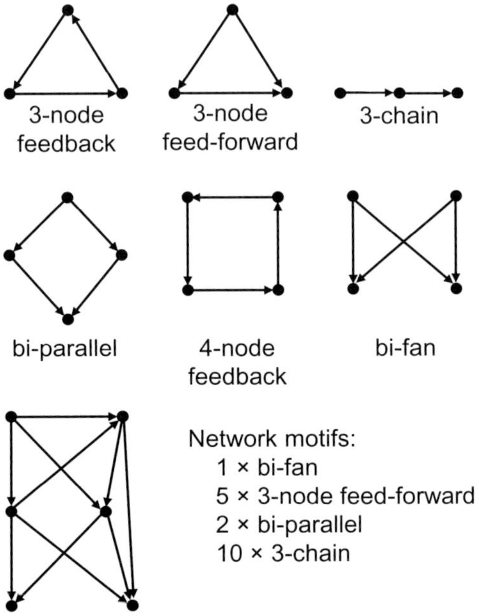

Figure 2.3 Digraph motifs, as labeled. The example has the following network motifs: 1 × bi-fan, 5 × 3-node feed-forward, 5 × bi-parallel, and 10 × 3-chain.

Baiser et al. (2015) looked at spatial scale by comparing the motif representations in local (one pitcher), site, and continental networks.

The usual manner of motif counting is unrestricted: instances that share nodes or edges are still counted separately. Schwöbbermeyer (2008) referred to this as counting concept F_1; counting concept F_2 allows separate motif counts if nodes are shared but not if edges are shared; and concept F_3 insists that neither nodes nor edges can be shared between counted motifs, giving what is called the motif's "uniqueness value." In general, the least restrictive counting scheme, F_1, is usually the rule in biological network analysis (Schwöbbermeyer 2008), and we suggest that practice should include ecological networks too.

The significance of a motif's frequency (call it motif q) can be evaluated by a Z score based on the mean and variance of its frequency in random networks:

$$Z_q = \frac{f_{\text{obs}}(q) - \bar{f}_{\text{rand}}(q)}{\sqrt{\text{var}(f_{\text{rand}}(q))}}. \tag{2.18}$$

Different networks have different characteristic motif profiles; for example, transcription networks in yeast tend to have feed-forward loops and bi-fans as the most common motifs, whereas trophic networks are characterised by 3-chain and bi-parallel motifs (Milo et al. 2002).

One suggestion is to take the Z values for the full set of motifs and create a vector of them, called a *significance profile* of the network (see Costa et al. 2007; Schwöbbermeyer 2008). How is the order to be determined, in the absence of a truly natural, functionally relevant, sequence of these motifs? The order should be related to a sequence of increasing motif complexity or directionality. The lack of independence among the motifs suggests that the interpretation of such a profile would have to proceed with caution. (Similar concerns arise in the interpretation of the graphlets described later in this section.) Randomisation methods seem to be the most promising for evaluating the implications of motif counts (see Schwöbbermeyer 2008).

Graphlets are induced connected subgraphs of $n = 2, 3, 4$ or (sometimes) 5 nodes. The term "induced" means that any pair of nodes in the graphlet must retain the edge between them if there is a corresponding edge in the original graph.

There is an impressive range of ways for graphlets to be used in network structural analysis, from simple frequency counts of defined graphlets in a whole network (like motifs) or counts of nodes in topologically distinct positions in a specific graphlet, to summaries and comparisons of related characteristics (Table 2.2). The key procedure is to count the occurrence of nodes in the distinguishable positions on each graphlet; nodes that are not distinguishable map onto each other by an automorphism and so they belong to the same orbit. These are the automorphism orbits *of the graphlet* (in the following section we will discuss the automorphism orbits *of the whole graph*) and are related to the symmetries of the graphlet. Although a node can belong to only one orbit of the whole graph, it can belong to more than one graphlet orbit because it occurs in more than one graphlet.

The first graphlet, g0, is two nodes joined by an edge; the nodes both belong to orbit $\gamma 0$ (Figure 2.4); the designations of "g" for graphlets and γ for the orbits is to

Table 2.2 A checklist of methods based on graphlets

Graphlet Frequencies	Counts of the 29 $n = 2, 3, 4, 5$ graphlets (show local structure)
	Compare networks by simple similarity or distance measure
Graphlet Orbit Signature	73 orbits in $n = 2, 3, 4, 5$ graphlets give each node 73 coordinates
	Measure node local similarity by comparing signatures
Graphlet Orbit Distribution	Counts of nodes touching each orbit
	Measure network agreement by comparing normalised GODs
Graphlet Correlation	Matrices and distances based on reduced list of orbits (11)
	Correlations of orbits provide clear structural profile

Modified from Dale (2017).

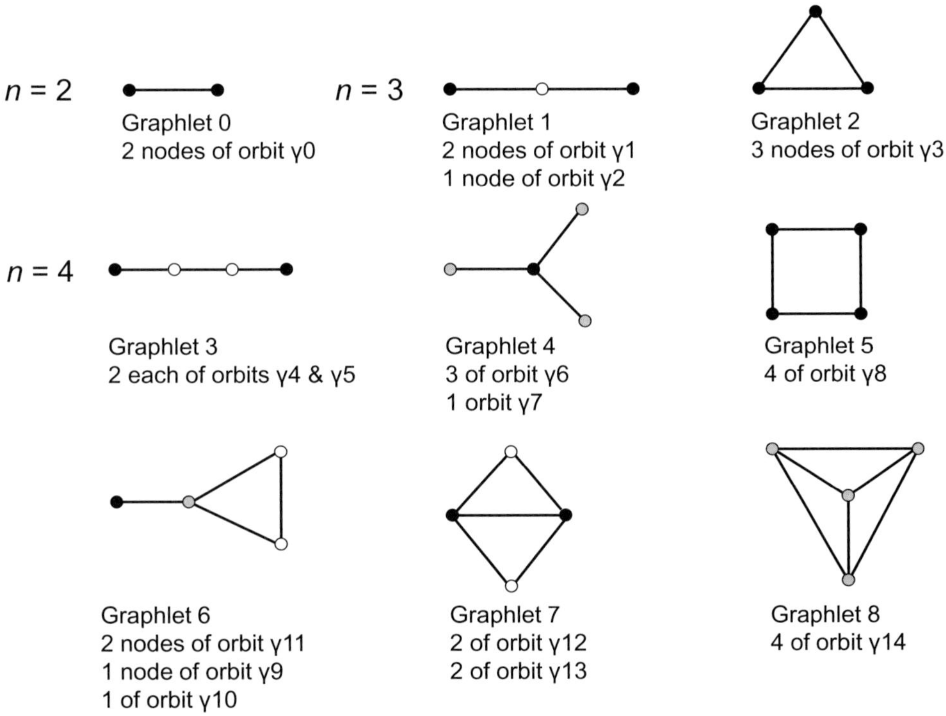

Figure 2.4 Basic graphlets and node orbits for graphs with edges that have no sign and no direction: $n = 2, 3, 4$. Each "orbit" identifies nodes that occur in equivalent positions in a graphlet (also called automorphism orbits). The prefix γ indicates undirected unsigned edges.

distinguish these from graphlets and orbits with directed or signed edges, which will be introduced later. There are two graphlets possible for three nodes: graphlet g1 is a two-edge path graph with two end-nodes in orbit $\gamma 1$ and a centre node in orbit 2. Graphlet g2 is a triangle with all nodes in orbit $\gamma 3$ (Figure 2.4). For $n = 4$, there are six graphlets, giving orbits $\gamma 4$ to $\gamma 14$ (Figure 2.4). The frequencies of whole graphlets in the entire network graph are indicative of the overall properties of local structure.

Any node may occur in several graphlet orbits (for example, the nodes in g3 also are in g0's orbit $\gamma 0$ and in orbit $\gamma 1$ or $\gamma 2$ of g1; see Figure 2.4) and the frequencies of the graphlet orbits in which a node occurs is its *graphlet orbit signature* (originally called "graphlet degree signature"; Pržulj 2007). (The reason for the suggested change from "degree" to "orbit" is to avoid possible confusion with "degree" as the number of edges at any node.) This signature indicates the node's topological position in the network structure, and the structural characteristics of the whole graph can be summarised by combining all the graphlet orbit signatures into an overall *graphlet orbit distribution* (originally called "graphlet degree distribution"; Milenković & Pržulj 2008). The orbits included in all 15 graphlet orbits for 2–4 nodes are not independent and it is recommended practice to use the 11 orbits of the 15 that are non-redundant. Having eliminated this redundancy, any correlations observed in the orbit frequencies must be due to the structure of the network graph itself (Yaveroğlu et al. 2014). These correlations are usefully displayed in a heat map format (an 11×11 *graphlet correlation matrix* [GCM]) and the structures of two network graphs can be compared using the Euclidean distance between corresponding entries in the GCM's, referred to as the *graphlet correlation distance* (GCD).

2.3 Methods of Network Analysis

2.3.1 Partitioning

One approach to studying a structure made up of units joined in pairs is to examine the ways in which those units can be divided into subsets or partitions. There are many good treatments of how to partition networks including Kolaczyk (2009), Newman (2010), and Estrada (2012). The two main methods for partitioning are clustering and spectral graph theory, and we will discuss these two in some detail. In addition to clustering and spectral methods, several other approaches or modifications to the basic ones have been explored (Kolaczyk 2009). These include the iterative removal of edges with the choice based on *edge betweenness* described earlier (see Girvan & Newman 2002).

Partitioning a network is closely related to the familiar clustering approach to ecological data in which the individual units are amalgamated one step at a time into successively larger groupings. In fact, Kolaczyk's (2009) treatment of partitioning begins with a description of the various approaches to hierarchical clustering. These can be *divisive*, starting with the whole network as a single large grouping and successively dividing it into smaller and smaller subclusters, or *agglomerative*, starting with single nodes and joining them into clusters that get larger at each step (see Legendre & Legendre [2012] for the ecological versions of these techniques). These methods usually produce a complete hierarchy of groupings with nested partitions, often displayed as a dendrogram. The basics of the methods require a measure of similarity or difference between any two groupings in the network (e.g. Euclidean distance) and a rule to recalculate that measure whenever a group is formed or divided (e.g. the maximum, minimum or average of the measure prior to this step of the process). These techniques are well covered in the ecological literature and we will not repeat the details here.

The second main approach to partitioning networks described by Kolaczyk (2009) is probably not familiar to ecologists. It is *spectral partitioning*, and it uses the relationship between characteristics of a network's connectivity and the eigenvalues of its adjacency or Laplacian matrix. We will delay a detailed description of this approach until we have covered the basic concept and techniques of spectral graph theory more fully.

2.3.2 Spectral Graph Theory

Spectral graph theory is the study of graphs and their characteristics using the eigenvalues and eigenvectors of matrices of the graph, such as the adjacency matrix or the Laplacian (Chung 1997; Brouwer & Haemers 2012). For an undirected graph, the adjacency matrix, **A,** has 0's on the main diagonal and elements $a_{ij} = 1$ where nodes *i* and *j* are adjacent and 0 otherwise. The Laplacian, **L**, has the nodes' degrees on the main diagonal and elements -1 where nodes i and j are adjacent and 0 otherwise. If the degree matrix is the diagonal matrix, **D**, the Laplacian is $\mathbf{L} = \mathbf{D} - \mathbf{A}$.

The eigenvectors of these matrices preserve an impressive amount of the information about the original graph as we can demonstrate with spectral graph drawing.

We might joke that, for the analysis of ecological networks, there is no such thing as "naïve graph theory" as something to be demonstrated with a few dots on paper with some lines joining them in pairs. The data sets to be analysed are often large and usually complex, and these must be first rendered as matrices, either of adjacencies or of weights. These matrices can be eigen-analysed, producing for each one a "spectrum" of the associated eigenvalues. As with any dimension-reduction approach, the eigenvalues can retain much of the useful information in the original data, but usually not it all. The eigenvalues can, however, be interpreted for several important network characteristics (Chung 1997; Brouwer & Haemers 2012). The properties accessible through spectral methods include graph diameter, numbers of cycles, connectivity, and bipartivity (Costa et al. 2007). Spectral analysis methods can also be used to detect modules or clusters within a network or to determine its partitions (Costa et al. 2007).

Some of the technical details of how spectral methods work are necessary, and we can begin with the concept of the *characteristic polynomial* of a graph, call it Γ, based on its adjacency matrix **A**. Γ is a determinant denoted $\det(\lambda\mathbf{I} - \mathbf{A})$ with **I** being the $n \times n$ identity matrix. For example, the triangular graph that is graphlet 2 (Figure 2.4) has the characteristic polynomial as follows:

$$\Gamma_{\Delta} = \lambda^3 - 3\lambda - 2. \qquad (2.19)$$

The roots of the characteristic polynomial are the eigenvalues. The same value can appear more than once; for example, the eigenvalues for the triangle graph are -1,

-1, and 2, expressed as $\{(-1)^2, 2\}$; the *multiplicity* of -1 is 2. The spectrum of the graph is the set of eigenvalues and the number of times they appear as roots of the characteristic polynomial (their multiplicities). The eigenvalues determine the eigenvectors, which are orthogonal. The eigenvalues of a graph, for either the adjacency matrix or the Laplacian matrix, are strongly associated with its key characteristics and thus are highly informative about its properties. Note that the eigenvalues for the two different matrices are not going to be the same (usually) although there is obviously a relationship between them; the Laplacian eigenvalues for the triangle graph (graphlet 2 in Figure 2.4) are $\{0, 2^2\}$. As a matter of convention, the Laplacian eigenvalues are given in increasing order, but the adjacency (or ordinary) eigenvalues are often given in decreasing order (Brouwer & Haemers 2012).

As a specific example, consider how the structure of a graph is contained in its spectral properties; this is called *spectral graph drawing*. Spielman (2012) demonstrated spectral graph drawing, beginning with the spatial graph of randomly placed nodes in the plane and the edges that form the Delaunay triangulation (see Gallier 2013). The Laplacian is calculated. The spatial information of the nodes' locations is discarded and the graph is redrawn with the nodes' positions determined by the eigenvectors of the second and third (smallest) eigenvalues of the Laplacian. This process provides a clear picture of the graph's structure even though the nodes' positions are based only on structural relationships, not on the original spatial data. Figure 2.5 illustrates the same procedure for a simple aspatial graph, following Gallier (2013). The Laplacian matrix for this graph is

$$\begin{matrix} \mathbf{2} & -1 & -1 & 0 & 0 \\ -1 & \mathbf{4} & -1 & -1 & -1 \\ -1 & -1 & \mathbf{3} & -1 & 0 \\ 0 & -1 & -1 & \mathbf{3} & -1 \\ 0 & -1 & 0 & -1 & \mathbf{2} \end{matrix}$$

The spectral graph drawing makes visually obvious the structural symmetry present in the original graph, but which was less obvious from its original drawing. In addition, while not planar, the drawing is clear, in part because there is less variation in line length (no very long edges on the periphery), with the edges between adjacent nodes kept short.

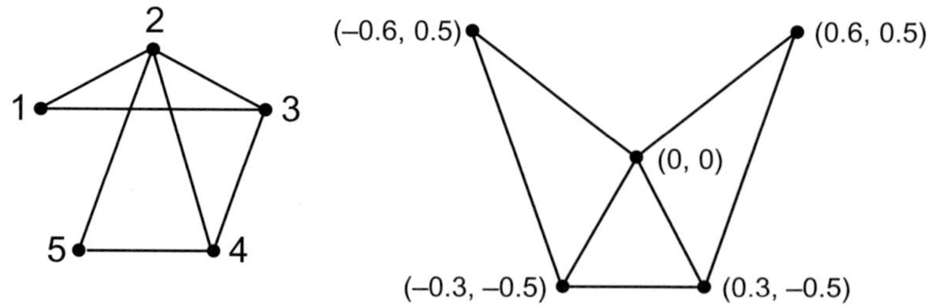

Figure 2.5 Original graph and its spectral graph drawing based on two eigenvectors of Laplacian matrix. Node 2 is at the origin (0, 0).

There are many other graph properties that can be evaluated using spectral characteristics; here are a few:

- The *spectral radius* of the graph is the largest eigenvalue of the adjacency matrix; larger values indicate a more strongly connected network. Staniczenko et al. (2012) used spectral radius as a measure of nestedness in bipartite mutualistic networks with both binary and quantitative data and advocated the spectral graph approach as a link to understanding local dynamical stability. We will say more about dynamics and stability of networks in a Section 3.4 of this book.
- *Zeros*: the number of times 0 is an eigenvalue of the Laplacian (its multiplicity) is the number of connected components in the graph.
- *Spanning trees*: for a connected network graph, the number of spanning trees, κ, is approximately the geometric mean of the $n-1$ largest eigenvalues of the Laplacian (including their multiplicities):

$$\kappa(G) = \frac{1}{n}\prod_{j=2}^{n} \gamma_j. \tag{2.20}$$

When we, as ecologists, are interested in the "sum of all paths" in an ecological network, the number of all spanning trees will be a useful fact. There are several algorithms available to generate all spanning trees (Gabow & Myers 1978; Matsui 1993), but in order to know the number by counting, they probably have to all be generated, possibly a very large task.

- The *algebraic connectivity* is the *second smallest* eigenvalue of the Laplacian matrix. It is positive for a connected graph (0 means not connected) and larger values are interpreted as greater structural resilience or robustness. Its maximum value is the node connectivity of the graph (the number of nodes that need to be removed to disconnect it). It is an important element in the process of spectral partitioning, the aim of which is to divide the network into two subnetworks of specified sizes, say n_1 and n_2. Division of a network into more parts is accomplished by iterating the division-into-two method.
- *Spectral partitioning*. The rationale and development of the spectral approach to partitioning a graph are well-described by Newman (2010). Apart from the algebraic connectivity, the other key concept for this process is the "cut size," which is just the number of edges between nodes that are in different sides of the partition. While the derivation is not easy, the resulting algorithm for our use is clearly explained (remember, n_1 has been chosen in advance):
    1. Start with the Laplacian matrix of the network graph and calculate the eigenvector associated with the second smallest eigenvalue.
    2. Order its elements from largest to smallest.
    3. Find the nodes corresponding to the n_1 **largest** elements and put them in partition 1; the remainder go in partition 2. Calculate the cut size.
    4. Find the nodes corresponding to the n_1 **smallest** elements and put them in partition 1; the remainder go in partition 2. Calculate the cut size.

5. Of these two, choose the partition that gives the smaller cut size, because it is the better alternative.

That is it!

- The *spectral gap* is the difference between the two largest eigenvalues of the adjacency matrix. Higher values are associated with structures that are less vulnerable to disruption or disconnection (Figure 2.6a). Kolaczyk (2009) has provided a helpful explanation of the term: given a graph that consists mainly of two d-regular graphs with only a few nodes joining them together, the two largest eigenvalues will be approximately d and the others will be about $d^{1/2}$ or less (see Figure 2.6b). There is therefore a gap in the spectrum of eigenvalues. The eigenvectors of the two largest eigenvalues will have large positive values for the nodes in one part of the network and large negative values in the other, thus separating the two parts in an eigenvector space. For dynamic networks, spectral gap is also a measure of how quickly a process converges to a steady state equilibrium (again, we will defer the discussion of dynamics until Section 3.4). The related concept of the *eigen-ratio* is the ratio of the largest to the second smallest eigenvalue of the Laplacian matrix; smaller ratios indicate greater stability of the synchronised state.
- *Spectral similarity:* Given two graphs on the same nodes but with edges based on independent data, their similarity can be measured by eigenfunction decomposition. For networks with undirected edges, use Moran's eigenvector maps ('MEM'

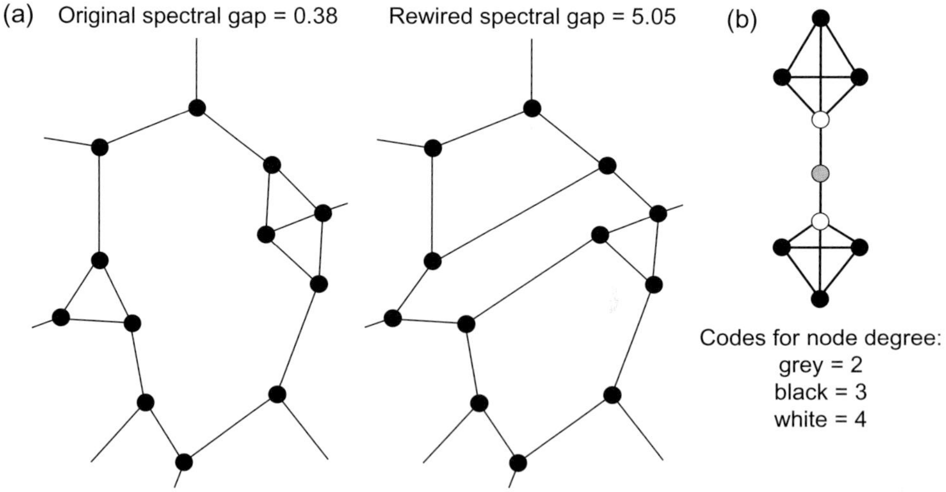

Figure 2.6 Spectral gap examples. (a) Spectral gap: A rewiring that conserves degree connectivity but decreases vulnerability to disconnection (Dale 2017; see Donetti et al. 2006). Rewiring increases the spectral gap from 0.38 to 5.05. [From Donetti, L., Neri, F., & Munoz, M.A. (2006). Optimal network topologies: Expanders, cages, Ramanujan graphs, entangled networks and all that. *Journal of Statistical Mechanics*, **8**, P08007. arXiv:cond-mat/0605565v2.] (b) Spectral gap: adjacency matrix eigenvalues are 3.17 and 3.00, then 0.68. The second eigenvector neatly splits the two groups of 4 nodes, at about +0.35 and −0.35, with the centre node at 0.

eigenfunctions; Dray et al. 2006). For digraphs, use asymmetric eigenvector maps ("AEM" eigenfunctions; Blanchet et al. 2008). The graphs begin as standard adjacency matrices but these are converted into distance matrices with an upper threshold. The eigenvectors of these distance matrices are then compared by a generalised Pearson coefficient to evaluate the similarity of the original graphs (for details see Borcard et al. 2004; Dray et al. 2006, 2012; Blanchet et al. 2008, 2011).

- *Bipartivity:* A bipartite graph is one in which the nodes belong to two categories (such as plants and their pollinators) with edges occurring only between the parts and none within. A graph is bipartite if and only if it has no odd cycles. One spectral property of bipartite adjacency matrices is that if λ is an eigenvalue, so is $-\lambda$. For example, the square graph of four nodes, graphlet 5 (Figure 2.4) has eigenvalues $\{-2, 0^2, 2\}$. In fact, the spectrum of a bipartite graph is symmetric about 0. For a graph that is not absolutely bipartite, the spectral measure of how close to bipartite it is depends on the ratio of even closed walks in the total of all closed walks (Estrada 2012). That measure is

$$B_s(G) = \frac{\sum_{j=1}^{n} \cosh(\lambda_j)}{\sum_{j=1}^{n} \exp(\lambda_j)} = \frac{1}{2}\left(1 + \frac{\sum_{j=1}^{n} \exp(-\lambda_j)}{\sum_{j=1}^{n} \exp(\lambda_j)}\right). \tag{2.21}$$

- For the Petersen graph (Figure 2.7), which is not bipartite, but not far off (it can be two-coloured with only 3 of the 15 edges within the same subset of nodes), the value is 0.96. Estrada (2012) has provided some values for this measure from real and artificial networks; real-world networks tend to have low bipartitivity because of the prevalence of triangles in these complex networks.
- *Spectra of some simple graphs* (see Brouwer & Haemers 2012):
  - ♦ *Star graph* on n nodes has adjacency spectrum $\{-\sqrt{(n-1)}, 0^{n-2}, \sqrt{(n-1)}\}$. Note that this graph is bipartite (1 central hub vs. $n-1$ spokes) and so the adjacency spectrum is symmetric to 0. Its Laplacian spectrum is $\{0, 1^{n-2}, n)\}$.
  - ♦ *Path graph* on n nodes has adjacency spectrum $\{2 \cos(\pi j / (n+1))$ for $j = 1, 2, \ldots n\}$. Its Laplacian spectrum is $\{2 - 2 \cos(\pi j/(n+1))$ for $j = 1, 2, \ldots n\}$.
  - ♦ *Cycle graph* on n nodes has adjacency spectrum $\{2 \cos(2\pi j/n)$ for $j = 1, 2, \ldots n-1\}$. Its Laplacian spectrum is $\{2 - 2 \cos(2\pi j/n))$ for $j = 1, 2, \ldots n\}$.
  - ♦ *Petersen graph* (5-pointed star within pentangle; Figure 2.7):
    Adjacency spectrum is $\{-2^4, 1^5, 3\}$.
    Laplacian spectrum is $\{0, 2^5, 5^4\}$.

  The number of spanning trees (based on the product of the $n-1$ largest Laplacian eigenvalues) for this graph is $2^5 \times 5^4/10 = 2000$ (Aldous & Wilson 2000).
  - ♦ *Sun* (or *sunlet*) *graph* (a cycle of n nodes, each with a leaf node attached). Bipartite for $n = 4$, but not for all n. The $n = 3$ sun graph is not bipartite.

The fact that two graphs have the same spectrum does not guarantee that the graphs are identical (isomorphic); Figure 2.7 shows two different graphs of five nodes with the

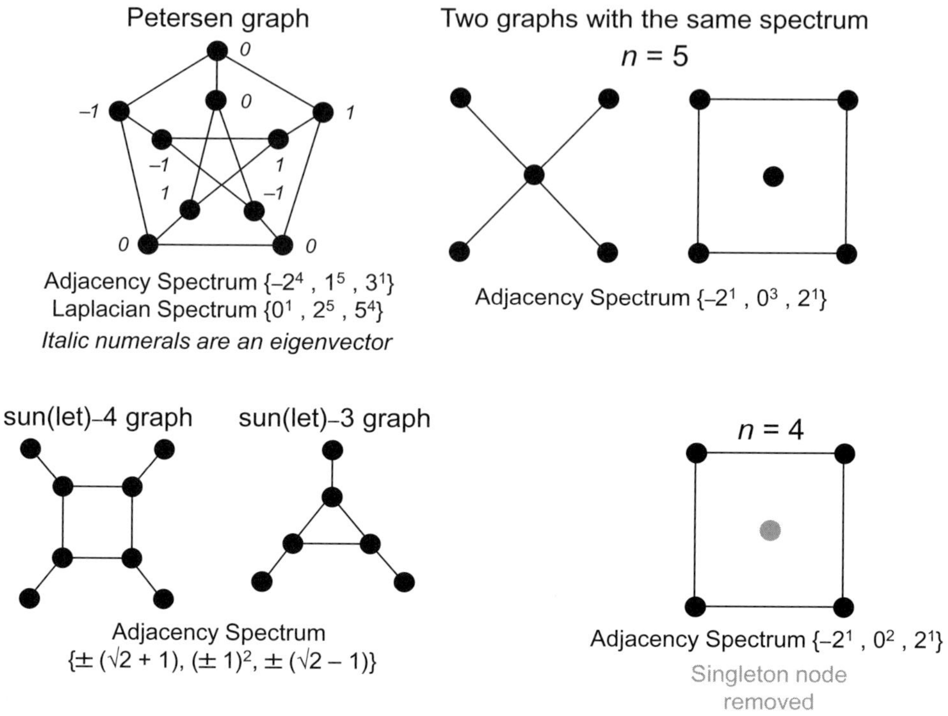

Figure 2.7 Spectral graph examples: Petersen graph, two graphs with the same spectrum, sun graphs, the effect of the removal of the central singleton node.

same adjacency spectrum. Notice that when the isolated node disappears from the "square-plus-singleton" graph, the adjacency spectrum changes only by losing a 0 (the 4-node graph is included in the figure for comparison). We have included the sun graph for $n = 4$, as an aside, because it looks somehow like an amalgamation of the square (4-cycle) graph with the star, but the spectrum is not clearly related to the spectra of the apparent components: $\{(\sqrt{2} + 1), 1^2, (\sqrt{2} - 1), -(\sqrt{2} - 1), (-1)^2, -(\sqrt{2} + 1)\}$. Similarly, the $n = 3$ sun looks to be based on the familiar triangle (3-cycle) graph which has spectrum $\{(-1)^2, 2\}$ compounded by three versions of graphlet 0 with spectrum $\{(-1)^2\}$. It looks equally simple to many other small graphs, but its spectrum unexpectedly complicated: $\{(1 + \sqrt{2}), (\frac{1}{2}(1 + \sqrt{5}))^2, (1 - \sqrt{2}), (-\frac{1}{2}(1 + \sqrt{5}))^2\}$.

2.3.2.1 Spectral Graph Theory and Ecological Applications

Why is this material important for the analysis of ecological networks? Spectral graph theory provides a route to evaluating or clarifying several important characteristics of ecological network graphs. In addition to specific data structure clarification (see Lee et al. 2010 for a genetic analogy), these include

- Objective portrayal of structure (spectral drawing)
- Measures of connectedness and components

- Related to measures of network complexity (see Section 2.4.1)
- Measures of structural resilience and vulnerability
- Objective partitioning
- Measure of network similarity
- Measure of bipartivity
- ... and there are probably many more, some of which have yet to be developed

The important features of this spectral graph theory are that they are direct and based on the matrices that must be created for any kind of analysis; in addition, they can be used for the very large and complex networks that are to be expected in ecological research. Although we have described the approach initially for simple graphs, they can be adapted for graphs with weights, edge directions, or edge signs. We will now examine the last of these cases and focus on the analysis of signed network graphs.

2.3.2.2 Spectral Graph Theory: Why Does It Work?

Spectral graph theory may seem like foreign territory to many ecologists, even those who may be comfortable with eigen-analysis techniques such as Principal Components Analysis (PCA; Legendre & Legendre 2012) and with graph theory conceptual approaches like random walks, at the intersection of which spectral graph theory lies. Part of the story is that a graph creates its own "space" and the edges, or whatever matrix is used to code the edges, define the structure within that multidimensional space. We also need to think of graphs in terms of processes and operators as well as merely static structures. "If we have an operator that is naturally associated with graph G, then the properties of that operator, and therefore of the graph, will be revealed by its eigenvalues and eigenvectors" (Spielman 2012). This is the adjacency operator realised by the adjacency matrix, such that a real vector $\mathbf{x}$ of length n is viewed as a function from the graph nodes to the real numbers. The *adjacency operator* (Godsil & Royle 2001; Spielman 2012) acts on such a function to produce the value at node a that is the sum of the values at the nodes that are its neighbours:

$$(A_G f)(a) = \sum_{b|A(a,b)=1} f(b). \tag{2.22}$$

If f is an eigenvector with eigenvalue θ, $\mathbf{A}f = \theta f$ and $\theta f\ (a) = \Sigma\ \theta f\ (b)$ over all neighbours of a. For example, the italic numerals on the Peterson graph in Figure 2.7 give an eigenvector with eigenvalue 1; each node has a label that is the sum of its neighbour labels (Godsil & Royle 2001).

The adjacency operator in question is most easily conceived of in terms of a process like diffusion; the adjacency matrix shows which neighbours of a given node can receive diffusion from that source. The process of diffusion is obviously closely related in this context to random walks on the graph itself. Evidence of that close relationship is the fact that the elements of the kth power of the adjacency matrix, $\mathbf{A}^k$, $a_{ij}^{(k)}$ gives the number of walks of length k between nodes i and j. The

Laplacian operator used in mathematical physics is also related to diffusion or to the flow of heat or fluid, and the name of the Laplacian matrix is no coincidence. The Laplacian matrix is also used for models of networks of springs or of the flow of current through a network of resistors (Spielman 2012). Studying a graph structure by spectral methods is analogous to studying some potential flow through the edges of the network.

As ecologists, any familiarity we might have with eigen-analysis comes from what we learned about linear transformations in linear algebra or from multivariate analysis methods for ordination such as PCA. As with PCA, one aspect of the spectral graph methods is dimension reduction, here referred to as *embedding*, either for unweighted or weighted network graphs (Bonald et al. 2018). The aim is to reduce the data structure from n dimensions to something fewer, while maintaining, as best as possible, the "distances" between nodes, especially those that are near each other. For example, based on the adjacency eigen-analysis, the structure of the 9-node example in Figure 2.6b, could be summarised in one dimension as: two clusters of four nodes with one node halfway between them. Laplacian eigenvectors with low eigenvalues have similar values for adjacent nodes, and so this suggests or allows using Laplace eigenvectors as coordinates for the nodes, since adjacent nodes will be kept close together (Bonald et al. 2018). This means that we are using structure as a surrogate for a process like diffusion and that we are using a surrogate process like diffusion to understand the structure. In addition, deriving coordinates for nodes based on structure can be seen as a complement to abstracting structure from locational data. In determining neighbour networks in spatial graphs, the locations of the nodes provide coordinates and then their proximities to define adjacencies based on some rule like nearest neighbour; in spectral graph drawing, we use the adjacencies to define the nodes' proximities and those determine the nodes' coordinates in eigenspace of reduced dimension (see Figure 2.6b).

2.3.3 Analysis of Signed Networks

Signed graphs arise in a number of circumstances, one obvious case being social networks, which distinguish both likes (positive) and dislikes (negative) from neutral relations (zeros). The second is in correlation networks, which record both the positive and the negative correlations among variables. In either case, any pair of nodes in the network graph can have a positive edge, a negative edge, or no edge at all. A third source of graphs with signed edges is any functional interaction network, such as a food web, although the edges tend to be directed or asymmetric, as well as signed.

A key concept for signed social networks is that of balance, but it is not the only one! A graph is balanced when all cycles have an even number of negative edges (we will expand on this brief description later in this section). The analysis of signed graphs requires the modification or adjustment of the methods for unsigned graphs, and we will describe some of these, starting with spectral methods for signed graphs (following Kunegis et al. 2010; see also Kunegis 2014).

The adjustment to the Laplacian matrix starts with the signed adjacency matrix, $\mathbf{A}_{\pm}$, and defines the "signed degree" of node i as its total number of edges whether positive or negative, more logically called the "absolute degree":

$$d_s(i) = \sum \mid a_{ij} \mid . \tag{2.23}$$

The modified degree matrix is then a diagonal matrix of these degrees, $\mathbf{D}_s$, and the "signed Laplacian" matrix is then $\mathbf{L}_s = \mathbf{D}_s - \mathbf{A}_{\pm}$. The naming of the matrices can be very confusing, because there is also, in the literature, something called the "signless Laplacian," where the standard Laplacian is $\mathbf{L} = \mathbf{D} - \mathbf{A}$ and the signless Laplacian is $\mathbf{Q} = \mathbf{D} + \mathbf{A}$, so that all the entries are positive (Brouwer & Haemers 2012). We will not discuss this variant further.

If the graph is not balanced and the smallest eigenvalues are greater than 0, then the eigenvectors of the k smallest eigenvalues, for example, λ_1 and λ_2, used as coordinate axes give an orthogonal basis for a graph drawing of "minimum energy" (approximately least distortion or least information loss; see Kunegis et al. 2010). Figure 2.8a gives an example in which the cluster structure is clearly portrayed with negative edges between the five clusters. Using λ_1 as a measure of balance is described in more detail later in this section.

If the graph is balanced and the smallest eigenvalue is $\lambda_1 = 0$, the next k smallest eigenvalues provide a minimum energy drawing, for example, λ_2 and λ_3.

If G is balanced, a graph drawing of a bipartite graph can be obtained using the eigenvectors of the eigenvalues $\lambda_1 = 0$ and λ_2, whatever its value. The resulting diagram has the nodes in each cluster aligned vertically, thus losing any intra-cluster details. A combined approach that relies on a linear combination of eigenvalues and their eigenvectors provides a compromise (Kunegis et al. 2010).

The concept of balance in signed graphs originated in the field of sociology (Harary 1953, 1959), based on questions such as "Are my enemies' friends also my enemies?" In undirected signed graphs, for any three nodes with three edges, the triangle's balance depends on the number of negative edges; the triplet is balanced if that number is 0 or 2. For more than three nodes, the graph is balanced if all cycles have an even number of negative edges. This occurs if all edges are positive or if the nodes can be partitioned into two clusters with only positive edges within and only negative between. Usually, the pattern is imperfect, and we look for a measure of partial balance (Aref & Wilson 2017), often based on the proportion of cycles that are positive, having an even number of negative edges. (The graph is completely balanced if all cycles are positive). Other measures of balance would be the number or proportion of edges that must be deleted or switched in sign to achieve perfect balance (see Wasserman & Faust 1994); this is a kind of "edit" measure that counts the number of changes required for a specific outcome. Large networks from real-world examples are almost never balanced (Kunegis 2014) and random signed graphs are almost certainly only weakly balanced (El Maftouhi et al. 2015). Aref & Wilson (2017) provided a summary of the measures available and some advice. Here is an extract of their summary, beginning with some of the measures:

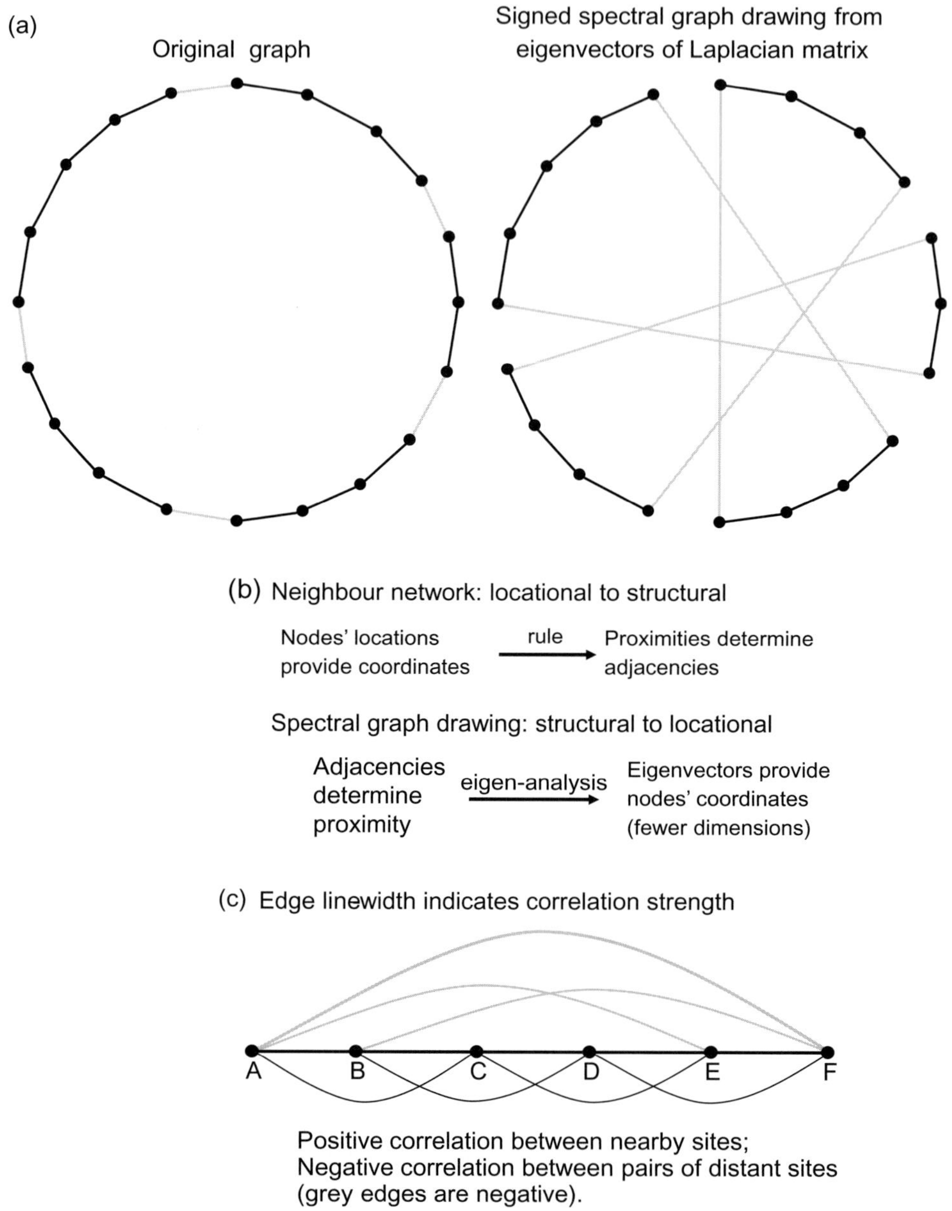

Figure 2.8 Spectral graph drawing. (a) Spectral redrawing of a signed graph: negative edges are elongated to show more clearly the +/− structure. (b) Relationship between locational and structural information: neighbour networks and spectral graph drawings. (c) Communities at sites on a strong environmental gradient: strong positive correlation between adjacent sites; negative correlation between pairs of distant sites (grey edges are negative). The relationships are consistent, but the graph is not balanced.

1. $D(G)$ is the proportion of positive cycles in graph G, based on O_k^+ positive cycles of length k and O_k in total:

$$D(G) = \frac{\sum_{k=3}^{n} O_k^+}{\sum_{k=3}^{n} O_k}. \tag{2.24}$$

2. $D_k(G)$ restricts the measure to cycles of a certain length, for example, for 3, it is $D_3(G) = O_3^+ / O_3$. ($D_3(G)$ is also called $T(G)$ from "triangle").
3. The smallest eigenvalue of the Laplacian is $\lambda_1(G)$, called the "algebraic conflict." It is 0 for a balanced graph. Where $S_{\max}$ is the maximum average degree total for the nodes joined by any edge in G, then the algebraic conflict is normalised as

$$A(G) = 1 - \frac{\lambda_1(G)}{S_{\max} - 1}. \tag{2.25}$$

4. A *deletion set* is a set of edges that are all removed, and no others, in a single act before the outcome of their removal is evaluated. Consider the deletion set of edges that produce a balanced graph; call it set E^*. A minimal deletion set has no proper subset that can accomplish the same outcome. The "frustration index," $L(G)$, is the size of the minimal deletion set:

$$L(G) = \min_{E*} \mid E* \mid. \tag{2.26}$$

   This is normalised and converted to a positive balance measure as

$$F(G) = 1 - \frac{L(G)}{m/2}. \tag{2.27}$$

Aref and Wilson (2017) have evaluated all these measures, and others, using Erdös–Rényi model graphs and specially structured random graphs, as well as some well-known data sets from sociology and biology. They found substantial differences among the results from cycle-based, eigenvalue-based, and frustration-based measures. Overall, the advice is that $F(G)$ is the best measure, but $L(G)$ is difficult to calculate (perhaps impractical) for large networks. The best alternatives are $A(G)$ and $D_k(G)$. Other measures, but including $D(G)$ and some not detailed here, are not recommended as indices of partial balance in ecological networks.

For ecologists, the question is not just whether the network graph is balanced, but whether there are forces that will increase or decrease balance over time, and whether there are predictable differences in balance as a function of scale (e.g. triplets of nodes vs. the whole graph). For example, Ilany et al. (2013) have found increasing balance in the social organisation of wild rock hyrax (*Procavia capensis*) populations. As to scale, very large human social networks tend to be nearly balanced, in part because most edges are positive, but also because negative edges cluster at very few nodes (Facchetti et al. 2011). These "unpopular" nodes affect the whole network but little, and it will be interesting to see how commonly this structural feature occurs in non-human networks of social structure or its equivalent.

Kunegis (2014) has described how the signed Laplacian can determine the best two clusters in a signed graph. If the graph is completely balanced, the smallest eigenvalue is 0, and the eigenvalues of +1 or −1 give the two internally positive clusters separated by negative edges. If the graph is not completely balanced, the signed cut approach minimises the proportion of positive edges between the two clusters while it maximises the proportion of positive edges within them (see Kunegis 2014 for the details). The message for ecologists is that there are direct techniques to detect the "best" clusters in signed graphs, even when they are not balanced. This will be especially useful when the number of clusters that will make the most sense for the circumstances is more than two and in working with graphs based on correlation networks which are not expected to be balanced. Kunegis (2014) has supplied an anthropological example with three higher order clusters of tribal groups in New Guinea, clustered using the eigenvectors of the Laplacian matrix (Kunegis 2014), and Gallier (2017) has provided details on *K*-way clustering of signed graphs (Kunegis 2014).

Having focused mainly on social networks in which balance can be important, we should emphasise that another major source of signed networks in ecology is correlation networks, in which balance is not a property of interest. More detailed discussion of correlation networks follows later but consider an ecological example: the plant communities on a strong monotonic environmental gradient. Adjacent or nearby sites are positively correlated but more distant pairs are negatively correlated (Figure 2.8c). This produces signed graphs that are not expected to be balanced, but the relative positions of the edges of different signs are consistent and correctly reflect the underlying structure. While "my enemy's enemy" may be my friend, the same logic does not apply to dissimilarity: when *A* and *B* are dissimilar and *B* and *C* are dissimilar, it does not follow that *A* and *C* are similar. By the same argument, while my friend's friend may tend to be my friend, the similarity of *A* and *B* and of *B* and *C* does not guarantee that the similarity extends to *A* and C, and less so to *D* or *E* (Figure 2.8c).

The third set of ecological networks with signed edges comes from interaction networks or subnetworks, but those edges tend to have directions as well as signs and so the concept of balance does not have the same role in evaluating their stability or dynamics. More detailed discussion of these signed networks (usually signed digraphs) occurs further on but consider the case of predation. Any two nodes that represent a predator species and a prey species may be joined by two edges: a positive edge from prey to predator showing the positive effect of prey on predator populations and a negative edge from predator to prey showing the negative effect of predator on prey. (We note that some conventions use a single edge for predation that is designated as being positive in one direction and negative in the other.) This creates a two-node two-edge cycle, which must be negative because it has an odd number of negative edges (one). Therefore, this subnetwork is not balanced. In addition, any larger cycles of which these two nodes become part can never be balanced, whatever kinds of interactions are now included in the network, because the two edges of opposite signs between predator and prey, will create both an odd and an even number of negative edges in the larger cycle. Therefore: not balanced. We will discuss this set of interaction networks further, but this consideration demonstrates why balance is not a key characteristic in interaction networks as it is in social networks.

Interaction networks are not usually symmetric and so the matrices that encode them are also asymmetric in the formal sense. The result of the asymmetry is that their eigenvalues tend to be complex (with real and imaginary parts) rather than real-valued, although they occur in conjugate pairs because the matrices themselves are real-valued. This means that although the spectral graph theory methods described earlier can still be useful, some adjustment is required for application and interpretation (see Brualdi 2010; Li et al. 2016).

It is clear that signed networks have much more potential for useful application in ecological studies than we have actually seen so far. The analysis of these networks can follow the methods of unsigned networks for many characteristics, but there are others for which the methods must be modified for signed edges or for which the methods are distinct. The analysis of signed networks is one area in which we are confident there are great new methods to be developed and exciting discoveries to be made.

2.3.4 Information and Entropy

Obviously, the graph of a network contains a great deal of information in the informal sense, from the identities of the structural units that are its nodes, from the pairs of entities joined by its edges (their interactions), and by its structure or topology. Here, *information* refers to the reduction of uncertainty. For example, designating a single element from a finite set can be seen as providing information proportional to the log of the number of elements in the set. The information in a graph can come from the distribution of its units (nodes or edges or subgraphs) into categories and their proportions; this gives *compositional* or *category information*. Information also comes from the structural characteristics of the graph; this is *topological information*. The two may be confounded if the categories are structure based.

Measures of information are often based on Shannon's entropy, related to the uncertainty of outcomes, but there are many ways to designate the information content of a network. In physics and thermodynamics, the entropy of a system is a measure of its disorderliness and it increases through time. In information theory, high entropy and high information content go together (Legendre & Legendre 2012). When the network's nodes belong to several categories of any kind, call them S_1 to S_Z, the *category information* of the nodes is like the familiar measure of species diversity and is contained in the proportions of the nodes in the Z categories, the p_i:

$$I_C(G) = -\sum_{i=1}^{Z} p_i \log_2 p_i. \tag{2.28}$$

This index reaches a maximum value when all the categories are equal. The categories can represent almost anything, including a graph theoretical characteristic of the nodes, such as degree (see later in this section), and this kind of information precedes knowledge of which nodes are attached to which. In particular, there has been much

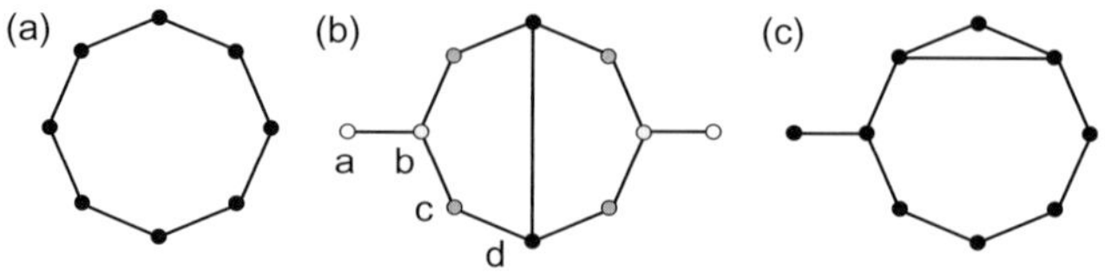

Figure 2.9 Graph symmetries and automorphism orbits. (a) The eight nodes are indistinguishable; all are in a single automorphism orbit. The graph has many symmetries. (b) The ten nodes occur in four distinct topological positions: two in each of *a*, *b*, and *d*, and four in *c*: the four orbits are indicated by shading. Nodes of the same shade can be mapped onto each other by automorphisms. (c) The eight nodes are all distinct, each in their own automorphism orbit. The graph has no symmetries. (A black and white version of this figure will appear in some formats. For the colour version, please refer to the plate section.)

interest in putting the network nodes into classes of structural equivalence: the *automorphism orbits* of the graph. Note that these are the automorphism orbits *of the whole graph*, not the automorphism orbits of the graphlets (small subgraphs) described earlier. Figure 2.9 illustrates the concept. In the simple ring graph on the left, the eight nodes are indistinguishable and therefore all are in a single automorphism orbit. The graph has many symmetries, both by rotation and by reflection. In the middle graph, ten nodes occur in four distinct topological positions: two in each of *a*, *b*, and *d*, and four in *c*. The four orbits are indicated by shading, nodes of the same colour can be mapped onto each other by an automorphism. There are fewer symmetries, but there are some, for example, rotation by 180° (note the nodes' degrees). In the third graph on the right, the eight nodes are all distinct, each in their own automorphism orbit. The graph has no symmetries.

The information associated with classifying nodes into automorphism orbits is the structural or topological information, but we will call it *graph orbit information*, I_o, because there are many possible kinds of information in a graph related to its topology. In fact, another suggestion for the information contained in a graph is based on the nodes' degrees, call it *node degree information*:

$$I_D(G) = \sum_{i=1}^{n} a_i \log_2 a_i. \tag{2.29}$$

Here, a_i is the row or column sum of the adjacency matrix, and so it is greater than 1, and there is no negative for the sum that produces the index. It is the amount of degree entropy by which a graph falls short of the maximum possible for the given total adjacency (degree sum or trace of the adjacency matrix), A, which is $A \log_2 A$ (Bonchev & Buck 2007).

Figure 2.10 illustrates the determination of orbits in small graphs, and how the directions of edges in digraphs affect the distinguishability of nodes and therefore the count of orbits. Figure 2.11 illustrates the conversion of node and orbit counts to measures of topological information for graphs with increasing numbers of orbits. The values for the node degree information are also included for comparison. It is obvious

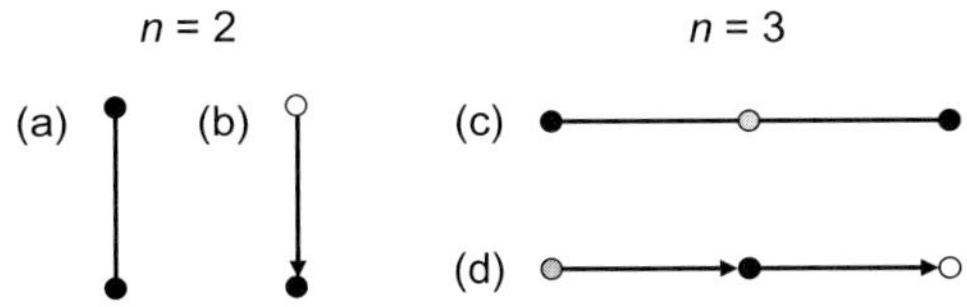

Figure 2.10 Information in graphs and digraphs. The information from orbit proportions is $I = -\Sigma p_i \log_2 p_i = \Sigma p_i \log_2 1/p_i$. (a) In the undirected graph with $n = 2$: two nodes in one orbit, $I = (2/2 \log_2 2/2) = 0$. (b) In the directed graph the digraph has two nodes in two orbits: $I = 2\ (1/2 \log_2 2/1) = \log_2 2 = 1$ bit. (c) Undirected graph with $n = 3$: two orbits with two nodes and one node $I = (1/3 \log_2 3/1) + (2/3 \log_2 3/2) = 0.92$. (d) Directed graph with $n = 3$: three orbits with one node each, $I = 3\ (1/3 \log_2 3/1) = 1.59$.

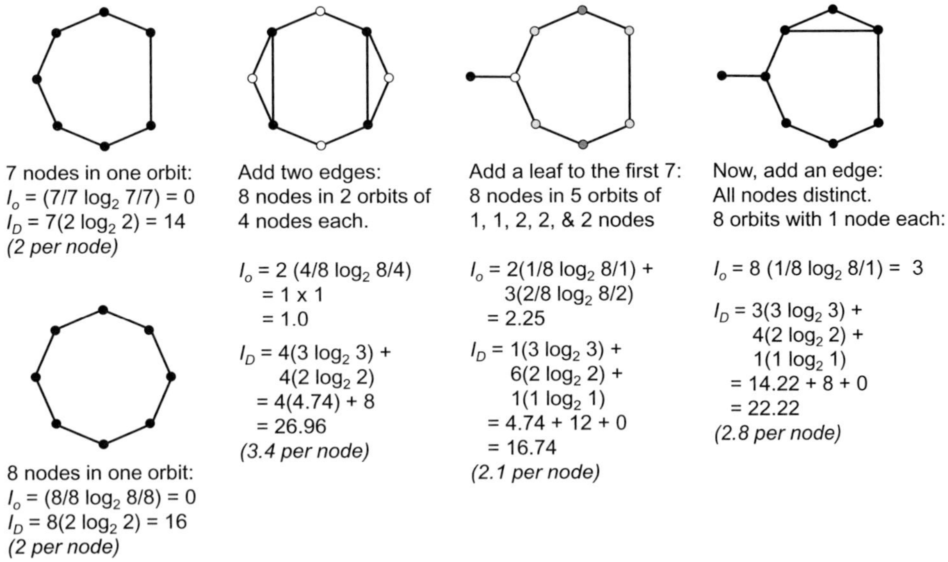

Figure 2.11 Information in graphs, a series illustrating I_o, the structural information measure based on orbits, and I_D, the structural information measure based on node degrees.

that the two measures are responding to different characteristics because they can change in different directions: one increasing when the other decreases.

Many networks in ecology and in other areas of science are not fully connected but consist of a number of components of various sizes, from very large (the "giant component") to isolated pairs of nodes with a single edge ("toothpick") and single unattached nodes ("singletons"); see Figure 1.9. When a network consists of a number of connected components, an important question is whether the information in the whole graph is the sum of information from all its components. The toothpicks and singletons add no orbit topological information and the leaf nodes in a single fan are all in the same orbit, but such networks are expected to have a high proportion of nodes that are topologically distinct and in their own orbits. Thus, they have high

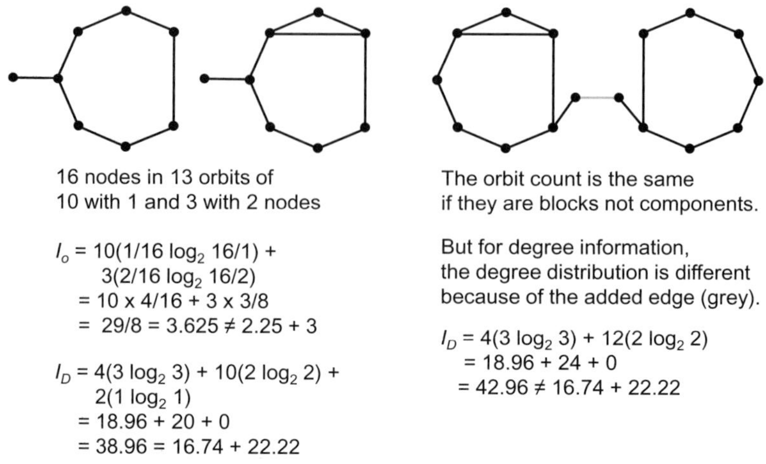

Figure 2.12 Information in components and blocks using I_o, the structural information measure based on orbits, and I_D, the structural information measure based on node degrees.

graph orbit information content. However, a few examples can illustrate another important difference between the orbit-based information measure and that based on the node degrees. Figure 2.12 shows that while the node degree information is additive over components, the graph orbit information is not. In fact, the high information content of one component can be "diluted" by the inclusion of another component with low information, so that the total is less than the sum of the parts. This has clear implications for real networks with many small components which may actually reduce the orbit information in the entire network. Figure 2.12, right panel, illustrates the effect of rejoining two separate components: orbit information is not affected, but (of course) the addition of an edge changes the node degree information. Lastly, having duplicate components has no effect on the orbit-based measure (because the orbit proportions are unchanged) and merely doubles the degree-based measure, summing the contributions of the two identical components.

2.3.4.1 Graphlet Information Measures

As indicated in the previous section, graphlets can be used in network analysis in a number of very helpful ways. They can also be used as the basis for an evaluation of the information contained or conveyed. We will describe two approaches: the first based on frequency counts of the graphlets themselves (*graphlet frequency information*) and the second based on the counts of nodes in all possible graphlet automorphism orbits (*graphlet orbit distribution information*). In either version the counts (of whole graphlets or of nodes in orbits) are converted to proportions and then evaluated by the Shannon diversity measure to give the average information content:

$$H_\gamma(G) = -\sum_{i=1}^{\max} p_i \log_2 p_i \text{ (subscript } \gamma \text{ can be } GF \text{ or } GO). \quad (2.30)$$

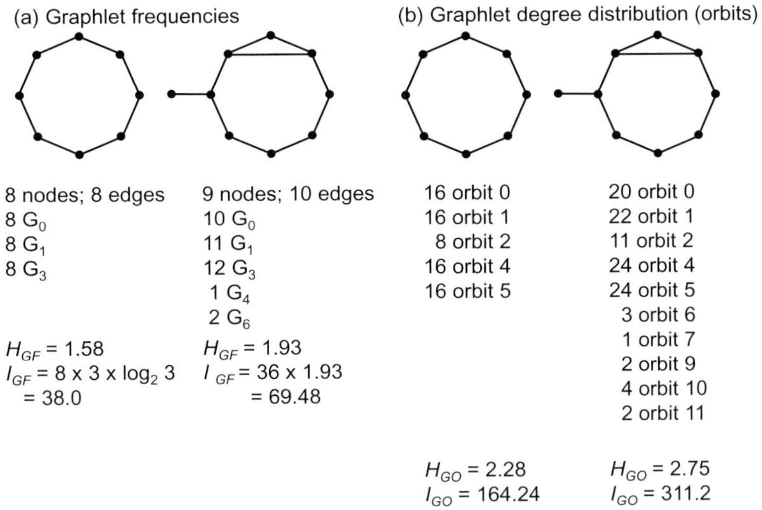

Figure 2.13 Information based on graphlets: examples showing the measures based on whole (a) graphlet frequencies (*GF*) and on the orbit (b) frequency distributions (*GO*).

The total information from the graphlet frequencies or the graphlet orbit distribution is that value multiplied by the number of graphlets or the number of graphlet orbits that contribute to the measure. Figure 2.13a compares the results for a simple ring graph of eight nodes with the same graph augmented by a leaf node and a short-cut edge, which cause all nodes to be distinct. The basic ring graph has $H_{GF} = 1.58$ giving a total information content of 38; compared to 1.93 for the augmented graph with a total of 69. The values for simple graphs are illustrated in Figure 2.14. (A ring graph of six nodes has the same information per node and a smaller total, but a ring graph of four nodes is quite different because it is small enough to be a graphlet itself: graphlet 4 with graphlet orbit 8). Using graphlet orbits as the basis for the measure gives values of $H_{GO} = 2.28$ and 2.75 for the same two examples, with totals of 164 and 311. While this approach requires further investigation, it seems that the graphlet orbit measure provides a good reflection of the relative complexity of the structures being considered, based on an intuitive assessment.

2.3.4.2 Search Information and Related Entropies

This next set of methods and measures focuses on the flow through a network, particularly the flow of information. Where p_{ab} is the shortest path from node a to node b, the probability that the path is followed in a random walk is

$$P(p_{ab}) = \frac{1}{a_i} \prod_{j \in p_{ab}} \frac{1}{a_j - 1}. \tag{2.31}$$

The *search information* (Costa et al. 2007) is the information required to identify one of the shortest paths between a and b, and it is then

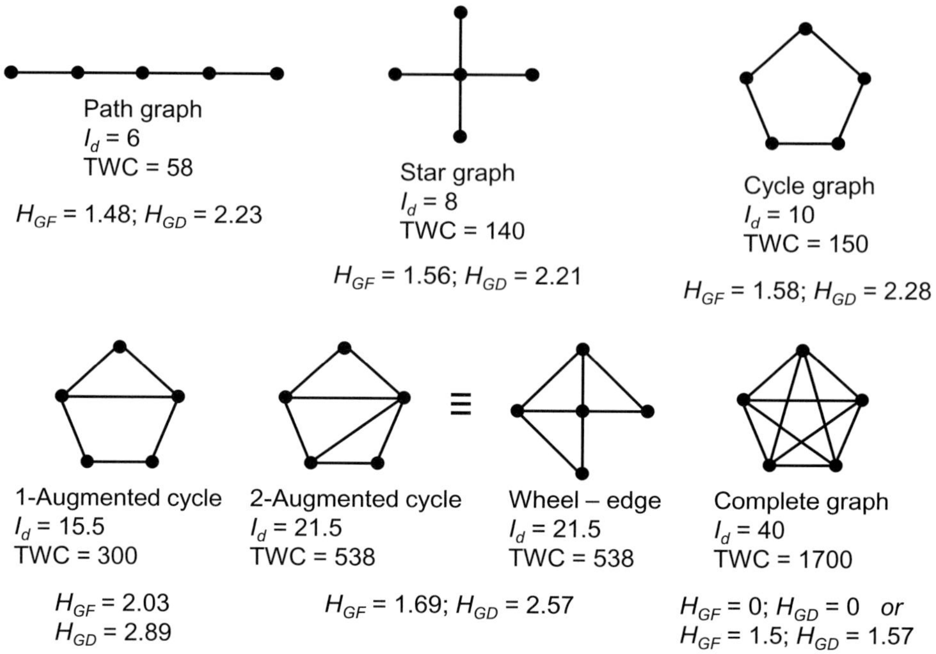

Figure 2.14 Simple graphs and their complexity measures based on whole graphlet frequencies (*GF*) and on the orbit frequency distributions (*GO*). TWC is the total walk count, described in Section 2.4.1.

$$S_{ab} = -\log_2 \sum_{\{p_{ab}\}} P(p_{ab}). \tag{2.32}$$

Continuing to think about paths from node a to node b, the *average search information* characterises the ease of navigation through the network, and it is

$$S = \frac{1}{n^2} \sum_{ab} S_{ab}. \tag{2.33}$$

The difficulty of locating other nodes from node a is the *access information:*

$$A_a = \frac{1}{n} \sum_{b} S_{ab}. \tag{2.34}$$

The difficulty of finding node b from any other node in the network is its *hide information:*

$$H_b = \frac{1}{n} \sum_{a} S_{ab}. \tag{2.35}$$

What is called *entropy* in relation to information, is a measure of how unexpected an outcome is, or the average uncertainty associated with data. In communication networks, the emphasis is often on the flow of information or messages through the available paths and the obvious parallel in trophic networks is the flow of energy. On the dubious assumption that only the shortest paths are used and that all these paths carry the "messages" at the same rate, two entropies can be defined based on the fraction of messages targeted to node k that go through node h, call it c_{kh}, and on the fraction of messages passing through node k that originated at node h, call it b_{kh}. These are (see Costa et al. 2007) the *target entropy* of node k:

$$T_k = -\sum_k \sum_h a_{hk} c_{kh} \log 2c_{kh}. \tag{2.36}$$

and the *road entropy* of node k:

$$R_k = -\sum_k \sum_h a_{hk} b_{kh} \log {}_2 b_{kh}. \tag{2.37}$$

Two general measures of the flow through the network are then the averages of these entropies over all nodes, $T.$ and $R.$; low values of $T.$ are associated with star structures and low values of $R.$ are associated hubs connected in a string (see Costa et al. 2007, and references therein). It would be worth exploring the properties of these entropies for ecological networks, but the assumption of the primacy of shortest paths is in sharp contrast to the growing emphasis for the analysis of ecological networks by considering the sum of all paths (see McRae et al. 2008).

2.3.4.3 Information Theory and Ecological Applications

Why is this material important for the analysis of ecological networks? While graphlets and motifs focus on the local structure of networks, we can argue that measures of information or entropy respond to structural features of the whole network graph. For applications to ecological networks, these measures tend to respond with characteristics that include

- Holistic response to features of the whole network graph.
- Sensitive to structural diversity.
- Depending on measure, especially affected by symmetries.
- Strongly related to the actual and potential dynamics of the network.
- Each provides one approach to measuring network complexity.

Discussions of networks and their statistical analysis in general and descriptions of ecological networks in particular often use the descriptor "complex" (e.g. the title of Estrada's 2012 book is *The Structure of Complex Networks*); just do a quick search of the bibliography for references to "complex networks"! This common description of networks must mean something more than just "large"; for example, Estrada (2012) explicitly contrasted "complex" with "regular" in the discussion of eigenvalue spectra. That seems to suggest that measuring the complexity of networks is an important endeavour, and many complexity measures will include concepts from information theory and entropy. It also brings us to questions such as "If not mere size, what is it that makes a network *complex*?" and "What factors determine network complexity?."

Complexity is therefore the first of the network properties we will discuss because knowing the complexity of the network you are studying is crucial for understanding the choices for analysis and for interpreting results.

2.4 Properties Evaluated

2.4.1 Complexity

Complexity is itself a complex topic and, yes, a complicated property of systems, networks, and the graphs that depict them, and the term takes on a variety of meanings based on the kind of network, for example, chemical versus biological versus ecological (see Bonchev & Rouvray 2007). The complexity of a network, which is made up of elements or units (the nodes) and the interactions between them (the edges), is related to the amount of information needed to describe the system, and that is often related, also, to the number of nodes and the number of different paths between them. The complexity of a network that is a functioning system can also be defined by the amount of information required to make good predictions about its future structural state or dynamics (see Shalizi 2006). In addition, characteristics related to the complexity of such a network may also include emergent properties and self-organisation.

Now it is usual to distinguish between structural complexity, which is a property of the topology of the graph, and compositional complexity determined by the different identities (e.g. atoms, molecules, or species) of the nodes of the network. Some early thoughts on network complexity measures were based on the idea that the complexity of an empty graph of n nodes and no edges should be the same as the complexity of a complete graph of n nodes and all $n(n-1)/2$ edges. The reasoning was that in both graphs all the nodes were indistinguishable (i.e. all in a single automorphism orbit). For most of the networks we may consider in ecological studies, this suggestion makes no sense because the interactions that are the edges are not all the same, and there is so much more that can be occurring in a graph of many edges than in a network of none (see Bonchev & Buck 2007).

There are many measures of network structural complexity, and these have been reviewed by several authors (Bonchev & Buck 2007; Mowshowitz & Dehmer 2012; Kamisinski et al. 2015). These efforts have been hampered by inconsistencies in the naming of the measures and by a lack of clear and consistent notation. For example, the terms "graph vertex complexity" is used by different authors to refer to different defined measures (Kamisinski et al. 2015; the same is true of "topological complexity"), and the many acronyms can be ambiguous (e.g. *OC* could be "overall complexity," "orbit complexity," or even "off-diagonal complexity"). We will not give a complete review of all available measures of network complexity (there are so many!), but will concentrate on four that are recommended for general use; these are based on adjacencies and distances, on entropy measures, and on paths or walks. We will also include several others based on different approaches: on eigenvalues, on subgraphs and on automorphism orbits. We will resist the temptation to rename these measures in a more consistent and logical scheme.

To review some of the notation, G is a graph of n nodes and m edges, and $\mathbf{A}$ is an adjacency matrix with elements a_{ij}. Its row or column totals are a_i which give the degrees of the nodes. $\mathbf{D}$ is the matrix of internodes distances, d_{ij}, with row and column totals $d_{i.}$; $e(v_i)$ is the eccentricity of the node, which is its maximum distance to any other node.

The first complexity measure is the "Vertex Degree Information" measure:

$$I_d(G) = \sum_{i=1}^{n} a_i \log_2 a_i. \tag{2.38}$$

The second is "Complexity B":

$$B(G) = \sum_{i=1}^{n} \frac{a_i}{d_i}. \tag{2.39}$$

The responsiveness of any of these measures can be evaluated by examining their values for familiar graphs of known structure. For example, B is n for a complete graph of n nodes (the graph K_n); about 3/2 for a star graph of any size; and about $8/n$ for a 1-cycle or path graph of n nodes (Bonchev & Buck 2007).

These first two measures are rather simple in their formulation; the next is not. It is the "Graph Distance Complexity." With $n_k(v_i)$ being the number of nodes within distance k of node v_i, the weighted sum for those nodes is

$$R(v_i) = \sum_{i=1}^{e(v_i)} k \times n_k(v_i). \text{ Then, } R(G) = \sum R(v_i). \tag{2.40}$$

The Vertex Distance Complexity is

$$V_d(v_i) = \sum_{i=1}^{e(v_i)} n_k(v_i) \frac{k}{R(v_i)} \log_2 \frac{k}{R(v_i)}. \tag{2.41}$$

The Graph Distance Complexity is

$$C_D(G) = \sum_{i=1}^{n} \frac{R(v_i)}{R(G)} V_d(v_i). \tag{2.42}$$

The last of the four measures is the "Total Walk Count," which is simply, as the name suggests, a count of all walks in the graph of length k from 1 to K. The computation might be daunting, but for simple graphs the number of walks of length k between any two nodes is the corresponding element in $\mathbf{A}^k$, such that

$$C_{TWC}(G) = \sum_{k=1}^{K} \sum_{i=1}^{n} \sum_{j=1}^{n} \left(A^k\right)_{ij}. \tag{2.43}$$

For example,

Matrix 2.1 Adjacency matrix $\mathbf{A}_1$

$$\begin{bmatrix} \mathbf{0} & 1 & 1 & 0 & 1 & 0 \\ 1 & \mathbf{0} & 1 & 1 & 0 & 0 \\ 1 & 1 & \mathbf{0} & 0 & 1 & 0 \\ 0 & 1 & 0 & \mathbf{0} & 0 & 0 \\ 1 & 0 & 1 & 0 & \mathbf{0} & 0 \\ 0 & 0 & 0 & 0 & 0 & \mathbf{0} \end{bmatrix}$$

Matrix 2.2 Self-product of adjacency matrix $\mathbf{A}_1$: $\mathbf{A}_1{}^2$

$$\begin{bmatrix} \mathbf{3} & 1 & 2 & 1 & 1 & 0 \\ 1 & \mathbf{3} & 1 & 0 & 2 & 0 \\ 2 & 1 & \mathbf{3} & 1 & 1 & 0 \\ 1 & 0 & 1 & \mathbf{1} & 0 & 0 \\ 1 & 2 & 1 & 0 & \mathbf{2} & 0 \\ 0 & 0 & 0 & 0 & 0 & \mathbf{0} \end{bmatrix}$$

For $k = 1$, the sum is 12, and for $k = 2$, the sum is 20, and so $C_{TWC}\,(G \mid K = 2) = 32$.

Graph Distance Complexity and Total Walk Count are the two most highly recommended of the four, according to Kamisinski's (2015) evaluations based on "different elementary, specifically designed, random and real network topologies."

Three more measures are worth mentioning. The first is "Topological Information Complexity," based on the proportion of nodes on each automorphism orbit of the entire graph, $p(o_z)$:

$$I_T(G) = -\sum_{z=1}^{Z} p(o_z) \log{}_2 p(o_z). \tag{2.44}$$

This is an interesting concept and sensitive to the details of network structure, but not very practical given the difficulty of determining orbits (and counting their nodes) in large networks. No polynomial time algorithm is currently available (Kamisinski et al. 2015; see also Mowshowitz & Mitsou 2009). It is easy to imagine circumstances in which large networks tend to have very many nodes that are distinct in their topological positions, resulting in a very large number of automorphism orbits.

The second is "Graph Index Complexity" derived from spectral graph theory and should be considered because spectral approaches have proved successful in other evaluations. This measure is based on the normalised value of the largest eigenvalue of **A**. It is

$$C_I(G) = 4c(1-c), \text{where } c = (\lambda - \lambda_{\min})/(\lambda_{\max} - \lambda_{\min}). \tag{2.45}$$

Normalising makes comparisons of different networks more straightforward but the index did not perform as well as others in the evaluation of Kamisinski et al. (2015).

Lastly, and without going into details, there are a number of complexity measures based on versions of enumerating subgraphs, for example based on counting all three-edge subgraphs (path, triangle, and star; see Bonchev & Buck 2007) or on spanning trees of the graph, perhaps with one or more edges deleted (see Kamisinski et al. 2015). There are obvious computational advantages for evaluating the complexity of very large networks by enumerating and characterising small subgraphs; the only requirement is to show that small-scale localised measures are able to capture and evaluate large-scale global characteristics.

A variant of this approach, described earlier, is to count the nodes that occur in each of the defined non-redundant graphlet orbits, creating the "graphlet orbit distribution" (Yaveroğlu et al. 2014). The standard diversity index of the values in that distribution can then be used as a measure of the network's structural complexity. This has advantages compared to other options based on automorphism orbits, because these need to be calculated only for small graphlets, e.g. only those with two, three, or four nodes. For example, the ring graph on the left of Figure 2.9 contains only graphlets G_0, G_1, and G_3, and so its only orbits are 0, 1, 2, 4, and 5. The augmented graph on the right of Figure 2.9 includes more graphlets than the ring graph (now adding G_2, G_4, and G_6), giving more orbits: 0, 1, 2, 3, 4, 5, 6, 7, 9, 10, and 11. As described earlier, the graphlet orbit distribution is a promising basis for measures of complexity. In essence, it is very similar to the Total Walk Count approach because it involves breaking the graph down into small multinode and multiedge pieces; the difference is that the graphlet measures are sensitive to the shapes of the pieces thus created.

Our recommendation for measuring network complexity is to accept Kamisinski's (2015) evaluations from a range of networks of different sources and topologies and to rely on Graph Distance Complexity and Total Walk Count. Specific interests may require other measures in addition to the first two, such as Vertex Degree Information Content and Complexity *B*, which are easy to calculate and performed well in test cases (Kamisinski et al. 2015). Graphlet counts or orbit distribution methods seem promising and are worth further investigation. These recommendations are for networks analysed as simple graphs, and so the evaluation of variants such as graphs with signed or directed edges or with loops or multiple adjacencies may require caution (see Kamisinski et al. 2015), although there are possibilities for graphlet methods based on the research of Sarajlić et al. (2016) and others.

A final comment on complexity is to repeat Shalizi's injunction (2006) that whatever measure of complexity we choose, it must be relevant to important aspects of the system we are studying. It needs to tell us something about both the structure and the functioning of the network of interest: its response to changes or disruption, its stability, or how easily controlled. Any measure of complexity, however excellent it may be in other situations, needs to be judged, and then chosen, based on its relevance to the system of interest.

2.4.2 Modularity

One property that can contribute to complexity is the uneven distribution of edges among nodes or among regions of the network. *Modularity* measures the non-homogeneous distribution of edges in regions of the network and can refer specifically to two different manifestations of non-uniformity. These are (a) the tendency of nodes to be joined more frequently to nodes of similar characteristics (assortativity) and (b) the existence of clusters of nodes with higher edge density within clusters and lower densities between them. In signed graphs, the distinguishing feature is the preponderance of positive edges within modules and negative edges between (Doreian & Mrvar 2009). Similarly, modularity measures should be modified for networks that are strictly bipartite (see Thébault 2013), particularly when the objective is to distinguish best-matching modules in the two parts of the graph. For an unsigned network of several partitions, one version of modularity is the sum over partitions of the difference between the observed proportion of edges within the partition and the number expected in a random graph with the same degree distribution (Estrada 2012). If the nodes belong to categories like those determined by species or gender, modularity can also emphasise the frequency of edges joining "like pairs" of nodes, as opposed to "unlike pairs" (Newman 2010). This emphasis on more homogeneous modules is directly related to the join count statistics that are familiar to ecologists from other applications, and the topic will be discussed further below in considering the independence of different measures.

In ecological studies, modularity is usually based on an interaction matrix, which records the frequency or intensity of the interaction between any two species. The most commonly used general measure of modularity for a network graph, G, consisting of K modules is denoted Q (see Dormann et al. 2017):

$$Q(G|K) = \frac{1}{2K}\sum_{i,j}^{K} \delta_{ij}\left(A_{ij} - P_{ij}\right), \tag{2.46}$$

$$Q(G|K) = \frac{1}{2K}\sum_{i,j}^{K} \delta_{ij}\left(A_{ij} - P_{ij}\right).$$

Here, A_{ij} is the observed value in the interaction matrix, P_{ij} is the expected value based on the marginal totals (i.e. the proportional null model), and δ_{ij} is an indicator function, being 1 when nodes i and j are in the same module and 0 otherwise.

Summarising the large body of information on ecological interaction networks, Dormann et al. (2017) have provided two interesting tables, one giving possible causes of the observed modularity in these networks and the other giving possible consequences. Plausible causes include competition, trait matching, one-sided adaptation and coevolution. Possible consequences include robustness, fragility, species richness and coevolution, but these are all speculative with little convincing evidence to date (Dormann et al. 2017).

From a study of four species of different dispersal abilities, Fletcher et al. (2013) concluded that modularity reveals critical spatial scales in metapopulation structure and therefore should be included in assessments of connectivity and population viability. As a second example, Gilarranz & Bascompte (2012) used an artificial modular habitat to show how modularity can buffer the spread of perturbation through a metapopulation. Clearly modularity is an important characteristic in ecological networks and of great interest, with more to be done!

2.4.3 Keystoneness

The character called *keystoneness* is the importance of an individual node to the function of the network and how that importance can be related to the node's position (broadly defined) in its network. In ecological networks such as food webs (trophic networks), a keystone species is one that has a disproportionately large effect on the system within which it occurs. For network analysis the important question is to what extent can keystoneness be determined from network topology (Benedek et al. 2007; Estrada 2007; Estrada 2012; Jiang & Zhang 2015). Based on a range of studies including the analysis of many real trophic networks and dynamic modelling experiments (Jordán et al. 1999; Libralato et al. 2006; Jordán 2009), the answer is that topology is indeed important and betweenness centrality is one recommended indicator. Other characteristics, however, need to be considered, especially the ecological interactions sometimes summarised as "mixed trophic impact" (Libralato et al. 2006; see Dale 2017), which combines biomass and interaction strengths for both direct and indirect interactions. Benedek et al. (2007) arrived at a parallel conclusion: considering more than a single species and thus using a multispecies approach is recommended for greater insight into node importance. It is a "sum of all paths" approach using all the information available and shows (again) the importance of complementary sources of knowledge (topological and ecological) to understand the structure and dynamics of ecological networks. The details of this method for food web models are in Libralato et al. (2006) and summarised with an example in Dale (2017). We will not repeat that material but suggest that the same kind of approach will prove useful in the analysis of other kinds of networks with suitable modifications to the meanings of the sum-of-all-paths within the network structure.

2.4.4 Hierarchies

Like keystoneness, the interest in hierarchical structures in network graphs stems from the expected relationship between these structures and network function. An obvious example arises in studies of competitive interactions, in which it may be possible to declare a "winner" in each pairwise test. If all pairs are tested, the resulting network structure is a tournament, as in sports, and each competitor can be given a score based on wins minus losses. In this case, a hierarchy of competitive dominance may result, although the hierarchy may be irregular with the possibility of "reversals," when the competitor with a poor record nevertheless prevails over a more successful one (see Dale 2017). Competitive hierarchies in ecological systems can be expected to affect population and community functional dynamics (Maynard et al. 2017) and thus species evolution (Chave 2009).

Many treatments of hierarchies in networks are based on one simple hierarchy that is centred on any single node; each has "first-order neighbours," those nodes that are adjacent to it (at graph distance 1), then "second-order neighbours," nodes adjacent to first-order neighbours but not to the focal node (at graph distance 2), and so on.

If the network has a well-defined origin, then the whole network is hierarchical based on path length from that origin. The most familiar example is the rooted tree in which the nodes fall naturally into a hierarchy of distances (measured by steps) from the root node; such structures arise in many ecological, genetic, and evolutionary studies. Figure 2.15a shows a rooted tree in which the gray nodes are all at a distance 2 from the root and therefore in the same level of the hierarchy. This also describes a partial order, since the order of some, but not all, pairs of nodes can be determined in the structure.

Extending the concept of distance-from-origin in a network from just a single node, allows any subgraph can perform this function. The *dilation* of a subgraph, g, is the subgraph itself together with all the nodes adjacent to the nodes of the subgraph (Costa et al. 2007). Figure 2.15b shows an example: the black nodes are the addition to the subgraph of grey nodes, forming the dilation. The converse is the *erosion* of the subgraph; it is the complement of the dilation of the complement of g (Costa et al. 2007). In other words, the erosion is the set of nodes in g that are not adjacent to any nodes outside g (dark grey nodes in Figure 2.15b). These two concepts come from mathematical morphology and image processing where the graphs under scrutiny are

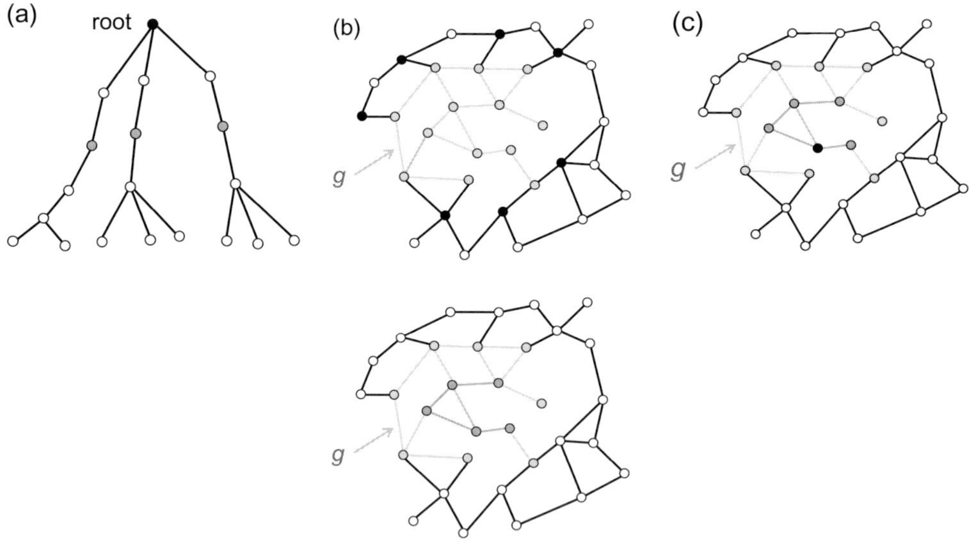

Figure 2.15 Hierarchies in networks. (a) Rooted tree. A rooted tree creates a hierarchy of nodes based on distance from the root. It is a partial order. Nodes in grey are all distance 2 from the root node (black). (b) Subgraph dilation and erosion. The dilation of the subgraph, g (green nodes), includes the black nodes. The erosion of the subgraph, g (all green nodes), includes only the darker blue. (c) Iterated dilation of a single node. The dilation of a single node (black) gives 1° neighbours (dark blue) nodes, 2° neighbours (light blue), and 3° neighbours (white). (A black and white version of this figure will appear in some formats. For the colour version, please refer to the plate section.)

lattice graphs. In spatial terms, the dilation of the subgraph annexes its periphery; the erosion is the "inside" or "core" of the set of nodes with no external connections. Repeated application of dilation produces rings of nodes that join the agglomeration at each step, thus producing a hierarchy of nodes based on path distance from the nearest node of g. In fact, if the subgraph in question is a single node, iterated dilation produces the hierarchy of first-order neighbours, then 2° neighbours, 3°, and so on (Figure 2.15c).

The same kind of hierarchy can develop in a non-spatial network. For example, in a trophic network, the root subgraph for dilation could be the primary producers. Speaking simplistically, subsequent levels of dilation would then include first the herbivores and omnivores, then the strict carnivores, and so on... (of course, real trophic networks are not likely to be quite so simple.) Whether spatial or non-spatial, temporal hierarchies develop during the spread of disease or information, or the transmission of messages (Newman 2010).

While networks often consist of modules or "communities," these may be overlapping sets of nodes, and smaller modules may combine recursively to form larger and larger groupings (Ahn et al. 2010). For biologists, the most familiar hierarchical network structure comes from a classification scheme forming a dendrogram of individual nodes representing taxa combining into pairs and then successively into larger and larger subgraphs (Legendre & Legendre 2012; Figure 2.16). Clauset et al. (2007) provided a statistical model for creating random graphs with this sort of hierarchical structure, which can generate useful null models for hypothesis testing. Hierarchical structure can also occur in networks that are basically scale free but with embedded modularity (see Estrada 2012). Although we generally think of hierarchical structure as being determined by groupings of the *nodes* of a network, Ahn et al.

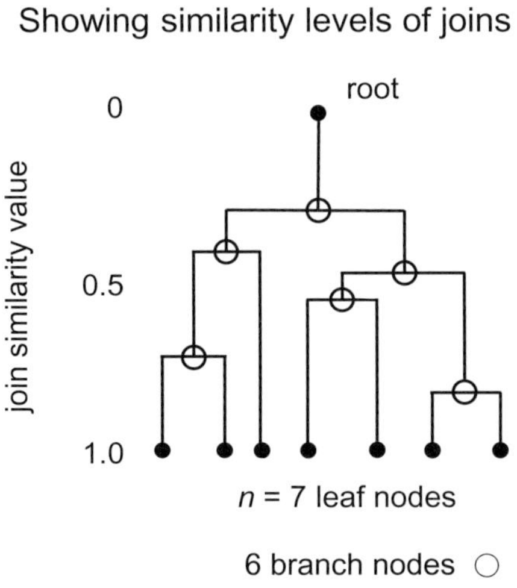

Figure 2.16 Hierarchical classification shown as a dendrogram. Similarity value shown by the level of the join. Root node, seven leaf nodes, and six branch nodes.

(2010) have found hierarchical groupings of *edges* in many networks, both biological (metabolic and protein–protein) and social (human), showing how modular overlap and the hierarchical organisation of such networks are two aspects of the same phenomenon.

2.4.5 Interdependence and Choice of Properties and Measures

With so many properties and with so many measures, it is no surprise that there is interdependence among them, no matter how the network is determined and no matter its size. The difficult question is then to find the best choice of measures: how many and which ones. Costa et al. (2007) presented an interesting table of the correlations among commonly used measures in familiar reference networks, ER (Erdös–Rényi), BA (Barabási–Albert) and the spatial random model, GN (Girvan–Newman) which contains modules or "community structure." They generated from 1000 realisations of each network for $n = 1000$ and an edge density to give an average node degree of 4. The table contains many large correlation values, showing that the measures taken together are at least partially redundant. Most important, however, is the fact that the correlations vary markedly between network models, providing further information on network characteristics (Costa et al. 2007). For example, consider three of the measures investigated by Costa et al. (2007): r, the correlation of the degrees of adjacent nodes; L, the average shortest geodesic distance; and C_I, the average triad completion for neighbours of i, standardised to the maximum possible triad completion for neighbours of i. Measures r and L are strongly positively correlated in the BA model, but not in ER or GN/ Measures L and C_I are strongly negatively correlated in the BA model, but not in ER or GN. Obviously for an observed network graph, the correlations of different measures cannot be assumed but nor can it be dismissed, the lack of independence among measures must be taken into account.

An example developed by Park and Barabási (2007), which we have already mentioned, provides a good illustration. Consider networks in which all nodes have binary labels, say 0 and 1, based on any of several criteria (e.g. performing a specific function or not). Consider the join count statistics for the edges: m_{00} is the number of edges between two "0" nodes; m_{11} is the number of edges between two "1" nodes; and m_{01} is the number of edges between nodes with different labels. The expected values of these counts under random attachment can be determined from the basic characteristics of the network and then two possible measures are

$$D = m_{11}/E(m_{11}), \text{"dyadicity" (assortative for 1's, and thus for 0's too)} \tag{2.47}$$

and

$$H = m_{01}/E(m_{01}), \text{"heterophilicity" (disassortative).} \tag{2.48}$$

The values of these two measures are constrained both by the properties of the network and by the fact that they are not fully independent. The suggestion is that while one measure is not sufficient, two can provide definitive and consistent information on network configuration. It will be interesting to determine whether ecological networks exhibit similar simple and consistent differences in their structural characteristics based on these two join-count measures or on any other pair of potentially complementary indices.

Rather than starting with only two measures, a logical procedure might be to begin with several or many measures of the networks of interest and then, considering each measure as a dimension, reduce the dimensionality and select the best measures (Costa et al. 2007). Real and simulated networks have intrinsic statistical variability (as well as other kinds) in their properties and in their measures (Costa et al. 2007). For example, 1000 realisations of the Barabási–Albert model produced a cloud of observations in the three-dimensional space defined by the three measures already mentioned: shortest geodesic path length, a clustering coefficient (standardised neighbour triad completion) and the correlation of adjacent node degrees. This creates a mathematical situation that is very familiar to ecologists: several variables, many observations, and a need to reduce the number of dimensions while preserving as much of the information as possible. The recommendation is to use Principal Components Analysis and Canonical Variable Analysis for this purpose, combined with Bayesian decision theory and Costa et al. (2007) provided convincing examples of the success of that approach. Applying this approach in different situations will provide some insight into the general relationships among these measures, but also into the factors that can affect those relationships, and thus the ecologist's best choice of combined methods. More important, perhaps, is to understand the ecological and network characteristics to which the measures are most sensitive and to make some initial choices for measures to be included, at least for initial screening, based on those sensitivities and the questions being asked.

One application of the various measures and combinations of measures of network structural properties is to compare networks. In the final section of this chapter, we examine a number of different approaches to compare networks, focussing on their structural properties.

2.5 Comparing Networks

2.5.1 Probability Calculations

In concept, an easy approach to comparing networks is to calculate the probability that two network graphs share as many nodes as they do, or, given the same sets of nodes, the probability that they share as many edges as they do (see Dale 1985). Calculations become more complicated, of course, as we move from basic graphs to digraphs or signed graphs, or to signed digraphs.

Computational complexity is not the only problem to face, of course; we have commented elsewhere on the difficulties of dealing with the lack of independence in data that involves spatial or temporal relationships (e.g. spatial autocorrelation, Dale & Fortin 2014). Many kinds of networks have similar interdependence structures, even when the autocorrelation is not explicit; after all, it is the interdependence that makes the network interesting and useful! For example, in studying Protein–Protein Interaction (PPI) networks, the dependence among interacting proteins may be at least partially the result of coevolution (Kelly et al. 2010), but knowledge of the sources of interdependence will help predict something about function from knowledge of structure. Remember that network analysis needs to be combined with other forms of information in order to provide "truly meaningful results" (Kelly et al. 2010). The interdependence in network graphs cannot be ignored, and frequently, the best approach is to use randomisations that respect the form of interdependence in the data while allowing the freedom to generate comparable reference "data." In some fields of science (e.g. physics), this is expressed as the concept of a statistical ensemble, being the set of all networks with the same constraints but possibly different measured characteristics (Kelly et al. 2010). An example already described is the set of all networks with a particular degree sequence, which can be generated by the random pairing of "stubs," consisting of nodes with unpaired edge ends that match the original degree sequence (see *Configuration Model* in the glossary). Similar procedures, giving similar results, are the random rewiring of networks or the conditional rewiring of networks, both of which are essentially constrained randomisations (see Kelly et al. 2010). Again, these can generate a reference set of comparable network graphs from the same ensemble with which any observed similarity measures for networks with the same constraints, can be compared. For probability calculations, sometimes the problem is that almost any amount of similarity turns out to be highly significant because so many different configurations are possible, and the data sets are so large. It gives some pause for thought as to what apparent statistical significance provides in such circumstances.

While providing some clarity on what comparison of network graph structure may now imply, considering the various forms of interdependence or constraints on randomisation may not make the calculations of improbable similarity any easier. Randomisation may be the only answer, and even that may provide some conceptual and computational challenges.

2.5.2 Graphlets

Graphlets have been recommended as providing insights into the structure and function of network graphs, and so an obvious suggestion for comparing networks is to involve graphlets. For example, that is the intent of comparing network GCMs (Yaveroğlu et al. 2014) using a graphlet correlation distance as described above, and all versions of graphlets can be adopted for network comparisons, from basic graphlets to directed and rooted graphlets (e.g. Aparício et al. 2015). The relative graphlet frequency vectors for two networks can be compared by the difference between the logarithms of the vectors, and the similarity of the graphlet ("degree") orbit distributions of two networks can also be calculated (see Newaz & Milenković 2019). Most comparisons are for two networks at a time, but more flexible methods for the simultaneous comparison of several networks using various of the elaborated graphlets are being developed (e.g. GRAFENE, "graphlet-based alignment-free network [comparison]"; see Newaz & Milenković 2019 and references therein).

In molecular biology, a major goal is to determine regions of greatest similarity in genomic sequences. This has given rise to a special set of methods for network comparison called "network alignment," which also make use of graphlets and related approaches (Newaz & Milenković 2019). This is a specialised area of genomic analysis, and we will not describe it here, but can refer the interested researcher to Malod-Dognin and Pržulj (2019) for a summary. That being said, however, a recent study of food webs used an alignment approach to identify a common "backbone" of interactions across more than 400 ecosystems (Mora et al. 2018), and so it is obvious that network alignment methods will have increasing application in ecological research. Alignment methods do not necessarily measure the similarity of whole networks but identify those parts of networks that match each other (align) best. While these methods for network comparison are both useful and flexible, several other methods are available to compare whole networks or network structures and configurations, although the basis and interpretations may differ, as we will describe.

2.5.3 Kernel Function Methods

2.5.3.1 "Kernel" versus "Kernel"

The term *kernel* can be a source of confusion if not used carefully because it has several distinct meanings in statistical applications. For ecologists, it may be most familiar describing a smoothing function, a window function, or as a method of modelling or estimating a variable's probability density function. In applications to machine learning, however, a *kernel function* is one that transforms input data into some desired form, as described in the text that follows. These two meanings of "kernel" are certainly different, but they share the property of being related to the measurement of similarity.

Kernel functions can compare network structures and we will start with a summary given in Dale (2017), and then provide some further details on their application to network analysis (see a useful technical review of graph kernels by Ghosh et al. 2017). This approach is popular in machine learning and related fields, and methods can be based on random walks, on paths, or on small subgraphs such as motifs, graphlets, or subtrees (Shervashidze et al. 2011; Ghosh et al. 2017). The advantage of kernels is the ability to map objects of interest, x (a node, subgraph, or a whole graph), from its own data space into a "feature space," which simplifies the analysis, using some mapping function, $\varphi(x)$ (Figure 2.17a). For example, two categories of data may be linearly separable in the feature space when they are not in the original data space (Figure 2.17a); see Alavarez-Vega (2011) as one of several good summaries. Comparing two objects, x and x', does not require converting each one explicitly with function φ (which may be computationally demanding), but can be accomplished by a kernel function, k, acting on the cross product for which $k(x, x') = <\varphi(x), \varphi(x')>$. This

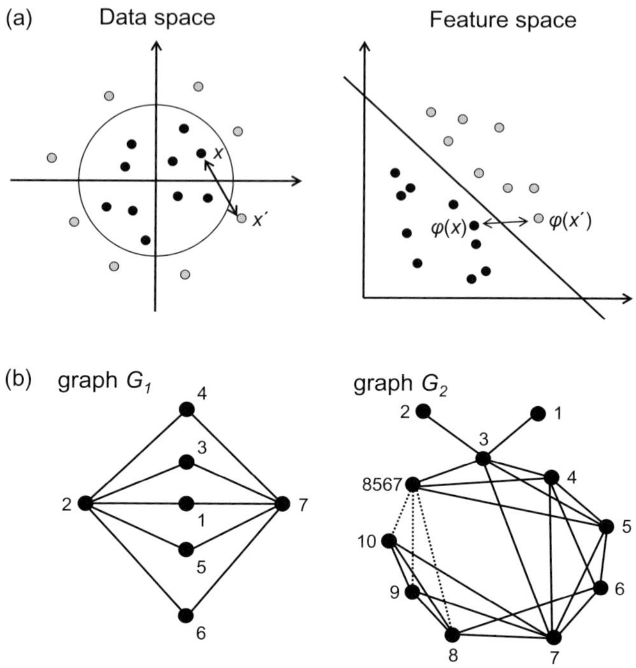

Figure 2.17 Graph kernels for comparing graphs. (a) Kernel trick: from data space to feature space (from many similar sources, e.g. Alvarez-Vega 2011). Compare x and x' from the data space by mapping the pair to the feature space using the function φ and then compare φ(x) and φ(x'). The two categories are not linearly separable in the data space, but they are in the feature space. The kernel trick is to replace explicit calculation of φ(x) with the kernel function, k, acting on the cross product in the data space. [From Alvarez-Vega, M. (2011). *Graph Kernels and Applications in Bioinformatics.* All Graduate Theses and Dissertations. 1185. https://digitalcommons.usu.edu/etd/1185]
(b) Comparing nodes in a graph using graph kernels based on random walks (the numbers are just node labels). Left panel: Random walks from node 2 or node 7 will be very similar. They also belong to the same graphlet orbits. Right panel: Graph with a large multiring subgraph: Random walks from nodes 1 or node 2 will be very similar and they belong to the same orbits. But random walks can become trapped in long cycles in the large multiring subgraph with 8567 nodes.

direct step is the advantage of this approach and its function measures the similarity of x and x', because it takes higher values the more features the objects share. Two networks can be compared directly by measuring the similarity of their graphs using these graph kernels (Vishwanathan et al. 2010; Shervashidze et al. 2011). These kernels have played an important role in studying the "large" graphs from molecular biology and chemo-informatics networks, as well as from information and social networks (Borgwardt et al. 2007; Vishwanathan et al. 2010; see also Schölkopf et al. 2004 and Gärtner 2008).

Although our main aim is to compare whole networks, we will start by showing how kernel methods evaluate the similarity of two nodes in the same graph, here based on random walks (see Kondor & Lafferty 2002 and also helpful talks by Vishwanathen, for example, www.stat.purdue.edu/~vishy/talks/Graphs.pdf or https://archive.org/details/Microsoft_Research_Video_103902). Given two nodes in one graph that "look the same" (similar topological positions and graphlet orbit signatures), the random walks that start from either node should have the same properties (Figure 2.17b, left panel). To compare, determine the number and lengths of random walks between the two nodes i and j. This simple method is not ideal because the random walks can stall in repeated cycles or in "tottering" back and forth between two nodes (Figure 2.17b, right panel; see commentary in Dale 2017 on random walks getting trapped). A solution is to down-weight walks of length k by $\lambda^k/k!$, where λ is an appropriate parameter between 0 and 1. Given $\mathbf{A}$, the graph's adjacency matrix, $\mathbf{A}^k$ enumerates the walks of length k. Therefore, the kernel that measures the similarity of the nodes i and j becomes

$$K(i,j) = \left[\sum_{k=1} \frac{\lambda^k}{k!} A^k\right]_{ij} = \left[e^{\lambda A}\right]_{ij}. \tag{2.49}$$

This is a simple solution for comparing nodes' positions within a graph, and it seems possible that some easy modification will allow us to compare whole networks.

For entire graphs, the kernel is a function of the two graphs, G and G', that measures their similarity and so we need a kernel that "captures the semantics inherent in the graph structure" (i.e. topological relationships; Vishwanathan et al. 2010). Random walks can evaluate structural characteristics of graphs and so these kernels can be related to the number of walks that match. One version is to take advantage of the properties of the *graph direct product:* $G_\times$ is the direct product of G and G' when its nodes are pairs of nodes from G and from G', and its edges join $G_\times$ nodes if and only if both pairs are joined in G and G'. Figure 2.18 gives an example. If $\mathbf{A}$ and $\mathbf{A}'$ are

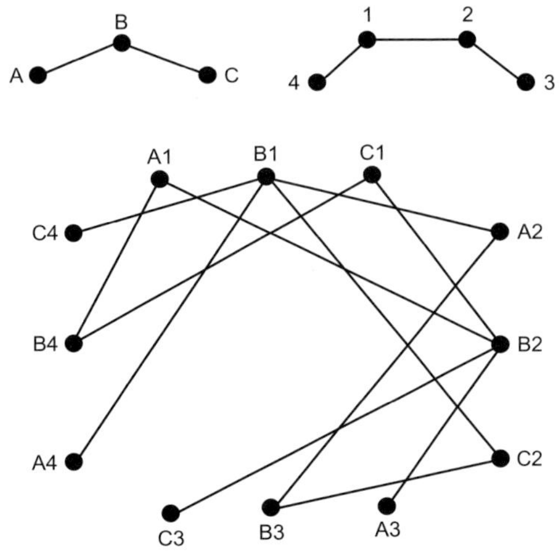

Figure 2.18 Two small graphs and their direct product, based on Vishwanathan et al. (2010).

the adjacency matrices of the original graphs, the adjacency matrix of $G_\times$ is their Kronecker product (each "1" in **A** is replaced by a copy of **A**'): $\mathbf{A}_\times = \mathbf{A} \otimes \mathbf{A}'$. Running a random walk on $G_\times$ is equivalent to running independent random walks on the original graphs (Gärtner 2008). To create the kernel, a weight matrix, $\mathbf{W}_\times$ is associated with $G_\times$, and with appropriate definitions, $\mathbf{W}_\times = \mathbf{A}_\times$. The *k*th power of any adjacency matrix counts walks of length *k*. The kernel is then based on simultaneous random walks on both G and G' between corresponding pairs of nodes and both of length *k*. It is essentially a sum over all walk lengths but with a length-dependent coefficient, $\mu(k)$, to reduce the effects of repeated cycles by down-weighting long walks (Vishwanathan et al. 2010). The last requirement is for initial and stopping probability distributions, $p_\times$ and $q_\times$. Then, the kernel function is (Vishwanathan et al. 2010; Ghosh et al. 2017):

$$K(G, G') = \sum_{k=0}^{\infty} \mu(k) q_\times^T W_\times^k p_\times. \tag{2.50}$$

That is the kernel function needed to compare the similarity of graphs; looks simple! This algorithm compares the two graphs but allows some flexibility in the actual measure, through different versions of $\mu(k)$ weighting walks of different lengths. The details are provided in Vishwanathan et al. (2010), and it looks like a promising method for applications in ecological studies.

A related concept in the comparison of networks is the "graph edit distance" (Neuhaus & Bunke, 2006; Shervashidze et al. 2009); this is just a weighted count of the primary operations needed to transfer one graph into the other by node addition or deletion, edge addition or deletion, and change of label where included. Another related concept uses the fact that if two network graphs, G and G', are identical, then there is an isomorphism that preserves the adjacency of any pair of adjacent nodes. If they are not absolutely identical but only similar, the minimum (over all bijections, φ) of the difference between the distance $d(u,v)$ and the distance $d(\varphi(u),\varphi(v))$ measures their similarity (for details see Baur & Benkert 2005).

An alternative for comparing two graphs, G and G',, is the graphlet frequency kernel. Where $F_G(i)$ is the vector of graphlet frequencies in graph G, let the proportions vector be $D_G(i)$: $D_G(i) = F_G(i) / \Sigma_j F_G(j)$. Then the graphlet kernel for G and G' is defined by $k_\gamma(G, G') = D_G^{\mathrm{T}} D_{G'}$ (Shervashidze et al. 2009; Ghosh et al. 2017). For example, comparing the graph in Figure 2.19 (graphlet frequency vector = (10, 11, 4, 0, 2, 4, 4, 4, 0)) with a derived graph shown in Figure 2.19, (graphlet frequency vector = (9, 10, 4, 0, 2, 4, 2, 1, 0)) the graphlet kernel is 0.1987, close to 0.1900 and 0.2168 as values for self-comparison measures (98% of their average). We could extend this concept to comparisons of the orbit signatures instead of the graphlet frequency distributions... The original graph in Figure 2.19 has an orbit distribution of (20, 22, 11, 12, 0, 0, 6, 2, 16, 4, 8, 4, 8, 8, 0), whereas the modified version has (18, 20, 10, 12, 0, 0, 6, 2, 16, 2, 4, 2, 2, 2, 0). The graphlet orbit kernel for the same two closely related graphs is 0.1231, close to 0.1140 and 0.1406 for self-comparison measures (97% of their average). As an aside, we suggest that future versions of this approach

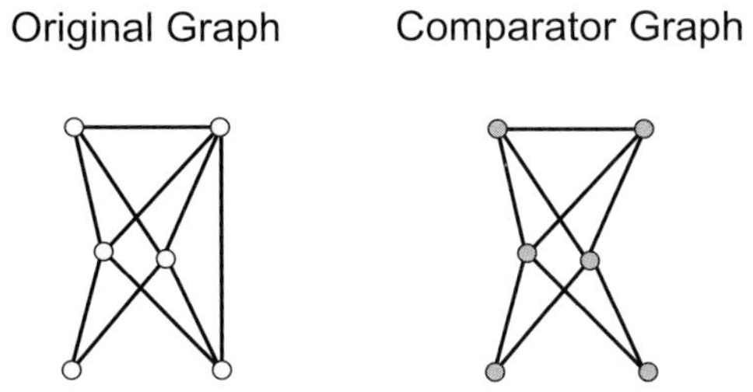

Figure 2.19 Counting graphlets and orbits: graphlet kernels. The original graph has graphlet frequency vector: (10, 11, 4, 0, 2, 4, 4, 4, 0) and orbit distribution: (20, 22, 11, 12, 0, 0, 6, 2, 16, 4, 8, 4, 8, 8, 0). The comparator graph has one edge deleted; graphlet frequency vector: (9, 10, 4, 0, 2, 4, 2, 1, 0) and orbit distribution: (18, 20, 10, 12, 0, 0, 6, 2, 16, 2, 4, 2, 2, 2, 0).

may include "graphlets" that are not connected (see Shervashidze 2009; Lugo-Martinez & Radivojac 2014; among several).

An important detail is that these graphlets are unlabelled, which does not make sense for chemical structures in which the labels on the structural nodes make a big difference (e.g. C vs. N vs. O), and so many chemistry applications focus in kernels that can use graphlets with labelled nodes (Gaüzère et al. 2015). The nodes may also be labelled for graphlets in protein function analysis determined from chemical properties or from their inclusion of a particular amino acid (Vacic et al. 2010).

Looking for further development and refinement of these kernel methods, we can ask what level of detail they can achieve. Given the overall measure of similarity of two large graphs, can the method identify parts of the graphs that are most or least similar, characteristics of the graphs that are most or least similar, or how similarity changes with scale? The localised graphlet kernel approach seems promising to help ecologists in future versions of analysis, although this is yet to be done explicitly (see Shervashidze et al. 2009; Lugo-Martinez & Radivojac 2014). More specifically, work by Lugo-Martinez and Radivojac (2014) to generalise graphlet kernels includes a technique that may be referred to as an "edit distance graphlet kernel" which combines the desirable features of both versions of kernel functions.

2.5.4 Spectral Methods

The similarity between kernel methods and spectral methods is clearly seen when we consider the relationships of kernel functions and the network graph matrices through the properties of the associated random walks. Brouwer & Haemers (2012) provided a useful guide to the current state of knowledge in this fascinating field of graph theory.

We described one approach to comparing two graphs and determining their similarity using spectral analysis, provided that the two graphs have the same set of nodes. Given two graphs with the same nodes but possibly different edges, one similarity measure is based on eigenfunction decomposition, using Moran's eigenvector maps (MEM eigenfunctions; Dray et al. 2006) for undirected edges and asymmetric eigenvector maps (AEM eigenfunctions; Blanchet et al. 2008) for directed edges. The adjacency matrices are converted into distance matrices with an upper threshold and their eigenvectors are then compared by a generalised Pearson coefficient measuring the similarity of the original graphs (for details see Borcard et al. 2004; Dray et al. 2006, 2012; Blanchet et al. 2008, 2011).

This interesting approach aside, spectral graph theory is not generally the basis for evaluating network similarity. There are good reasons for this: (1) More than one graph can have the same spectrum as already mentioned. (2) The spectrum of a graph can change quite markedly in response to seemingly small structural changes. For example, while it is possible for a bipartite graph and a non-bipartite graph to have same spectrum (Brouwer & Haemers 2012), adding a single line to a bipartite graph that makes it non-bipartite can have a significant effect on the spectrum (see Brouwer & Haemers 2012). The simplest example is the addition of a single edge that turns a 3-node path graph (basic graphlet g1, bipartite) into a 3-node triangle (basic graphlet g2, not bipartite), which changes the adjacency eigenvalues from $(\sqrt{2}, 0, -\sqrt{2})$ to $(2, -1, -1)$. This represents a large difference in the structural evaluation by the spectral method, but perhaps it matches an intuitive evaluation that the path is qualitatively different from a triangle.

There is evidence that the approach may well be worth pursuing in future research (see Wilson & Zhu 2008; Gera et al. 2018). Part of the solution may be to use the Laplacian matrix or a derivative of it, instead of the adjacency matrix which has proved to be inferior (Wilson & Zhu 2008). For example, our two 3-node graphlets cited earlier, g1 and g2, have Laplacian spectra of (0, 1, 3) and (0, 3, 3): not quite so different. The twin graphs of the saltire have different spectra for matrices other than the adjacency matrix (e.g. Laplacian or signless Laplacian; Brouwer & Haemers 2012), responding to the obvious difference between them.

For assessing a large network derived from data from a series of time-slices, Gera et al. (2018) provided some advice on appropriate analysis procedures. They recommend spectral analysis, with sequential time-slice analysis of the adjacency, Laplacian, and normalised Laplacian matrices. They suggested that this sequential approach is a promising method for these particular circumstances, but it is not clear how general is this recommendation. As ecological researchers, we will await more advice from the experts in this field, but it may be that new successful approaches apply only in narrow ranges of conditions and may not all be suitable for general application.

2.5.5 Dissimilarity

So far, our discussion of comparing networks has focused on measuring their *similarity*, but with very good reason, Poisot et al. (2012) have brought to our attention the importance of evaluating the *dissimilarity* of species interaction networks in ecological studies. The intent is to investigate the sources of variability among comparable networks, rather than to describe how interaction networks are different from other sorts. That approach would follow the path of Newman and Park (2003), who have provided a clear explanation of how social networks differ from other kinds of commonly studied networks (positive correlation of neighbour node degrees; high rates of triad completion). Instead, Poisot et al. (2012) have chosen to examine (and partition) the sources of dissimilarity, being differences in interactions between species common to both networks and differences due to species turnover. These sources are clearly analogous to the familiar partitioning of species diversity in different communities, and like community (or sample) β diversity, the dissimilarity of interactions is the sum of species turnover and in-common species interaction differences:

$$\beta_A = \beta_S + \beta_C, \tag{2.51}$$

with the three subscripts indicating all interactions (subscript A), species turnover differences (S), and in-common species' interactions (C). To measure these dissimilarities, β can use any of the usual indices, such as Jaccard's, but Poisot et al. (2012) have used Whittaker's index:

$$D_W(1,2) = [2(a+b+c)/(2a+b+c)] - 1, \tag{2.52}$$

where a is the count of items (species or interactions) shared by the two networks, b is the count of the same items in network 1, and c is the count in network 2. It is possible to calculate probabilities for the observed values using combinatorial considerations in order to evaluate "significance," but a restricted randomisation approach is probably the most practical and most generally applicable.

Other research on differences in changing ecological networks has tended to emphasise the details of the differences (e.g. Burkle et al. 2013) or the scales at which changes occur (Trøjelsgaard & Olesen 2016), as well as possible causes for change.

2.6 Concluding Comments

As with spatial statistics, network structural statistics can occur both in a global form, applied to the whole network, and in a local form, applied at the scale of a single node, a motif or graphlet, or a larger module. Understandings of the local characteristics of nodes and edges can combine in many ways to study a range of more global characteristics of networks. As described earlier, because the characteristics of interest are not independent, we need to understand those dependencies and include that knowledge in our interpretation. Similarly, the great diversity of approaches, properties, and measures offers the potential for important insights into network structure, but the key step in our analysis is to understand the relationships between structure and function (or topology and dynamics), thus providing answers to our ecological questions. The complementary direction is to determine how the graph or network properties respond to or are shaped by ecological factors, and whether those responses belong mainly to the realm of static or one-time effects, or they represent an ongoing evolution of the structure.

The other part of the effort is to compare different networks or different realisations of the same network for the properties of interest. Such comparisons will need to be quantitative, almost always, leading to the consideration of the statistics of network and graph properties and the evaluation of significance. These considerations lead directly into the next part of this book, to discuss the statistics of quantitative analysis and their application to understanding the dynamics of ecological networks.

3 Quantitative Analysis of Dynamic Networks

3.1 Introduction

Ecological networks that require analysis to answer ecological questions are many, as detailed through specific examples in Chapter 1. Some of the most important, with their associated data and processes, are

- Species association networks: frequencies of interspecific proximities or contacts
- Mutualistic networks: numbers, intensities or rates of interactions
- Trophic networks (food webs): interspecific predation rates or consumption amounts
- Host–parasite or host–parasitoid networks: abundances and interaction intensities
- Social networks: individuals' proximities and nature and strength of interactions
- Epidemiological networks: timing, distances, and modes of disease spread
- Ecosystem compartment and flow networks, with quantities and rates
- State and transition networks of landscape system evolution
- Phenology networks: temporal shifts in ecological processes
- Spatial networks: species' numbers in landscape patches and species movement between them

All these ecological networks have dynamic properties related to topological changes through time and to functional changes of the network, in addition to the structural properties reviewed in Chapter 2. We usually distinguish between the dynamics ***of*** networks and the dynamics ***on*** networks (Figure 3.1; Gross & Sayama 2009; Peterson et al. 2013; Dale 2017; Silk et al. 2017). The dynamics ***of*** networks includes the development or "evolution" of a network graph by random or non-random processes, as nodes or edges accrue or are lost. As a network evolves, its structural properties may change, and the trajectories of such changes can provide insights into the network or its network model. Those insights may be well beyond those available from the network's characteristics when static (Figure 3.1; Costa et al. 2007). The observed dynamics of a network may be initiated by perturbation or disturbance, whether internal to the network (a species goes extinct) or external (changing climate affects pollination interactions). Recording or modelling the network's response to such perturbations by looking at structural or functional characteristics provides an evaluation of its sensitivity to change (fragility) or its robustness and stability (see Costa et al. 2007). Changes in ecological networks often show up as

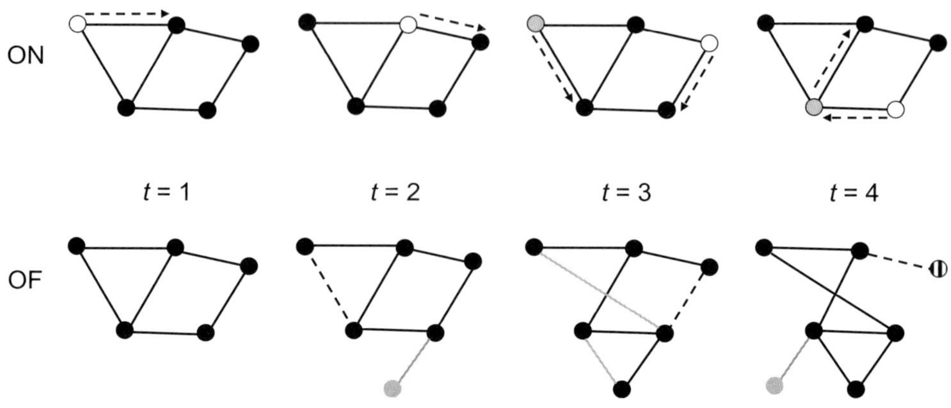

Figure 3.1 Dynamics **on** a network and dynamics **of** a network. Dynamics on a network: node state affects neighbour states (or not) as indicated; no structural change. Dynamics of a network: nodes and edges are lost (grey and then absent) or gained (green and then black) between time intervals. (A black and white version of this figure will appear in some formats. For the colour version, please refer to the plate section.)

turnover of species (e.g. beta-diversity) or of pairwise interactions with changing circumstances (see Kemp et al. 2017).

The dynamics ***on*** networks include the flow of energy, information, or disease, and how those are affected by the characteristics of the node and of the edges that join the node to neighbours. Technical approaches to network dynamics may use systems of coupled differential equations that describe the state of a node and the influence through time of the states of neighbours (Porter & Gleeson 2015). Whether the dynamics are *of* or *on* a network, the response to perturbation may include a period of transient dynamics before the system settles to a new equilibrium (see Hastings 2001, 2004; Hastings et al. 2018). Various kinds of feedback loops are at work in these network dynamics, and they also depend on local structural configuration and the structural characteristics of the neighbourhood. In general, networks of systems in which the topology and function evolve by reciprocal changes, thus representing a kind of coevolution, are called *adaptive networks* (Gross & Blasius 2008). With a related but not identical terminology, Moore et al. (2018) discussed *adaptive food webs* as dynamic structures and processes affected both by external drivers and changes in adaptions of the species within them.

The conundrum is the dual quandary of how network structure affects its dynamics, both *of* and *on*, and how function and dynamics affect network structure. Those reciprocal effects within a single network are also affected by external factors (many abiotic) acting on structure or on function and by the action of those external forces on the reciprocal interactions. In addition, there may be affects from other interaction networks (Figure 3.2), creating a complex dynamic structure for which predicting behaviour is very difficult. A simple model of these effects is to consider a state-and-transition model in which the nodes of the network exist in a few defined states (Phillips 2011; Daniel et al. 2016), which affect the network topology according to

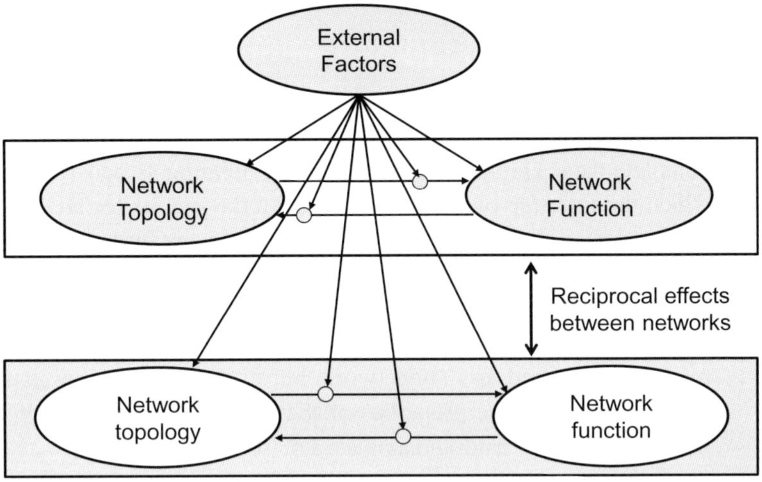

Figure 3.2 The interplay of structure and function in two interaction networks is affected by external factors and by each other. The shaded nodes were originally edges between the topology and function nodes, but now acting as nodes, affected by external factors.

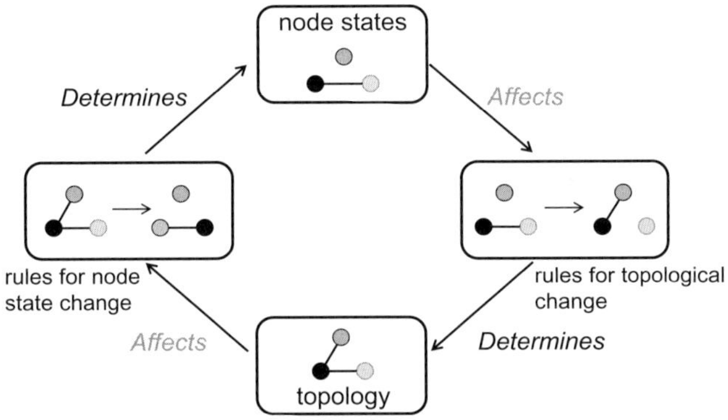

Figure 3.3 Adaptive network: the reciprocal effects of network topology and network functional dynamics that affect node states.
[Based on Gross, T., & Blasius, B. (2008). Adaptive coevolutionary networks: A review. *Journal of the Royal Society Interface*, 5, 259–271 and Dale, M. R. T. (2017). *Applying Graph Theory in Ecological Research.* Cambridge: Cambridge University Press, republished with permission.]

some rules, and in which the nodes change states according to rules based on topology and current states (Figures 3.2 and 3.3). Before we work directly on the reciprocal effects in ecological networks, we need the background of statistics and models for network applications as well as considering statistical inference for ecological networks, especially when our data are incomplete or lack independence (Naujokaitis-Lewis et al. 2013; de Aguiar et al. 2019). Let us start with inference when the information is incomplete.

3.2 Inference

3.2.1 Sampling Networks

Many ecological networks of interest must be created from samples rather than from complete data. Think of the difficulties of building an ecosystem's trophic network in which the first step of merely detecting all the species (or all the taxa) is not possible, far less documenting all the interactions that occur (Jordano 2016). The challenge is to make inferences about the large-scale or global characteristics of the network based on only some of the data (Bliss et al. 2014). Just as a subset of the nodes and their edges from a graph form a subgraph, a subset of the nodes and their edges from a network may be considered as a subnetwork, but the properties measured in subnetworks are unlikely to reflect the properties of the whole network (Stumpf et al. 2005), even if they are based on random samples. For "big data" networks, analysing the complete data set may present insurmountable practical difficulties, and so only samples of the entire network data are used for analysis. Therefore, we can have either networks that are created based on samples, or samples from existing networks, each presenting challenges for analysis and interpretation.

For sampled networks, several methods are more-or-less standard; here are five:

- *Node sampling*: q nodes are taken at random from all n nodes in the network graph; any or all the edges between nodes in that sample can be included as appropriate (Figure 3.4a). If all the edges joining nodes in the sample are included, this is an induced subgraph based on a random sample of nodes and the method can be called *induced subgraph sampling* (Kolaczyk 2009). If all the sample's nodes are included, but some of the edges are invisible or have been left out, this is the case of *link failure* (Bliss et al. 2014) in node-based sampling (Figure 3.4b, left panel).
- *Edge sampling*: w edges are taken at random from all m edges in the network graph; all nodes attached to those edges may form a secondary sample (Figure 3.4a, left panel), with source and sink nodes distinguished in a digraph (a graph with directed edges) if needed. If the edges' end-nodes are always included, this is called *incident subgraph sampling* (Kolaczyk 2009). When the edges represent relationships between objects or individuals, such as communication events, this is what Crane (2018) called relational sampling, creating a dyadic sample with two nodes in each. Sampling edges rather than nodes creates bias in the sample towards nodes of high degree (see Crane 2018) because with more edges attached, there is a greater chance that one of their edges is sampled. The quantitative data associated with an edge may itself be incomplete; for example, if we are counting pollinator visits as interaction events, we may observe only a portion of the visits that occur, for example, 13 out of the actual 22. This is essentially *interaction sampling* (Figure 3.4b, right panel).
- *Hyperedge sampling*: A "hyperedge" extends the concept in graph theory of an *edge* that by definition joins two nodes to a *hyperedge* that can join more than two nodes at a time. If hyperedges are the sample units, clusters of two to many nodes

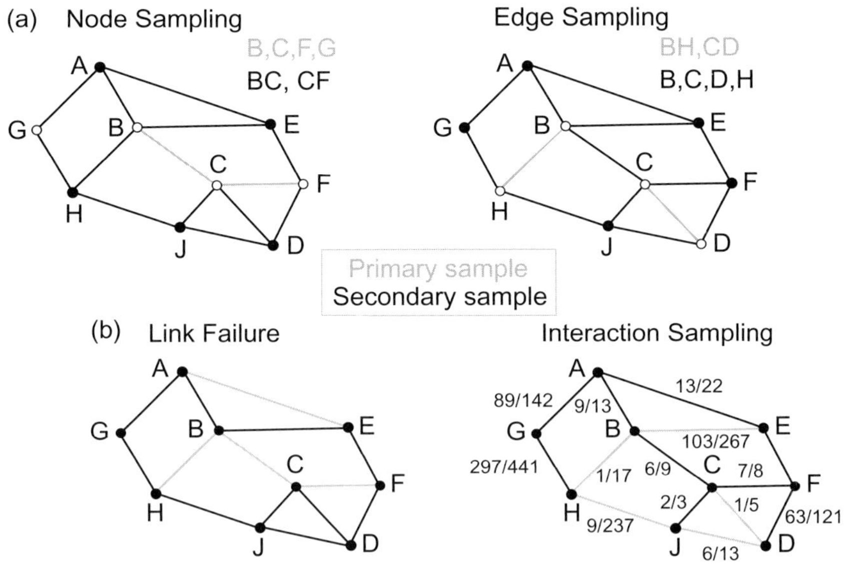

Figure 3.4 Network sampling schemes. (a) Node sampling (left). Including the nodes' edges gives an *induced subgraph* sample. Edge sampling (right). Including the edges' end-nodes gives an *incident subgraph* sample. (b) More about network sampling. The left panel shows "link failure": all nodes are detected but some edges are missed (shown in grey). Right panel: Interaction sampling. Only a proportion of interaction events are detected: partial data. Grey edges are those in which fewer than half the interaction events were detected.

are the result. Two familiar examples would be scientific papers which provide samples consisting of one to many authors, or movies, each of which provides a set of actors (Crane 2018). In such cases, despite the potential for long participant lists, the sample unit is the paper or the movie, and so the sample size is the number of productions, not the number of participants. Hyperedge sampling is closely related to the concept of an affiliation network, which is a bipartite graph consisting of two sets, for example, (1) the publications and (2) the authors, with simple edges joining each participant to each production in which they participated.

- *Path sampling*: A network such as the internet is studied based on samples of paths through the network from source S to target T, based on a list of the servers in the path (see "traceroute" example in Crane 2018). Although originally conceived of for information networks, the same sampling technique can be applied to trophic networks using individual food chains in the network (Figure 3.5). Again, the sample size is the number of paths, not the number of nodes or of edges.
- *Snowball sampling*: Starting with a random sample of q nodes, the sample associated with a single node is augmented iteratively by some nodes attached to it by paths of length 1 (first-order neighbours), then some nodes attached to it by paths of length 2 (second-order neighbours), and so on up to order k (see Figure 3.6, left panel). Snowball sampling tends to use a maximum of $k = 2$ or 3 neighbour layers,

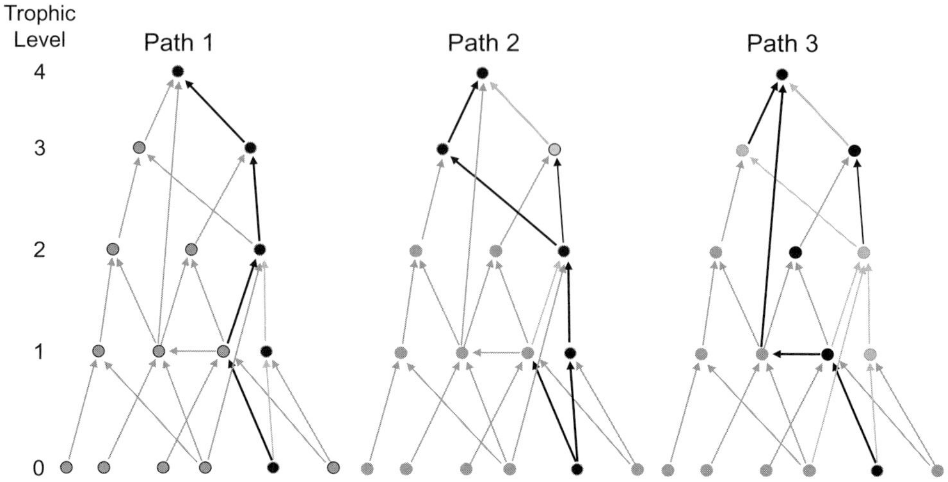

Figure 3.5 Path samples: food chains are path samples in trophic networks.

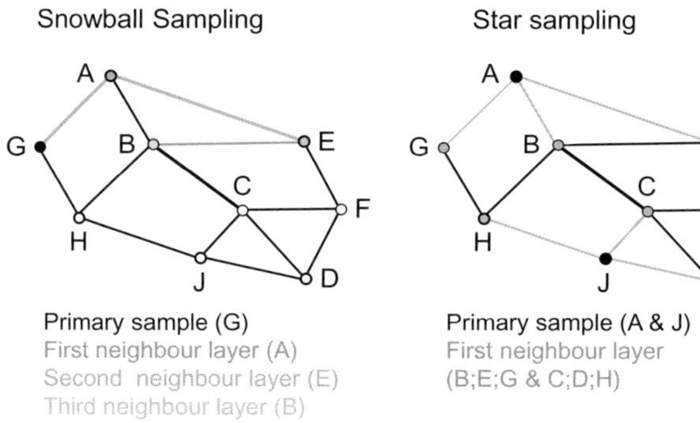

Figure 3.6 Network sampling schemes: snowball sampling and star sampling. Nodes coded by shading to show primary sample and subsequent neighbours. (A black and white version of this figure will appear in some formats. For the colour version, please refer to the plate section.)

but it could be greater for large networks. Depending on the application, the edges between the nodes of the neighbour order included may or may not be considered part of the sample. The sample size is the number of "snowballs." One common variant is to use all incident edges and all first-order neighbour nodes of a random subset of nodes; this is called *star* sampling (Kolaczyk 2009; see Figure 3.6, right panel). For a network digraph, the paths that determine the layers of snowball samples may be directed or not.

In all five cases, sample size is just the number of units in the sample, whether the units are nodes, edges, hyperedges, paths, or snowball subgraphs. Of course, some consideration is required when the samples, such as clusters of nodes, are allowed to

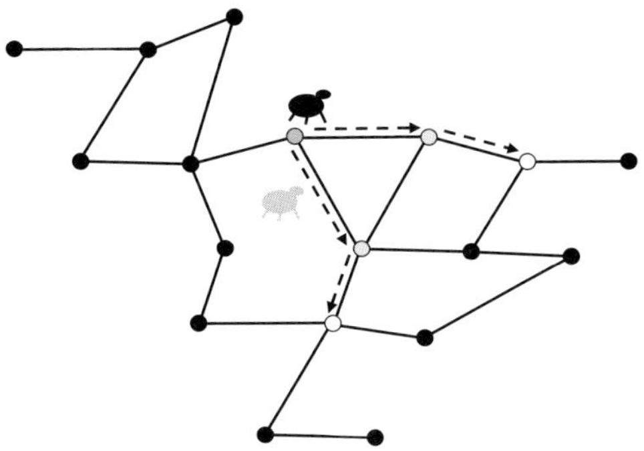

Figure 3.7 Crawler methods of sampling: a crawler starts at a random node and explores the network following its own rules. The crawler may replicate itself and follow many routes at once. (A black and white version of this figure will appear in some formats. For the colour version, please refer to the plate section.)

overlap and do. The lack of independence among such samples can reduce the effective sample size when probability calculations are to be used as a guide to interpretation.

Snowball sampling is just one of a family of "crawler" sampling methods, with many variants. Those variants may be based on the priority of neighbours to be included in the sample: breadth-first search, depth-first search, simple random walks, and so forth (see Newman 2010), any of which can be pictured as a crawler being dropped at random on the network and then exploring according to its own particular rules (Figure 3.7).

Lee et al. (2006) have focused on the statistical properties of samples taken from scale-free networks, using the three of the sampling methods just described: node sampling, link sampling (edges), and snowball sampling. They have investigated network characteristics such as node degree, betweenness centrality, path length, clustering coefficient, and assortativity by comparing the samples with the original networks. The networks considered were the Barabási–Albert scale-free model (BA), a protein interaction network (PIN), the autonomous systems level of the internet (AS), and the e-print authorship network from "arXiv.org." There are two findings to consider in interpreting studies of ecological networks. The first is that, for some characteristics, the estimates from samples are not close to whole network values, even with up to 50% sampling. The second is that the results of individual node or edge sampling are very different from those of snowball sampling, often overestimating the true values where snowball sampling underestimates, or vice versa. Remember that the properties of snowball sampling are related to those of a random walk; the sample remains within a single component, whatever the value of k, and if the component is small, so is the sample! Of course, the "all neighbours" snowball sampling approach is equivalent to following all possible (if short) random walks at the same time.

A biased random walk version of the snowball sampling is found in human social research in which current nodes (people) being studied (the "respondents") have a role in selecting the next layer to be sampled for study (their friends or contacts); this is called *respondent-driven sampling* (Heckathorn 1997). Some idea of the consequences that may result from this biased form of sampling are described by Malmros et al. (2013; see also Newman 2010), and a similar approach could provide efficient directed sampling in behavioural ecology. In sampling theory more generally, there is a similar approach for the detection of clusters of similar objects, called *adaptive sampling*, and it is worth considering for ecological networks.

Adaptive sampling is an iterative approach, in which the choice of the following layer of samples depends on the contents or results of the previous sampling layer. It is adaptive in the sense that the results to date affect how the sampling next proceeds, and we can consider both adaptive sampling in gathering data to create the network, and adaptive sampling to retrieve information from an existing network (see Thompson 2006, 2011). For example, as an oversimplification, given a spatial grid of samples intended to investigate a pair of species forming one potential link of a species co-occurrence network, say species A and B, the contents of a sample at random location (x, y) could determine the next sample location (e.g. ring-necked duck and blue-winged teal; Smith et al. 1995):

If A and B are both present: sample a contiguous location, say $(x + 1, y)$ or $(x, y + 1)$.
If only A or only B: sample a nearby location, say $(x + 1, y + 1)$ or $(x - 1, y - 1)$.
If neither A nor B: sample a new random location.

Another simplified example is sampling the network of a population of individuals (the social network nodes) to measure characteristics associated with a known communicable disease. Given a random sample of 10 individuals, those with no exposure to the disease can be omitted from further consideration, but those that have been exposed are then used as the source of subsequent snowball sampling of their contact neighbour nodes (via the network edges). This adaptive procedure increases the rate of sampling individuals of interest based on the network edges (Thompson 2011) but unlike respondent-driven or "chain-referral" sampling (Heckathorn 1997), it is driven by the researcher rather than by sample individuals.

In either form, adaptive sampling has the advantage of increased "return" on effort; there are more samples of the kind needed for the study than there would be from random sampling. The drawback, of course, is the increased autocorrelation in the sample: the samples are not fully independent due to their spatial association in the first case, or due to their circumstantial similarities and (probably) shared contacts in the second.

The same considerations of lack of independence apply when the mode of sampling itself results in clusters of samples, as in any snowball, adaptive, or "crawler" sampling scheme. Fish community data from fisheries research provide a good example of cluster sampling in collecting data for (correlation) networks (Nelson 2014). Because the samples are acquired as multiples rather than as individuals, for example, by seine nets, the standard (independent random sample) methods of statistical analysis do not apply. Clusters of samples are also collected by design in *adaptive cluster* sampling, a method used frequently in studies of ecological communities (e.g.

Roesch 1993 for forest insects; Smith et al. 1995 for waterfowl). Even if not explicitly converted into a network graph, any such compositional study can form the basis for a correlation network of the taxa encountered. The lack of independence in these cluster samples can be addressed in part, as in data with spatial autocorrelation, by invoking the concept of the effective sample size, which is the number of *independent* samples that would provide the same information content as the dependent set (Nelson 2014; see also Dale & Fortin 2014). This may be difficult to do, and Krivitsky and Kolaczyk (2015) have investigated the effective sample size of growing networks using "independent dyad" (such as mutual friends) exponential random graph models (ERGMs). Depending on the sparseness or density of edges per node, the effective sample size can range from $O(n)$ to $O(n^2)$, the latter meaning that computing time increases with the square of the number of nodes. Preliminary results for triadic (friend-of-a-friend) models are similar (Krivitsky & Kolaczyk 2015).

The lack of independence in samples is also a consideration in sampling existing networks rather than using samples to create networks. The implications of adaptive sampling or cluster sampling of any kind will depend on the nature of the network (what it represents) and whether it is assortative or disassortative: Will sampling first-neighbour nodes increase or decrease the proportion of samples with the desired characteristics? What are the implications for effective sample size? Although positive autocorrelation decreases the effective sample size, negative autocorrelation, where it exists, can actually *increase* the effective sample size beyond the nominal value (see Dale & Fortin 2014).

3.2.2 Incomplete Data

The problem of incomplete data is ubiquitous in ecology, as in a great range of scientific endeavours that involve statistical analysis. In classical statistics, missing data are usually part of a rectangular data matrix representing the outcomes of experiments or observations, and there are many ways of dealing with these missing bits, depending on the data structure (Little & Rubin 1987). One option is to discard the incomplete records, but that can be wasteful or lead to biased results, and a second option is to fill in the missing values with best estimates, calculated, for example, by regression on the other observed values, an "imputation-based" procedure (Little & Rubin 1987; Ellington et al. 2015). A third option is a weighting procedure using the inverses of the responses' probabilities, and a fourth is modelling using maximum likelihood to give estimates of parameters describing the responses (Little & Rubin 1987). We will discuss these options further in the context of the statistical and other analyses of ecological networks; for some, the options for dealing with missing data are very limited due to the nature of the system and the form of the data (e.g. the animal social networks of behavioural ecology; see Newman 2010).

The challenge for ecologists is to estimate as accurately as possible the important statistics that characterise the network of interest when the data are incomplete (too often the case!), whether missing nodes or interactions between nodes (Bliss et al. 2014). While some studies of missing data rely on the assumption that the proportion missing is known somehow (e.g. Bliss et al. 2014), in ecological networks, we will usually not know the proportion of the network that the observed subset represents. This gap will have direct implications for our interpretation of the estimates derived from the subset observed.

As a final area for consideration, Gera et al. (2018) have discussed the appropriate procedures for assessing a large network derived from data in a series of time-slices, each of which provides only partial information on the whole network. The samples of networks can be compared by probabilistic calculations involving counts of nodes or edges, or similarities may be compared with statistics such as degree distributions, clustering coefficients, path lengths, and so on. Gera et al. (2018) listed alternative approaches including kernel functions and graphlets, but describe in detail the usefulness of spectral analysis by the sequential examination of the adjacency, Laplacian, and normalised Laplacian matrices. Gera et al. (2018) have suggested that this sequential approach is a promising method for network data in time slices.

3.2.3 Network Inference

3.2.3.1 Estimation

From sampled networks, we can derive estimates of the network measures of interest. These would usually include the numbers of nodes and edges, the degree distribution, average degree, and the clustering coefficient. Kolaczyk (2009) added node betweenness centrality to this list, and Bliss et al. (2014) included the maximum node degree (possibly inherent in the degree distribution estimates) and the proportion of nodes in the giant component. The important concept for estimating node properties is to acknowledge that the nodes have unequal probabilities of being included in the sample and to include those probabilities as weights in forming the estimate, using what is called the *Horvitz–Thompson estimator* (Kolaczyk 2009; Bliss et al. 2014). For a set of sampled nodes, S, where node i has observed value y_i and probability p_i of inclusion, the estimate of the total, τ, given $\{p_i\}$ is

$$\hat{\tau}_p = \sum_{i \in S} y_i / p_i \tag{3.1}$$

Many of the properties of interest can be expressed as totals, and so versions of this Horvitz–Thompson estimator can be used to derive estimates for all of those, and the estimates derived in this way are unbiased (Kolaczyk 2009). The concerns that remain are (1) what is basis for sampling method (nodes? how chosen? edges? how chosen?) and (2) what is our knowledge of the probabilities of the elements that form the sample? For example:

- From node sampling, seeking the average degree (k-bar), y_i is taken as the degree of node i (see Kolaczyk 2009; Bliss et al. 2014) and

$$\text{est}(\bar{k}) = \hat{\tau}_p \frac{\hat{N}}{n} = \frac{\hat{N}}{n} \sum_{i \in S} y_i / p_i = \text{obs}(\bar{k}) \frac{\hat{M}}{m}. \tag{3.2}$$

We will not work through all combinations of sampling methods and properties to be estimated, but refer to the authors cited.

3.2.3.2 Significance and Randomisation

Bliss et al. (2014) have described the scaling effects on the estimates of network statistics based on four different sampling regimes:

- Random node-induced subnetworks (some nodes and all their edges)
- Random link-failure subnetworks (some nodes but some of their induced edges omitted; Figure 3.4b)
- Random edge-incidence subnetworks (some edges and all their end-nodes)
- Random interaction event subnetworks (some interaction events, for example, a subset of all competitor A versus competitor B events; e.g. 9/13 in Figure 3.4b).

Developing estimated values for the statistics of interest is sufficiently problematic as to indicate the difficulties of attempting to derive significance test for those estimates, and there are so many sources of lack-of-independence in the information. While Horvitz–Thompson methods can be used to derive variances and unbiased estimates of the variances of some of the statistics estimates, these can become very difficult to calculate depending on the sampling design (Kolaczyk 2009). Insight can be gained by compiling the results of iterated resampling of the same network using the same sampling method and sample size (see Kolaczyk 2009). Randomisations, in particular constrained or restricted randomisations, are often the best approach, and the comparison of results with the outcomes from known and familiar network models can be the most convincing route to evaluation.

3.3 Statistical Considerations

3.3.1 Statistics and Ecological Networks

Biological data of all organisational scales produce interesting networks for analysis. In ecological studies, these can range from the level of molecules, genes, and individuals, to populations and communities (Junker & Schreiber 2008; Stumpf & Wiuf 2010; Estrada 2012; Dormann et al. 2017). While the "raw" data, as collected, can sometimes be used to create a network of interest directly, we often either create or evaluate the ecological network using statistical inference, hoping to distinguish the most important results based on statistical significance. As is often the case in the analysis of ecological data, network analysis is affected by several sources of lack of independence, which all affect our interpretation.

As an illustrative example, let us consider how we study the co-occurrence patterns in communities of organisms such as plants or microbes (Williams et al. 2014; Weiss et al. 2016), based on calculations of correlation. Motivations for such studies include the determination of drivers in environmental ecology, the understanding of habitat niches or disease, or the generation of hypotheses on which interactions are worth further study (Weiss et al. 2016). The abundance (could be presence-absence data) of each of n taxa (or "operational taxonomic units") is determined in a sample design appropriate to the study goals, and all possible pairwise correlations are determined; there are n $(n - 1)/2$ of those. The network graph is then created by including edges with "large" or "significantly large" values, whether positive or negative; and choosing those with large absolute values is implicitly performing a statistical test with the null hypothesis that the correlation is not different from zero (Kolaczyk 2009). Weiss et al. (2016) have described difficulties inherent in this approach, some of which may be familiar from discussions of determining pairwise species associations in plant communities based on quadrat data and goodness-of-fit tests (see Dale 1999).

To begin with, microbial community data may include at least 5000 "features" and so there may be more than 12 million correlation coefficients (Weiss et al. 2016). If each choice of a "large" valued coefficient is equivalent to performing a statistical test, this is a multiple testing problem with millions of simultaneous tests. Second, the correlation pairs are not independent. A positive correlation of A and C is less surprising (and less informative) if there are positive correlations of A with B and B with C. Third, the relationship between taxa A and B may change magnitude or sign depending on the presence or absence of taxon C, or of taxa C and D! (See discussion for species associations in Dale 1999.) As Weiss et al. (2016) phrased it: because microbes live in communities (and they may function as consortia), we need to consider "three-feature interactions, four-feature, and more." The rarity of taxa or of combinations of features, missing data due to incomplete sampling, and non-linear responses add to the complications of statistical analysis (Weiss et al. 2016). These authors do not even mention two further sources of difficulty: the possible presence of spatial or temporal autocorrelation among the samples used to create the network (see Dale & Fortin 2014), although they do cite the possibility of creating correlation networks from time series data.

Several methods were successful in detecting three-member interactions on their tests, but Weiss et al. (2016) suggested that interactions with more taxa may be impossible to detect with any of the methods currently available. We will return to a further discussion of networks derived from these calculations of correlation in more detail, but this complex and sophisticated study illustrates the difficult relationship between "ordinary statistics" and their application in large and complex networks with multiple-feature interactions.

3.3.2 Correlation Networks

Correlation networks are widely used in biochemical, metabolic, genetic, and compositional studies (Friedman & Alm 2012; Lasserre et al. 2013; Batushansky et al. 2016). Correlation network analysis has several purposes, but two of the most common are these (Layeghifard et al. 2017):

1. Detecting and characterising clusters of the nodes, whatever they represent (e.g. micro-organism taxonomic units ("OTUs," Williams et al. 2014).

2. Identifying nodes that are most critical in the structure and function of the system being studied ("keystone" nodes, e.g. Dempsey & Ali [2011], who examined several centrality measures, although keystoneness should be more about network function whereas centrality is a structural measure).

Both these goals are the same as those of the familiar approach known as association analysis, whether in ecology or business. Correlation networks are formed by calculating a correlation coefficient (often Pearson's or Spearman's) or some other measure of similarity between each pair of "objects" in the study (Batushansky et al. 2016; Weiss et al. 2016). If there is sufficient information on causal structure in the system, partial correlation coefficients based on Pearson's may also be a good choice (see Kolaczyk 2009). Weiss et al. (2016) have investigated and compared eight similarity measures, including the familiar Bray–Curtis, Pearson's, and Spearman's, but also newer methods such as sparse correlations for compositional data (SparCC; Friedman & Alm 2012), local similarity analysis (LSA; Ruan et al. 2006), and maximal (mutual) information coefficient (MIC; Reshef et al. 2011). They also used what are called *copula functions* to generate random variables with specified covariance matrices, although the same approach could be used to quantify the dependence between variables of interest.

We will not repeat the full technical details of their investigation, but they found much variability among the results of different methods and the newer methods LSA, MIC, and SparCC performed well but under different conditions. Not surprisingly, applying more than a single method can improve precision. The main point, however, is that the standard measures have their own sensitivities to features such as the distribution shape (e.g. symmetric normal vs. skewed) or linear versus non-linear responses. The answer proposed is to choose copula functions to assess the dependencies between pairs of variables and then to consider modified approaches for the multivariate case (Weiss et al. 2016).

How do we create the network given all the pairwise correlations? How best to determine which edges are included? Having chosen a correlation or similarity measure, there are choices on how the edges are determined to create the network itself. These include (Dempsey & Ali 2011)

1. A single threshold value selects the edges with the largest positive values, creating an unweighted, unsigned network graph. The threshold may be determined for the correlation coefficient's magnitude, for its significance level (e.g. p value), or for a combination of the two.
2. Two threshold values distinguish edges with large negative correlations and large positive correlations, thus creating an unweighted but signed graph. The criteria are usually symmetric, for example, include edges with $|r| > 0.8$, but do not need to be. Again, thresholds may be correlation coefficient values, their p values or a combination of the two.
3. Retain all correlation values as weights for the edges to form a weighted network for further analysis, such as "weighted gene co-expression network analysis" (WGCNA, Zhang & Horvath 2005) which is very popular in genetic studies. More

generally it is "weighted correlation network analysis" (Horvath 2011) and it can be adapted for other kinds of data, such as community composition (Duran-Pinedo et al. 2011; Geng et al. 2016). This kind of analysis does not use a "hard" threshold such as absolute value or a set p value, but uses "soft" thresholds for the correlation coefficients themselves (Horvath 2011).

The first approach seems to be wasteful of the information available (see Costantini & Perugini 2014) whether the threshold is set low or high, but it is often followed to give precedence to positive associations (e.g. Dempsey & Ali 2011; Duran-Pinedo et al. 2011; among many). The third approach is popular, and weighted correlation analysis appears frequently in research in many areas from biochemistry and medicine to ecology and community studies.

The second approach, which produces both positive and negative edges without weights, is the one we will discuss in more detail. Even given the bias towards "positive" results in much of science (Kolaczyk 2009), we can still ask why anyone would discard the negative edges of a correlation network. This choice retains them, but allows some flexibility on treating positive and negative coefficients differently, by allowing asymmetry of the thresholds for inclusion.

In a subsequent section, we will describe the analysis of networks for structure–function relationships using small subgraphs called graphlets (Pržulj 2007 and subsequent publications), introduced in Chapter 2. Graphlets were originally used to investigate the relationship between structure and function in unsigned and unweighted networks, but they can be usefully modified for signed network graphs. Correlation networks that do not discard edges with negative weights have graphs to which signed graphlet analysis can be applied. These graphlets should provide key insights into the relationship between structure and function in these systems. In addition, most examples of the third approach, which produces signed weighted edges, can be converted to a signed but unweighted form by the use of suitable thresholds (e.g. Friedman & Alm 2012; Lasserre et al. 2013), allowing signed graphlet. There may be some real advantage to complementing WGCNA by a graphlet-based investigation of network structure and function.

In the analysis of signed networks, we are looking for clear groupings of the objects (species or products) or of the collections within which the objects occur, and we are trying to determine which nodes are most important for how the system operates (as with keystone species or transport hubs). As in the overall search to understand the relationship between topology and dynamics, it is easy to see that signed graphlets provide a promising method to identify the roles of individual nodes or classes of nodes in correlation networks. A main goal of studying signed correlation networks is to move from these undirected networks to directed acyclic networks (DAGs) that show causal relationships (see e.g. Opgen-Rhein & Strimmer 2007). In Section 3.5, we provide a detailed discussion of causal networks and their place in ecological research.

3.3.3 Methodological Comment

We are all familiar with coefficients used to quantify the mutual dependence or similarity of pairs of variables like Pearson's or Spearman's measures of correlation. As mentioned, these standard measures have their own sensitivities to features such as the distribution shape (e.g. symmetric normal vs. skewed) or linear versus nonlinear responses; and a more flexible general approach is offered by what are known as *copula functions* (Fisher 1997; Nelsen 1999). Copulas as measures of dependence have appeared seldom in ecological studies, but they are common in financial applications and show great promise for ecological research (Warton et al. 2015). Copulas provide scale-free measures of dependence between variables or of one variable on many, as well as providing the basis for designing and creating families of bivariate distributions (Fisher 1997; see also Nelsen 1992, 1999). Copulas are bi- or multivariate distribution functions with one-dimensional marginal distributions that are uniform on [0, 1]. When used to measure the dependence of one variable on another or to quantify multivariate dependence, they have the advantage of not assuming a linear response and not relying on Gaussian distributions for valid application (see Zhang & Shi 2017).

Where a multivariate copula might be required, it is often replaced by the hierarchical application of bivariate copulas, with an increasing number of variables as conditionals; in particular, the vine copula. A vine for n variables is a set of nested trees in which the edges of tree j are used as the nodes of tree $j + 1$ and joined (for example) if they share a node in tree j (Figure 3.8a). The edges at each iteration have as weights the partial correlation appropriate to their end "nodes," which may be the edges of the preceding level of tree Figure 3.8b and c. Of course, the vine needs a copula for each edge to create a complete vine describing the multivariate dependence for the full set of variables.

The basic concept of vines (Joe 1997; Bedord & Cooke 2004), as described here specifically for copula functions and correlation measures, provides the right format

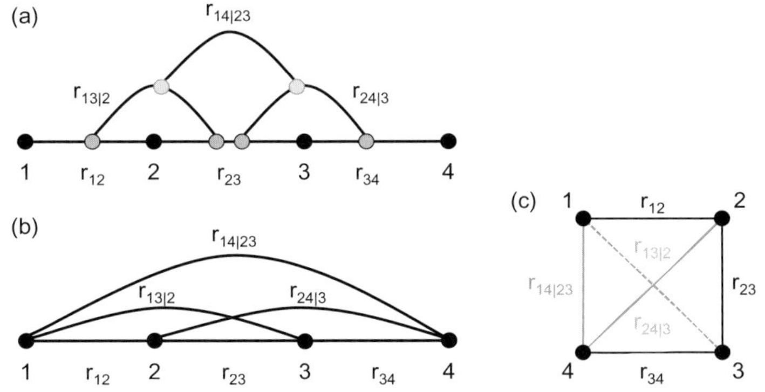

Figure 3.8 A regular vine on four nodes (variables 1–4): partial correlations as indicated. The vine (a) can be redrawn as a complete graph (b and c: two versions of how that looks).

for any discussion of "interactions among ecological interactions" (Golubski & Abrams 2011; Delmas et al. 2019). Consider the effects of two parasites on one host, and then the interaction between the two specific forms of parasitism. That creates a second-order edge between the two edges that represent parasitism. A second example is from Delmas et al. (2019; cf. Golubski & Abrams 2011), where a precursor infection on the host enables an opportunistic parasite to attack the host. Figure 3.8 shows a similar hierarchical layering of interactions between variables. The relationship of variables 1 and 2, r_{12}, and the relationship between variables 2 and 3, r_{23}, interact at the next level, designated as $r_{13|2}$, and so on. When we move on to discussions of multilayer and other possible complexities of multimode interactions, this vine-inspired approach to the relationships among interactions will definitely come into its own.

3.4 Structure and Function in Dynamic Networks

3.4.1 Dynamic Networks: Concepts and Attributes

At the beginning of this discussion, we distinguished between *dynamics on* a network, where the topological structure (nodes and edges) remains constant while other properties such as categories and weights change, and *dynamics of* a network, with attendant gains and losses of nodes or edges (Figure 3.1). In either case, the network is a *dynamic system*; its state is described by quantitative variables and changes through time according to rules or sets of equations (Newman 2010). While changes to topology and changes to function can sometimes be considered separately, in ecological networks we are often most interested in the reciprocal changes of structure and function (Figure 3.2), and such systems are usually considered to be *adaptive networks*, those in which topology and states coevolve (Gross & Sayama 2009; Sayama et al. 2013; Moore et al. 2018). Ecological adaptive networks have the additional feature to consider: they are affected by external factors or by the interaction with other different networks (Figure 3.3). Because dynamics must involve time, we need to consider the possibilities of discrete time intervals, with iterative change perhaps described by difference equations, or continuous time, often involving description by differential equations (Newman 2010), although one converges to the other as the discrete time interval, δt, shrinks to an infinitesimal that gives instantaneous transition probabilities (Kolaczyk 2009).

3.4.2 Stability, Resilience, Robustness, and so on

The list of terms describing the ability of a network to resist, absorb, or recover from disturbance is long: stability, resilience, robustness, recovery, persistence, restoration, sustainability, and so on; contrasting with terms such as fragility, vulnerability, catastrophic shift, failure cascades ... (see, among very many: Holling 1973; May 1973; Pimm 1984; McCann 2000; Gunderson 2000; Solé & Montoya 2001; Ives &

Carpenter 2007; Scheffer 2009; Gao et al. 2016). In general, for ecological networks, the property of greatest interest is probably resilience, which is the ability of the system to maintain its main and basic functions when disruptions, disturbances or errors have occurred, *even when they have resulted in changes to the network itself* (our emphasis) (e.g. Gao et al. 2016; see also Haegeman et al. 2016). Ecological studies have usually emphasised questions about the relationship between diversity (or complexity) and stability (or resilience) (May 1973; Pimm 1984; McCann 2000; Ives & Carpenter 2007; Gravel et al. 2016), but questions about local and global structure have also been considered intensively (Thébault & Fontaine 2010; Solé & Montoya 2001). We will not review all the literature on ecological network stability but will focus our commentary on the relationship between structure and function, especially as it brings insight to the concept and nature of resilience. Resilience is just one form (or version) of stability (Holling 1973) in ecological systems, and Thompson and Williams (2017) distinguished among nodal resilience (a reduced population recovers on its own), landscape resilience (an extinguished population is re-established by re-colonisation), and network resilience (interactions among organisms allow a reduced population to rebound), as illustrated in Figure 3.9.

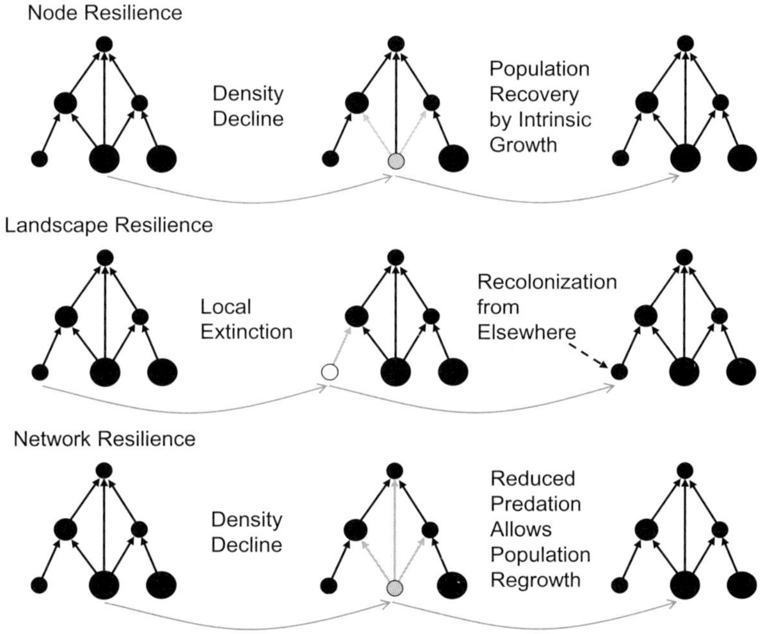

Figure 3.9 Population resilience in a food web after disturbance. Three modes are node resilience, landscape resilience, and network resilience.
[Redrawn with permission from Thompson, R. M., & Williams, R. (2017). Unpacking resilience in food webs: An emergent property or a sum of the parts. In J. C. Moore, P. C. de Ruiter, K. S. McCann, & V. Wolters (eds.), *Adaptive Food Webs: Stability and Transitions of Real and Model Ecosystems*, pp. 88–103. Cambridge: Cambridge University Press, republished with permission.]

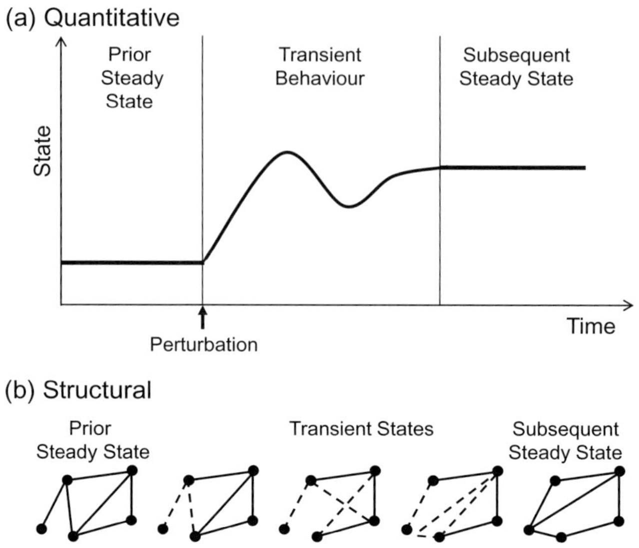

Figure 3.10 Transience: progression from a previous steady state through a period of transient change, following perturbation or disturbance, to a subsequent steady state established. (a) Quantitative. (b) Structural.

The property of resilience may be broad enough to allow that when a system is perturbed, the post-disturbance steady state that re-establishes may not be the same as the prior state. In some cases, ecological systems, like electrical circuits, may exhibit transient behaviour between the perturbation and the establishment of the new steady state (Figure 3.10, see also Hastings 2001, 2004; Hastings et al. 2018). This transience may take the form of a sequence of changes in the state of the system that may temporarily overshoot or undershoot the previous and subsequent states as in Figure 3.10a. Think of it as "bounce." Generally, transience describes changes to the quantitative function or dynamics of the system, such as flow, but it could also apply to system topology and structure with the loss and establishment of links (network edges) in the system as in Figure 3.10b.

Before discussing resilience and related concepts further, we need to remind ourselves that the topic should be network analysis and network statistics, not ecosystem modelling or experimental approaches to evaluating community stability. Therefore, our questions should be about what measures or what analyses provide the best information on network stability, on resilience, and on fragility. It is tempting to develop easy and informal equivalence between graph theory properties and ecological characteristics that are related to stability, but that could be misleading. There is a certain degree to which properties can be loosely matched, as laid out in Table 3.1 (Dale 2017), but it would be a mistake to suggest that any such pairs are fully equivalent. Thus, we might justifiably object to labelling a *graph theory* property as "resilience" or "fragility," but we know from spectral graph theory that the *algebraic connectivity,* the *second smallest* eigenvalue of the Laplacian matrix, indicates greater

Table 3.1 Matching ecological stability and graph theory properties of networks

For stability or coexistence	Graph theory property
Competition hierarchy reversals	Intransitivity, hierarchies with reversals
Association network dynamics	Dynamic balance and intransitivity, may prevent extinctions
Food web hierarchy	Blurred levels (omnivory) may aid persistence
Food web dimensionality	Graph intervality
Keystones species	Degree and neighbour degree distributions; centrality
Composition nestedness	Node degree relationships
Redundant pathways	Node connectivity, algebraic connectivity

Ecological characteristics indicating stability or coexistence or resilience may match some graph theory characteristics of their networks. In some cases, orderliness (see Dale 2017, table 12.2) can increase stability (e.g. modularity in food webs); in other cases, the opposite may occur (e.g. strict competition hierarchies may enhance extinction). The relationship between stability and structure may depend on context: hubs and keystone nodes may be stabilising under normal conditions but can be sites of network vulnerability under rare circumstances.
From Dale (2017).

structural resilience or robustness. The maximum value of that eigenvalue is the node connectivity of the graph, which is the number of nodes that need to be removed to disconnect it (Chapter 2). Obviously, this is a structural property *of* the network graph and is not the same as the conditions for functional resilience *on* an unchanged network structure.

In the general network literature, what is called the *robustness* of a network is its resilience to the loss of components or component failure so that it continues to function when nodes or edges are removed (Klau & Weiskercher 2005). The same literature usually makes the distinction between targeted attacks that focus on important components of the network, such as hubs, and random failures which are not targeted (Barrat et al. 2008). In addition to the connectivity metrics already described as measures of network cohesion, Klau and Weiskercher (2005) suggested two that are based on probability. The first of these is the *reliability polynomial*: for network graph, G, and the random failure of any edge with probability $(1 - p)$, the reliability polynomial $R(G, p)$ is the probability that the graph remains connected. The second is the *probabilistic resilience*: this is the largest number of node failures such that the graph remains connected with some probability $(1 - p)$.

By whatever means the robustness of ecological networks is evaluated, it is no surprise that it varies from network to network. Pocock et al. (2012) assessed several networks in the same agroecosystem and found that predicted variability in robustness, with networks that include pollinators being particularly fragile. Not only that, but there was also low covariance among network robustness, indicating that ecological restoration efforts that might benefit one functional group may not benefit the others (Pocock et al. 2012).

The statistics or methods of network analysis that will provide the best insight into network dynamic properties related to stability will depend strongly on the kind of network and the kind(s) of interactions it includes. The evaluation of these statistics may well rely on models.

Based on models, to understand the relationship between structure and local stability, we often consider networks in which the edges are governed by equations that determine their dynamics. Those dynamics can be related to a matrix of first derivatives of those functions, called a Jacobian matrix, and its properties indicate the stability of the system it describes. For example, a trophic network's dynamics can be based on the equations such as Lotka–Volterra, that describe predator–prey interactions. A qualitatively stable trophic system is one that is stable based on the signs of the Jacobian matrix (e.g. at least one negative on the main diagonal), and Allesina and Pascual (2008) used this concept for the basis of studying one version of system stability ("quasi sign"). In turn, Borrelli (2015) used the same approach to rank 13 possible configurations of three species that allowed double edges (motifs, really; see Borrelli 2015). The main result of this study was that empirical trophic networks tend to have an over-representation of the more stable 3-node subgraphs across marine, freshwater, and terrestrial systems.

Gravel et al. (2016) used Jacobian matrices of interspecific interactions and of dispersal to study stability and complexity in the context of modelling metaecosystems to investigate the role of spatial dynamics. The structure used was a spatial network of local interaction network (a network-of-networks or a metanetwork) including also the effects of intraspecific density dependence and the dispersal between patches. Their study confirmed that spatial dynamics can stabilise metaecosystems. These stabilising effects are proportional to dispersal rates when they are low and proportional to metacommunity size when the dispersal rates are high, and the effects are strongest at intermediate rates. In this kind of system, therefore, network connectivity and the distribution of diverse local environments have a critical influence on the species diversity of the whole.

3.4.3 Relating Structure and Function Using Graphlets

Network topological structure affects the dynamic processes ***of*** or ***on*** the network, and the dynamic functioning of a network can change both its non-structural dynamics (i.e. *in* or *on* but not *of* the network), and the structure itself (Figure 3.1). We can compare an ecological network with a railway network in which the physical elements and layout determines how machines can use the network and thus its traffic flow; but the network's machines have the ability to modify the locations and capacities of the network routes by building new tracks, new bridges, and so forth and demolishing previous routes and infrastructure (Dale 2017). This reciprocal and dynamic interplay is an essential characteristic of ecological networks, but it is difficult to determine and to quantify. The difficulties relate to familiar concerns about determining process from pattern and understanding fully the relationship between form and function. What can we know about the dynamic functioning of ecological systems from studying their structure? What can we understand about their structure by studying their dynamics? A key concept is that of *functional topology* (Pržulj et al. 2004), which indicates the confidence that some knowledge of function can be determined by understanding network structure. The dangerous implication is that, given sufficient contextual knowledge, function can be predicted from topological structure; the analogy is that, given enough background on how trains and railway tracks work, you can predict the function of the roundhouse turntable without seeing it in action (see Dale 2017). In ecological systems, this becomes even more difficult because usually more than one network with reciprocal affects between networks need to be considered (Figure 3.3). One promising approach is using graphlets for network analysis. Graphlets were introduced in the previous chapters for the discussion of measures of network and graph complexity. Much of the work on graphlets is due to Pržulj and her research group (e.g. Pržulj 2007; Milenković & Pržulj 2008; Yaveroğlu et al. 2014), and what follows is a summary of that approach and its implications.

3.4.3.1 Basic Graphlets (No Directions or Signs on Edges)

As introduced in Chapter 2, Section 2.1.7, a graphlet is an induced connected subgraph of the network graph of interest; *induced* means the subgraph must have *all* the edges of the original that join nodes in the subgraph. *Connected* means you can trace a path along nodes and edges from any node to any other. In theory, a graphlet could be relatively large (perhaps up to 10 nodes) but for practical reasons the graphlets investigated have only up to 4 or 5 nodes. The first structural evaluation is to count the occurrences of each graphlet in a given network as the basis for inference and comparison, just as is done with "motifs" (also subgraphs of usually three or four nodes; Milo et al. 2002). These counts are the *graphlet frequencies* (Milenković & Pržulj 2008) and differences between two instances of a network can be catalogued by the changes in these frequencies (see later in this section). The most common subgraphs, however, may not be the most important structurally, and so it is recommended to count the frequencies of nodes at uniquely identifiable positions in the graphlets, rather than just whole graphlets (Milenković & Pržulj 2008). Among the three nodes of graphlet g1 (Figure 3.11), the two end-nodes are structurally indistinguishable, with one edge each to the central node of degree 2, whereas that central node is structurally different. The two end-nodes therefore belong to the same automorphism orbit (they can be mapped onto each other without changing the structure), call it orbit 1, but the central node is distinct and belongs to a different automorphism orbit, orbit 2 (see Milenković & Pržulj 2008). Changing to counts of nodes in orbits from simple graphlet frequencies enhances the sensitivity of the analysis, as will be shown.

The first graphlet, g0, is two nodes joined by an edge; the nodes both belong to orbit $\gamma 0$ (Figure 3.11). (We have designated the orbits of the original basic graphlets with "γ" because other kinds of graphlets will be introduced with their own orbit designations: "δ" for directed, "σ" for signed, and "τ" for signed and directed.) There are two basic graphlets possible for three nodes: graphlet g1 is a two-edge path graph with two end-nodes in orbit $\gamma 1$ and a centre node in orbit $\gamma 2$. Graphlet g2 is a triangle with all nodes in $\gamma 3$ (Figure 3.11). For $n = 4$, there are six graphlets, giving orbits $\gamma 4$ to $\gamma 14$ (Figure 3.11). The frequencies of whole graphlets in the entire network graph are

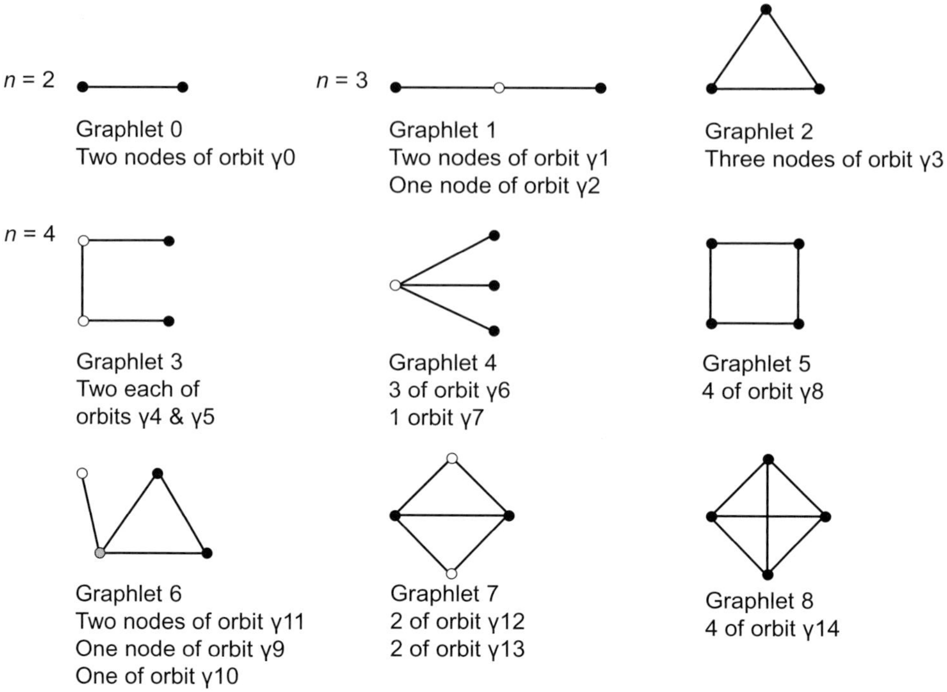

Figure 3.11 Graphlets for basic graphs with no signs or directions on the edges. $n = 2, 3$, and 4. Within any graphlet, different shades indicate different automorphism orbits.

indicative of the overall properties of local structure. Figure 3.11 gives a catalogue of all graphlets with two to four nodes, and a list of all their orbits (each distinguishable node position is a separate orbit). Some graphlets have nodes in three different orbits (g6 has four nodes in three orbits) whereas some have only one orbit (g0, g2, g5. . .) because all the nodes occupy equivalent positions. For 2- to 4-node graphlets, there are 8 graphlets and 15 orbits, and 29 and 73, if $n = 5$ is included. Each orbit provides its own "dimension" to describe a graph in 15-dimensional space using the totals or proportions of nodes in the full census of orbits (count illustrated in Figure 3.12). Similarly, any single node can be characterised by its frequency distribution in these orbits, called its *graphlet orbit signature* (originally called "graphlet degree signature"; Milenković & Pržulj 2008). The whole network graph is described by the frequency of its nodes in these orbits, which can be called its *graphlet orbit distribution* (originally called "graphlet degree distribution"; Milenković & Pržulj 2008). We suggest changing the terminology from "degree" to "orbit" to avoid possible confusion with "degree" as the number of edges at any node. Figure 3.12 illustrates how a graph can be disarticulated into component graphlets, producing graphlet frequencies of (8, 7, 4, 0, 0, 1, 0, 4, 0) and an orbit distribution of (16, 14, 7, 12, 0, 0, 0, 0, 0, 4, 0, 0, 0, 8, 8, 0). It also shows how node 1 has the orbit signature of (3, 3, 1, 2, 0, 0, 0, 0, 1, 0, 0, 0, 1, 2, 0).

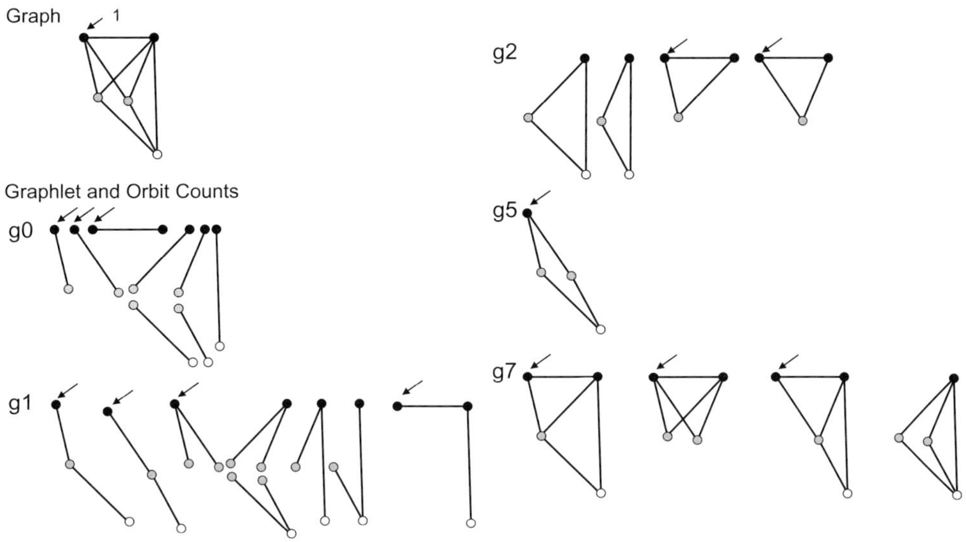

Figure 3.12 Counting graphlets and orbits. How a graph is decomposed into possibly overlapping graphlets and the node orbit occurrences are enumerated (node 1 indicated by arrow).

Although each orbit can be considered as a separate descriptive dimension, the orbits are not fully independent: any 4-node graphlet contains some 3-node graphlets; for example, any node in orbit $\gamma 4$ (in g3) is always also in orbit $\gamma 1$ (in g1) as shown in Figure 3.11. Comparing graphs based on their orbit distributions therefore must consider the effects of such redundancies Figure 3.13b. The 15 dimensions of all orbits can be reduced to 11 with the redundancies removed; having done so, any dependence that remains must be due to the topology of the particular network and is therefore the information we are looking for (Yaveroğlu et al. 2014). For a single network graph, the 11 dimensions and 121 correlations between the graphlet orbit signatures of all its nodes provide an 11×11 *graphlet correlation matrix* (GCM). Clearly there is a great advantage to be able to reduce even the largest of network to an 11×11 structural summary, which can be portrayed as a "heat map" (Yaveroğlu et al. 2014). This format is an 11×11 grid of coloured squares: the red end of the spectrum for large positive scores and the blue end for large negative scores. This visual presentation of the results may be helpful, but it is not the most important outcome. The important part is the ability to use the correlations to follow the coevolution of the network structure and the flow within it, as in the example of the world trade network observed from 1962 to 2010 detailed in Yaveroğlu et al. (2014).

In that example, the nodes of the network are countries and the edges are weighted by the yearly amount of trade between any two countries. The insights that the graphlet analysis provides for that adaptive network are the roles of particular nodes as "brokers" versus "peripheral" in both position and influence, based on their orbit signature. In that example, the correlation matrices show that the orbits fall into positively associated clusters reflecting brokers (orbits 2, 5, 7, 8, 10, and 11) versus

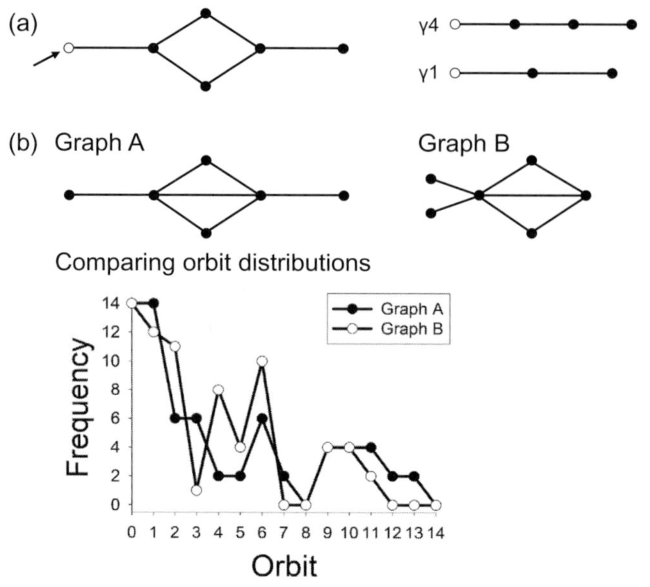

Figure 3.13 Orbit distributions and redundancies. (a) Any node in orbit 4 is also in orbit 1. (b) Comparing orbit distributions of two similar graphs.

peripheral (orbits 4, 6, and 9). Even more insight is available by following the changes to these signatures through time (e.g. an increase in China's broker score after 2000, coincident with a long decline in Britain's) which can be tracked in relation to crude oil prices and changes in the trade network itself. These changes can be evaluated by comparing instances of the world trading network from different times through their graphlet correlation matrices (GCMs). This example provides an excellent model for the analysis of the structure, function, and evolution through time of complex ecological networks.

To compare two networks, the Euclidean distance between the corresponding entries of two graphlet correlation matrices is the *graphlet correlation distance* (GCD). This distance is an index of structural dissimilarity, which responds clearly to graph characteristics, even with noise or random losses in the data (Yaveroğlu et al. 2014). Interpreting the results may benefit from an assessment of the relationship between graphlet orbits and measured node characteristics such as environmental factors. Familiar multivariate methods can do this, e.g. *Canonical Correspondence Analysis* (CCA; Legendre & Legendre 2012), which provides a simultaneous ordination of two kinds of data. Here, they are the graphlet orbit signatures and those node characteristic measures; characteristics external to the network graphs themselves are essential for interpreting the graphlet analysis to provide the full ecological context, and methods such as CCA can help to clarify the results. The basic graphlet methods are summarised in Table 3.2.

Table 3.2 A checklist of basic graphlet methods

Method (level)	Details and application
Graphlet frequencies (network)	Counts of the 29 $n = 2, 3, 4, 5$ graphlets (show local structure)
	Diversity in the vector measures network complexity
	Compare networks by simple similarity or distance measure
Graphlet Orbit Signature (node)	73 orbits in $n = 2, 3, 4, 5$ graphlets give each node 73 coordinates
	Diversity in the vector measures (node) local complexity
	Measure node local similarity by comparing signatures
Graphlet Orbit Distribution (network)	Counts of nodes touching each orbit
	Distribution of orbit frequencies
	Measure network agreement by comparing normalised GDDs
Graphlet Correlation	Matrices and distances based on reduced list of orbits (11)
	Correlations of orbits provide clear structural profile
	Heat maps of correlation matrices facilitate comparison

Modified from Dale (2017).

3.4.3.2 Directed Graphlets

Some ecological networks in which the interactions are asymmetric are best portrayed as digraphs, for example, trophic networks, and therefore suggest the development of analysis with directed graphlets (Sarajlić et al. 2016; Trpevski et al. 2016), perhaps to be called *digraphlets*. Where only one edge is allowed per pair of nodes (i.e. no reciprocal pairs of edges), $n = 2, 3,$ and 4, provide 39 directed graphlets and 128 directed orbits (Figure 3.14). The number of graphlets increases quickly when directions are included; the 6 basic graphlets for $n = 4$ produce 34 edge-directed graphlets. Taking directions into account will be especially important when it is necessary to distinguish between divergent and convergent processes. With directed edges, g4 becomes d10 to d13, with orbits δ29 to δ38 (Figure 3.15). There is divergence or convergence at the central node, and combining or splitting flow through it, and thus these dynamics should be detected by these digraphlets. Having directions on the edges also allows an assessment of small cyclic structures in digraphs (Figure 3.16); this sensitivity to cyclic structures in network graphs may be critical to understanding the relationship between the structure and dynamics in these systems.

Trpevski et al. (2016) have used a slightly different approach to study the effective connectivity networks in the brain: edges can be either unidirectional or reciprocal, and the directional edges also have a sign indicating inhibitory or excitatory function. In contrast, the approach described by Sarajlić et al. (2016) dealt with any possible reciprocal edges by counting them in two graphlets, one for each direction. The study by Trpevski et al. (2016) concentrated on $n = 3$, with two directed edges forming a connected "wedge" (including a pipe or a collider) or three edges forming a triangle (including cyclic or transitive). For each node, they created a *signature vector* of 16

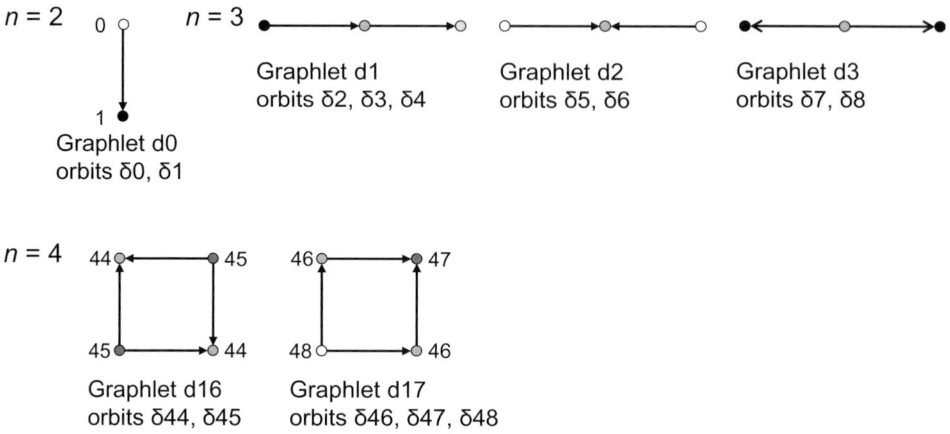

Figure 3.14 Graphlets for digraphs. For $n = 2, 3, 4$, and only one directed edge per pair of nodes, there are 39 d-graphlets and 128 δ-orbits.
[Based on Sarajlić, A., Malod-Dognin, N., Yaveroglu, O. N., & Pržulj, N. (2016). Graphlet-based characterization of directed networks. *Scientific Reports*, **6**, 35098. Copyright 2016, SpringerNature]

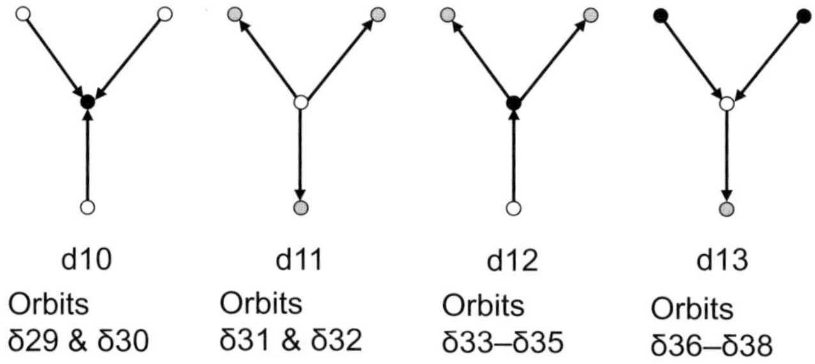

Figure 3.15 Directed graphlets for digraphs: with directed edges, graphlet g4 becomes di-graphlets d10 to d13, with orbits δ29 to δ38. These di-graphlets detect divergence versus convergence at the central node and combining versus splitting flow through it.
[Based on Sarajlić, A., Malod-Dognin, N., Yaveroglu, O. N., & Pržulj, N. (2016). Graphlet-based characterization of directed networks. *Scientific Reports*, **6**, 35098. Copyright 2016, SpringerNature]

coordinates, one for each of the 16 graphlets they used. They found that the signature vectors were dominant in the excitatory effective brain networks. Also, the excitatory networks showed stronger causal patterns than the inhibitory ones, as shown by positive and negative values in the graphlet correlation matrix. Although not an ecological study, this research does provide a good example of using directed graphlets to investigate function and causality.

For any version of directed graphlets, the most profitable applications to relate structure and function seem to be correlational, at least to begin: the graphlet

Figure 3.16 Directed graphlets. Here are 5 of the 11 d-graphlets with directed cycles. The sensitivity of these graphlets to the topology of cyclic structures in the network graphs is critical to understanding the relationship between structure and dynamics in these systems.
[Based on Sarajlić, A., Malod-Dognin, N., Yaveroglu, O. N., & Pržulj, N. (2016). Graphlet-based characterization of directed networks. *Scientific Reports*, **6**, 35098. Copyright 2016, SpringerNature]

correlation matrix (Trpevski et al. 2016), and the correlation of graphlet orbit signatures with other node attributes determined again by CCA (Yaveroğlu et al. 2014; Sarajlić et al. 2016). Sarajlić et al. (2016) have suggested that these methods allow sophisticated analysis and interpretation of topology for directed networks from any area of science. The potential for analysing ecological networks using correlation measures for directed graphlets seems to be both obvious and enormous. The important feature, of course, is that the effects are known to be directional, giving a crucial step towards convincing causal conclusions.

Other extensions of graphlet analysis are being pursued, for example, graphlets for temporal (time-only) networks (Hulovatyy et al. 2015). These dynamic graphlets investigate temporal networks and changes between observations in time slices. When we consider adding the dimension of space, for spatiotemporal analysis, explicitly spatiotemporal graphlets may not be necessary because the results from basic graphlets over several time periods may be sufficient. This is another area that is of particular interest to ecologists and deserves further detailed investigation.

3.4.3.3 Signed Graphlets

Just as considering networks with directed edges inspired directed graphlets (or digraphlets), many ecological networks are naturally represented by graphs in which the edges have signs. These would include, for example, social networks that include the equivalent of *dislike* as well as *like*, or networks of similarity or correlation measures that record negatives as well as positives. This consideration leads to creating graphlets with signed edges. To modify basic graphlets for signed graphs requires an enumeration of graphlets and their orbits with signed edges (Figures 3.17 and 3.18). There are 62 signed graphlets with 150 orbits for $n = 2$, 3, and 4. This is somewhat more than for the directed edges (40 graphlets; 128 orbits; see Sarajlić et al. 2016), which might seem counter-intuitive if we picture mapping left and right arrows

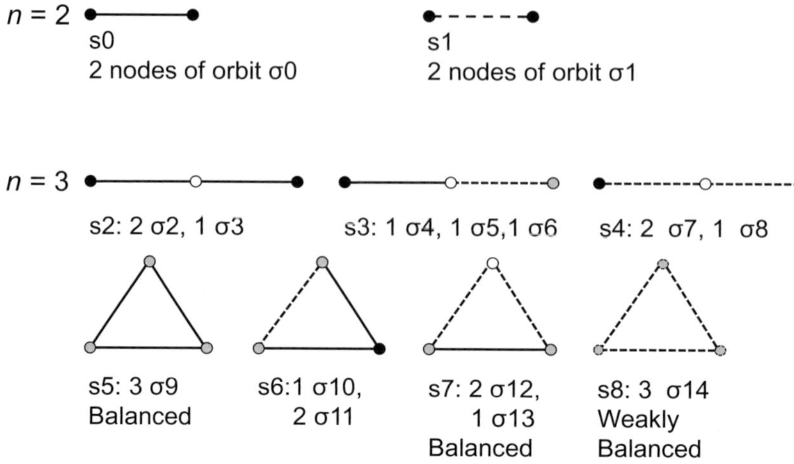

Figure 3.17 Signed graphlets (edges positive or negative) and their node orbits. Each orbit identifies nodes that occur in equivalent positions in a graphlet (automorphism orbits). Within any graphlet, different shades indicate different orbits. Graphlets s5 and s7 are balanced; s8 is weakly balanced (3-clusterable).

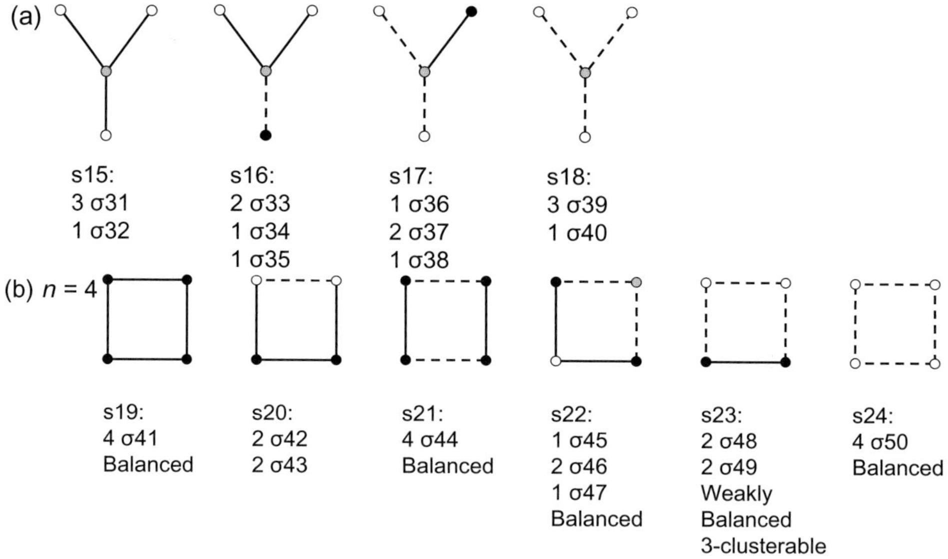

Figure 3.18 For $n = 4$, (a) the four signed graphlets derived from graphlet g4: s15 to s18. (b) Six signed graphlets derived from graphlet g8: s19 to s24. Graphlets s19, s21, s22, and s24 are balanced; s23 is weakly balanced (3-clusterable).

onto plus and minus edges, but that is not quite the case. The 4-node path graphlet with right-directed edges $= (\rightarrow\ ,\ \rightarrow\ ,\ \rightarrow)$ is essentially the same as one with left-directed edges $(\leftarrow\ ,\ \leftarrow\ ,\ \leftarrow)$ because of symmetry. The 4-node path graphlet with three positive edges (+ , + , +), however, is not the same as one with three negative edges (– , – , –). The kinds of symmetry explain why the accounting of signed graphlets is different from that of directed graphlets.

For any kind of graph, (basic, signed, directed), the full set of graphlets includes some redundancy, just due to their structural properties, for example, 3-node graphlets must contain some 2-node graphlets, and so on. For example, in the basic (unsigned, undirected) graphlets, C_0 is the number of times that any given node is in orbit 0; this is the degree of the node = number of edges = number of node neighbours. Selecting any two of its neighbour nodes, either those two have an edge between them, in which case the focal node is in orbit 3 and contributes to the count of orbit 3, C_3; or they have no edge between them, in which case the focal node is in orbit 2, contributing to the count of orbit 2, C_2. There are $\binom{C_0}{2}$ such pairs of neighbours each contributing either to C_2 or to C_3; therefore,

$$\binom{C_0}{2} = C_2 + C_3. \tag{3.3}$$

This is the first of four equations describing orbit redundancy that permit the reduction of necessary orbits (for n = 2–4) from 15 to 11 (see Yaveroğlu et al. 2014 for supplementary information). For signed graphs, determining these equations becomes complex and involves many equations (six or eight for the signed version of each for the ordinary graphlets) for many variables because there are so many orbits in the signed graphlets with the same topological structure as any one in ordinary graphlets. One suggestion is to follow Yaveroğlu et al. (2014) and omit the signed equivalents of the orbits they determined to be redundant: 3, 12, 13, and 14. For signed graphlets, we would therefore omit orbits s9 to s14, and all of s91 to s149. Removing redundancies in this way would greatly simplify the analysis, although there are still many more than the 11 of ordinary graphlet orbits (85!). Unfortunately, this approach is not correct because of the extra structural relationships due to the signs of the edges. For example, the signed graph equivalent of Equation (3.3) being one equation with three variables becomes 5 equations and 11 variables. An alternative is to follow Sarajlić et al. (2016) and to limit the signed graphlets to $n = 2$ and 3, rather than 2, 3, and 4, thereby reducing the number of orbits from 85 to 15. Even with the reduced range of n, there are several equations for many variables in the set of signed graphlet orbits. Their tests of digraphlets showed that the reduced size of the graphlets and of the number of orbits considered (13 vs. 129) gave superior performance as indicated by distance comparisons between digraphlet correlation matrices using the directed versions of a number of familiar null models and several data sets. This needs to be tested for signed graphlets.

3.4.3.4 Other Extensions of Graphlet Analysis

A next step might be to consider graphlets for signed digraphs. Limiting the size to $n = 2$ and 3, and to digraphs with no reciprocal edges, there are 23 graphlets and 59 orbits for this kind of graph (see Figure 3.19). If reciprocal or contradictory directed

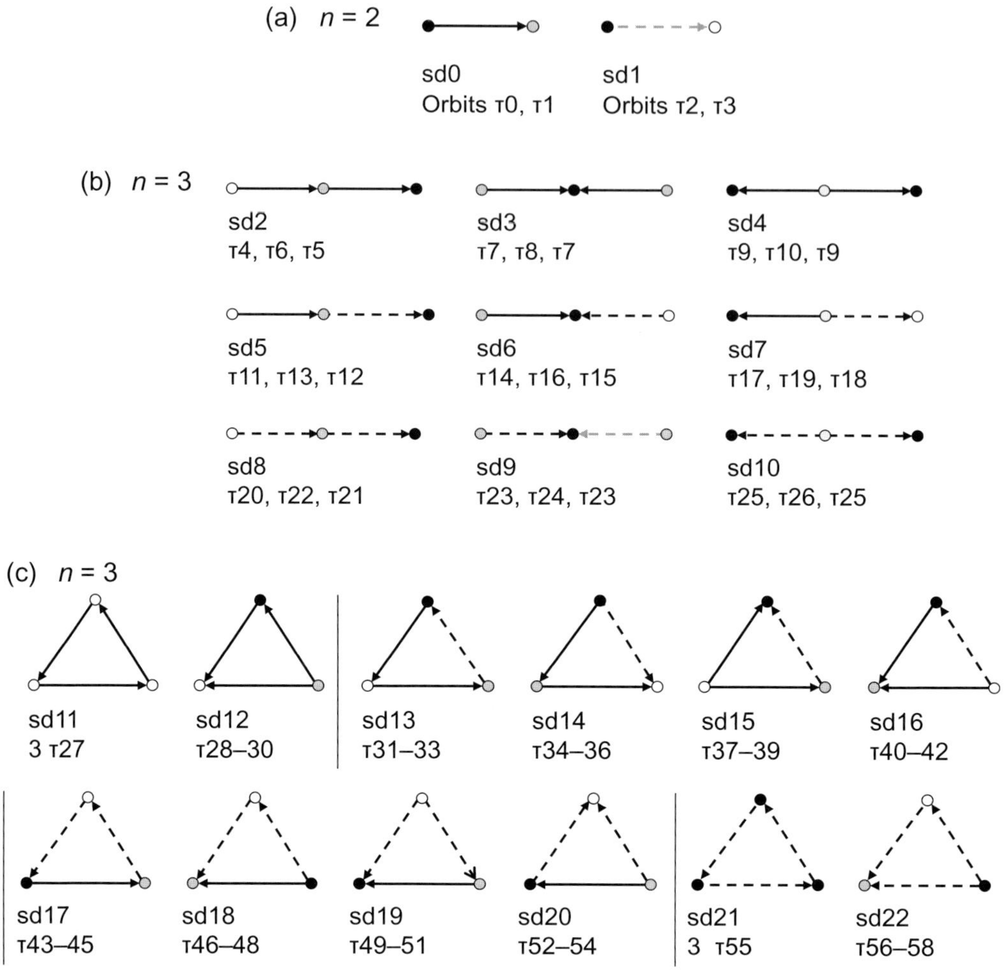

Figure 3.19 Signed digraphlets, edges positive or negative (dashed), and their node orbits. Each "orbit" identifies nodes that occur in equivalent positions in a graphlet (automorphism orbits). Within any one graphlet, different shading indicates nodes in different orbits. (a) $n = 2$, (b) $n = 3$, (c) $n = 3$, continued.

edges are allowed, a "count them twice" is one approach to analysis using these single-edge graphlets. The responses of these graphlets to both structure and functional roles require further investigation, and we await further developments, both theoretical and practical. Evaluating structure and configuration in ecological networks, such as food webs, in which the edges have both signs and direction seems to be an obvious but complex approach to understanding stability and resilience in such systems.

There are many further variations on the basic graphlets that have been proposed if not fully developed. One such is labelled graphlets (Newaz & Milenković 2019); the graphlets discussed so far have unlabelled nodes or nodes that are all of the same kind (i.e. homogeneous), which does not make sense for chemical structures in which the labels on the structural nodes identify the atomic composition (e.g. C vs. N vs. O). Many chemistry applications focus on heterogenous graphlets (and graphlet kernels) with labelled nodes (Gaüzère et al. 2015), where the label may refer to active or inactive compounds in a network of related proteins or chemical products (Lugo-Martinez 2011). The labels may also indicate their composition, such as the inclusion of a particular amino acid (see Vacic et al. 2010). Instead of giving the nodes labels ("colours"), the edges can be coloured or labelled (Gu et al. 2018).

The nodes may be labelled to show order, rather than type, so that the three node path graphlet g3 has three instantiations: 1—2—3, 1—3—2, and 2—1—3 (Newaz & Milenković 2019). For example, in protein structure networks (SN) which are frequently studied in bioinformatics, the nodes are amino acids and the positions of the amino acids can be imposed as their order onto the nodes (Newaz & Milenković 2019). As with many of these elaborations of the basic graphlets, the number of elaborated graphlets increases very quickly with n; for example, the 6 basic graphlets for $n = 4$, produce 38 node-ordered graphlets. It is worth noting that, as each protein has its own structure network of amino acids and proteins themselves are the nodes of a protein–protein interaction network (PPI), a PPI is a good example of a network-of-networks.

Noting that many important networks are dynamic, not static, Aparício et al. (2015) have extended graphlets' application to temporal (time step) networks, enumerating all 4-node graphlet orbits in each temporal snapshot of the network. The changes then build an orbit-transition matrix of the dynamics. The orbit-transition matrices of different dynamic networks can then be compared, and the authors provide examples with up to several thousand nodes and up to 16 time observations. Šourek et al. (2013) used timed snapshot observations for graphlets with directed edges to study trends in keywords used in Twitter. The approach of using graphlets for timed snapshot data is an interesting first step in what may be a promising avenue to explore for dynamic or time-only networks.

Another variant is to use graphlets that need not be connected (see Shervashidze et al. 2009; Gaüzère et al. 2015; among several). This modification has the advantage of helping to deal with disconnected networks, and to include nodes that are not in the "giant component" of a "typical" network structure. However, it is not clear how much functional information is coded in a graphlet of three unconnected nodes. Does that concern contradict the aim of trying to understand function from structure and network form from network dynamics?

A third modification is to consider rooted graphlets (Lugo-Martinez & Radivojac 2014). The approach has two versions, as follows:

1. The root indicates the source and thus the direction of flow or causal influence on a biochemical network (e.g. in a gene regulatory network, Aparício et al. 2015).
2. The root indicates the focal node for which a neighbourhood of nearby nodes is being defined (Lugo-Martinez 2011).

Again, this refinement may be unnecessary if the other variants of graphlets are as effective as might be hoped for. Further examples of application should help clarify the relationships among the different versions of graphlet analysis.

A more radical approach to using graphlets to study relationships between topology and function has been developed by Windels et al. (2018), which we will describe in some detail. That approach re-envisions the concept of node adjacency to create graphlet Laplacians. The revised concept is that any two nodes are *graphlet-adjacent* (in graphlet gX) with respect to the basic graphlet gX (g0 to g8; Figure 3.11) when they co-occur in an instance of that graphlet. This means that two nodes can be graphlet-adjacent even when there is no edge between them (e.g. in g3 or g6). For example, node 1 in Figure 3.12 is g1-adjacent to node 5, although there is no edge (1, 5). In fact, node 1 is g1-adjacent to node 5 three times: through (1, 2, 5), (1, 3, 5), and (1, 4, 5), so that the multiplicity of the g1-adjacency of nodes 1 and 5 is 3, written as $a_{15}(\mathrm{g1}) = 3$. This multiplicity gives rise to the concept of a graphlet adjacency matrix $\mathbf{A}_{\mathrm{gX}}$ with elements that are not just 0's and 1's; they are $a_{ij}(\mathrm{gX})$ for $i \neq j$ and 0 otherwise. The *graphlet degree* of any node is the sum of its multiplicities or the row or column sum of matrix $\mathbf{A}_{\mathrm{gX}} : d_i(\mathrm{gX}) = \Sigma_{\mathrm{j}}\left[a_{ij}(\mathrm{gX})\right] = \Sigma_{\mathrm{j}}\left[a_{ji}(\mathrm{gX})\right]$. The *graphlet degree matrix* is then $\mathbf{D}_{\mathrm{gX}}$ with elements $d_i(\mathrm{gX})$ on the main diagonal ($i = j$), and 0 elsewhere. This then allows us to define the *graphlet Laplacian* for graphlet gX:

$$\mathbf{L}_{\mathrm{gX}} = \mathbf{D}_{\mathrm{gX}} - \left[\mathbf{A}_{\mathrm{gX}} \div (|gX, -1)\right] \tag{3.4}$$

where |gX| is the number of nodes in gX (its *size*) and the division is element by element. Of course, $\mathbf{L}_{\mathrm{g0}}$ is the standard and familiar Laplacian (Windels et al. 2018).

The Laplacian is usually the preferred matrix for spectral analysis (Chung 1997) within the tradition of spectral graph theory, and in the graphlet context, it can now be used for spectral embedding and spectral clustering (following Windels et al. 2018). Spectral embedding is a dimension reduction so that a network graph is represented by points in fewer dimensions than originally for the nodes and the edges intersect only at the nodes (i.e. no crossings). For the graphlet Laplacian embedding, the object is to have pairs of nodes that are frequently graphlet-adjacent for gX in the original graph to be close in the embedding space. Windels et al. (2018) recommended a version of the Laplacian Eigenmap embedding for this purpose. They also performed normalised spectral clustering to create c clusters using k-means clustering, based on the symmetrically normalised Laplacian:

$$\mathcal{L}_{\mathrm{gX}} = \mathbf{D}_{\mathrm{gX}}{}^{-1/2}\mathbf{L}_{\mathrm{gX}}\mathbf{D}_{\mathrm{gX}}{}^{-1/2} \tag{3.5}$$

The example we wish to focus on is the yeast Genetic Interaction (GI) network in which the genes are classified into 14 groups by biological function (e.g. metabolism/ mitochondria, cell cycle/meiosis, signaling/stress response, and so on). Spectral embedding in two dimensions using the symmetrically normalised Laplacian for the graphlet g3 clearly clusters and separates the data by biological process, as shown by the colour coding in their figure 3 (Windels et al. 2018). Once again, the graphlet approach has produced convincing insight into the relationship between topological structure and system function.

3.4.3.5 Comment on Network Sampling and Graphlets

There is a clear, if somewhat inexplicit, relationship between some of the network sampling schemes and graphlets as a general approach, because they are both based on the concept of using local structure and the information therein to make inferences about the global characteristics of the whole graph. Equivalently, either local sampling or localised graphlet scores can be used to map explicitly local characteristics of the network structure onto its network graph. There is also a clear and explicit algorithmic relationship between network sampling with random walks or path sampling for counting graphlets; see for example, Bressan et al. (2017).

3.4.4 Analysis of Network Flow

The flow of energy, information, or material is often the key feature of interest in ecological networks and may be central to understanding the relationship between structure and function in them. Figure 3.20a illustrates a good example of how structure affects function in determining flow in a network. The edges in the network have both directions and weights that represent the capacity of each edge for the flow of whatever the network carries (often referred to generally as "traffic"), in some cases each edge also has a cost of the transport it represents (Ahuja et al. 1993; Bang-Jensen & Gutin 2009; Li et al. 2013). Two principles are that the actual flow in any edge cannot exceed its capacity (of course), and that flow is conserved: inflow equals outflow at any interior node and so there is no storage as seen in ecosystem models. Where e_{ij} is an existing edge between the two nodes indicated, we can let k_{ij} be the capacity of the edge for flow through it and c_{ij} is the cost; the actual flow on the edge x_{ij} cannot be more than its capacity. In some applications, there is also a minimum flow for each edge as well as capacity which is its maximum; see Bang-Jensen and Gutin (2009) or Marcus (2008). In the example shown in Figure 3.20a, the flow is from source, S, ("S" for "source" or "start") to sink, T, ("T" for "target" or "terminus") and the directions of all edges follow that trend. The flow assigned to particular paths depends on the order in which the paths are considered, but the example illustrates the general rule that the maximum flow through the network is the same as the minimum-weight cut-set of edges (here 15: edges SA and SB form the cut-set). As you might expect, once we allow and include edges with counter-flow directions, the calculation of flows becomes more complicated.

The analysis also becomes more complicated when edge costs are included. The cost associated with traffic flow can be any one of a number of factors: actual expense, distance, time taken or delay, and so on, and cost can be associated with an individual edge or can be calculated for an entire path (Kolaczyk 2009). A key assumption in this kind of analysis is that the cost of a path is equal to the sum of the costs of the edges in it (often, but not always, a reasonable assumption). Interpreting direct measurements of the costs of a subset of all paths, called *end-to-end* measurements, Kolaczyk (2009) provided guidance on (a) inference about the characteristics of edge costs and (b) inference about the characteristics of the full set of all possible paths. Because there is

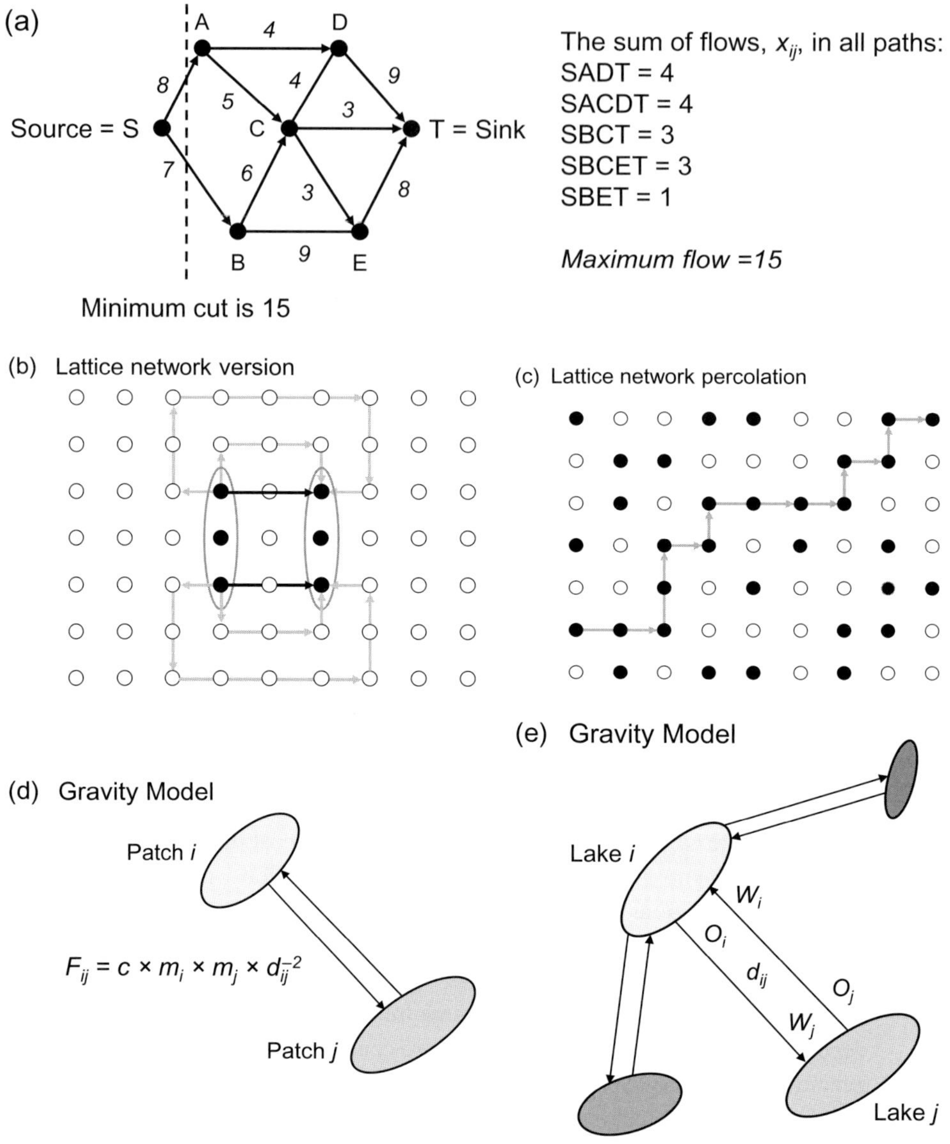

Figure 3.20 Network flow. (a) Minimum cut and maximum flow: network digraph with flow capacities, k_{ij}, on the edges. The sum of flows, x_{ij}, in all paths: SADT = 4, SACDT = 4, SBCT = 3, SBCET = 3, SBET = 1; *Maximum flow =15.* The minimum cut is SA and SB, with weight 15. For comparison, a cut of DT, CT, and BT is 20, and a cut of AD, AC, BC, and BE is 24. (b) If longer dispersal paths are more costly, dispersal between elongated patches is easier if they are side-by-side, not end-to-end. (c) Lattice network version of percolation. As "open" sites become more numerous, eventually a path from one side to the other is formed, the percolation cluster, like a giant component. (d) Gravity model of organism movement between patches i and j. The basics that underlie a gravity model of the movement of organisms between patches: flow is proportional to the product of patch "mass" (e.g., population density), divided by the square of the distance between them. $F_{ij} = c \times m_i \times m_j \times d_{ij}^{-2}$. (e) Gravity model: underlying this model of the movement of invasive exotic organisms between lakes are quantities O_i, the "push" from a lake or productivity, and W_i, the "pull" on propagules towards it or its attractiveness. The distance between lakes is d_{ij}.

much redundancy in the set of all paths in the form of re-used edges, there is no need to examine and measure all possible paths.

The cost-related concept that parallels the "maximum flow" goal of network analysis described earlier is "minimum cost," where the cost is in an appropriate "currency." Phillips et al. (2008) used network flow analysis as the basis for recommendations to optimise dispersal corridors for endemic plants (Proteaceae) in the Western Cape region South Africa. Dispersal was represented as chains (path or sequence) of suitable cells in a spatial grid in a series of years from 2000 to 2050. They used both the maximum flow (based on "maximum flow = minimum cut"; see Figure 3.20a) and minimum cost principles, with the latter related to the ease of dispersal by multiple paths between habitat patches determined by the patches' relative positions (Figure 3.20b). Formulating the problem in the format of network flow analysis allows ecologists to use both well-established theory and existing software from other fields such as computer science (e.g. see Ahuja et al. 1993). In relating flow and connectivity to population maintenance, the asymmetry of the connections may be important; Bode et al. (2008) have shown that asymmetric connectivity can act to decrease metapopulation persistence in patch occupancy models (Figure 3.20b).

For any flow at all to occur, there must be an open path through the network, and one approach to determining whether a path exists is percolation theory. The basic concept is most easily explained in reference to a two-dimensional lattice network of nodes that are either open to flow or closed. In this version, the edges all permit flow, and as the proportion of random nodes that are open increases from zero, eventually a critical number and arrangement of open nodes is reached that allows flow from one side of the lattice to the other (Figure 3.20c; Barrat et al. 2008). The nodes in that path form the *percolation cluster*. Different networks have different values for the proportion of open nodes that allows flow as their *percolation thresholds*, for example, square lattice versus triangular lattice, and the dimension of the network has its own effect, for example, square plane lattice versus three-dimensional cubic lattice. Where the nodes control the flow through the network, it is called *site percolation*, but it also possible to allow the edges to control the flow by being open or closed, called *bond percolation*. There is a clear and obvious conceptual similarity between bond percolation and a random walk on the network graph. In network theory focused on information flow, *inverse percolation* is a key concept related to the vulnerability or resilience of a connected network. Initially all the nodes are open, but as more and more of the nodes are closed or removed, at what stage does the flow become blocked? Knowing the properties of the degree distribution of the initial undamaged network, it is possible to predict those same properties when a fraction, f, of the nodes are removed at random. For further details and discussion, see Barrat et al. (2008).

Other aspects of network flow or movement are addressed by "gravity" models. These are based on the concept that the strength of many interactions may vary with the product of some measure of "mass" (e.g. patch area) of the interacting entities divided by the square of the distance (often geographic) between them, analogous to Newton's law of gravitational attraction. As a subset of network flow models, they use the same prescription (product of two measures of "mass" divided by distance squared) to predict the flow between the nodes of a network. For example, we might want to compare the observed flow of individuals between two subpopulations in a landscape to the product of their patches‘ densities divided by the square of their separation (Figure 3.20d). Gravity models are widely used to evaluate international trade flow in a deterministic form which are evaluated for fit with the data by standard statistical methods (Kepaptsoglou et al. 2010), but they have also been applied in both deterministic and stochastic forms to study and model patterns of lake invasions by exotic species (Schneider et al. 1998; Bossenbroek et al. 2001; Potapov et al. 2011).

That research focuses on the freshwater systems in North America, which are liable to invasions by a range of non-native species from fish to bacteria (zebra mussels and grass carp may be the most familiar examples). Because humans are often dispersal agents (think of diaspores attached to boat hulls), it is important to understand the factors affecting this kind of dispersal, and gravity models are recommended for that purpose. Previous work specifically on zebra mussel invasion used deterministic and probabilistic gravity models, as described in Bossenbroek et al. (2001). Schneider et al. (1998) and Potapov et al. (2011) have developed a stochastic gravity model for a similar purpose. The basic form of a gravity transport model consists of three parts:

1. *Source factors* related to the production or generation of the material being moved.
2. *Destination factors* that enable or enhance the transport, often termed "attractiveness."
3. A *separation function* that is related to the effects of distance or cost on the transport.

Gravity models can be constrained by the production part of the model or by both production and attractiveness. For example, in considering the potential movement of zebra mussels from one watershed to another by recreational boaters, the number of boats being moved from a lake (the source) is its production variable, where the number of boats being brought to a second lake (the target) is its attractiveness (Figure 3.20e).

In a deterministic version the gravity model for the propagule pressure related to transport from site i to site j, call it T_{ij}, is based on the outward "push" or production from the source, O_i; the "pull" (attractiveness) of the sink, W_j; and some function of the distance between them, $\phi(d_{ij})$:

$$T_{ij} = O_i \times W_j \times \phi(d_{ij}). \tag{3.6}$$

A truly "gravity" version would have $\phi(d_{ij}) = d_{ij}^{-2}$, but more general functions, $d_{ij}^{-\beta}$, are possible. The stochastic gravity model approach proceeds by finding the measurable covariates that determine a site's push and attractiveness (e.g. lake area, perimeter, elevation, location, and human population density). Surveys provide data on the movement of boats or boaters between sites: n_{ij}. Once the form of the models to be fitted is determined, estimates of its parameters are obtained by fitting the model to the data, and the fit is evaluated by AIC (Akaike information criterion) or BIC (Bayesian information criterion); for a comparison of these two measures, see

Burnham and Anderson (2004). Potapov et al. (2011) have given an explicit example describing boat movement between lakes in Ontario, and details of the model they have developed can be found there.

A second example of gravity models comes from a study of the landscape genetics of metapopulations of a high mountain frog, *Rana luteiventris* (Murphy et al. 2010). Gene flow was explained as a function of distance between sites, attraction or production at each site, and the resistance to flow between sites. Connectivity of the spatial network was found to be negatively correlated with distance, at-site presence of predatory fish, and topographic complexity; gene flow was positively correlated with site productivity and length of growing season. The authors concluded that gravity models provide a powerful new approach for answering a wide range of questions in landscape genetics, with obvious implications for applications in ecology.

3.5 Causal Networks

Just as the flow of energy, information, or material is a key feature of ecological networks, a very special kind of "flow," that of the causal relationship, is studied by networks of a different kind. These causal networks are important for understanding ecological systems; and the nodes are not sites or species, but rather the variables of interest themselves. Using graphical models is a key component of causal analysis, which has seen rapid and revolutionary developments in the last decade or two (Grace 2006; Pearl 2009; Shipley 2016; Pearl & MacKenzie 2018). A graphical model is a graph that represents joint probabilities of variables and functions describing their interdependence. Therefore, the nodes do not represent *things* (individuals, species, locations), but rather binary, categorical, or quantitative *variables*. The edges do not represent *interactions* among entities (predation, communication, friendship) but instead they indicate or quantify *conditional independence* or *conditional dependence* among the variables.

Probably the best place to begin to understand causal networks is by considering correlation networks, which we described in Section 3.3.2. The basic data in correlation networks are either observations or measurements of n variables at several times or several locations, from the correlations between all possible pairs are derived. For example, Vile et al. (2006) measured a number of vegetative, reproductive, and phenological traits of 34 plant species in eight old fields in the South of France. Given the correlations of all possible pairs of variables (i.e. pairwise correlations), it is easy to create a meaningful correlation network by discarding correlations (i.e. edges) of low value or low importance, and retaining network correlations and thus edges with large absolute values, positive or negative, with higher apparent significance.

Once a correlation network is obtained, we can develop and test hypotheses about causal structures within the data and thus causal mechanisms. In other words, based on prior knowledge and hypotheses, we can put directions on the edges between variables in the graph to reflect causal influences, whether unidirectional (preferred) or bidirectional, reflecting mutual influence. The task of selecting edges or finding directions for

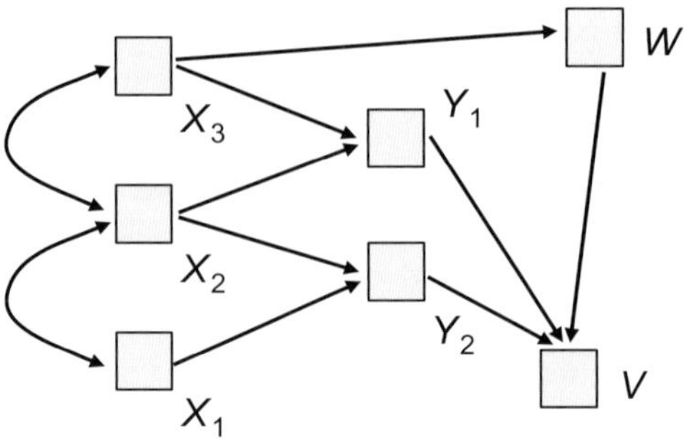

Figure 3.21 A hypothesised causal network for path analysis. The letters represent variables. The unidirectional edges represent causal effects and the bidirectional edges are unresolved correlations.

edges is equivalent to hypothesis generation. Usually, one causal hypothesis is proposed for each edge to omit or to include and this causal network assembly may use or may require detailed background knowledge, structural analysis, or temporal sequencing, and so forth (see Shipley 2016). Having directions on the edges, we will then want to determine the intensity and significance of the causal effect through the network. To carry out this assessment, a constellation of methods has been developed, including path analysis, factor analysis, structural equation modelling (SEM; Grace 2006), and several versions of causal analysis and causal modelling. We will now provide a quick guide to the range of these methods and some of the relationships among them.

The basic form of path analysis begins with the work of Sewall Wright in the 1920s (Shipley 2016), in part inspired by the complexities of the factors that determine the coat colour characteristics of guinea pigs (Pearl & MacKenzie 2018). The original version uses only observed variables and their correlations or linear regression coefficients. The paths under consideration are those between the variables, portrayed as the nodes of a graph, based on directed edges as hypothesised causal links and bidirectional edges representing non-causal correlation or unresolved relationships (Figure 3.21). The inclusion or exclusion of any possible edge in the path diagram represents a hypothesis, and edges should be omitted only when it is certain there is no causal effect.

Each directed edge goes from the source node known as the *parent* to the sink node known as the *child*. A distinction is usually made between *exogenous* variables in the digraph, which are those with no parent, and *endogenous* variables which do have at least one parent in the structure. Figure 3.22 illustrates the allowed composition of a path between two variables, based on the directions of the edges in the path: (1) a "common effect" variable (known as a *collider*) may not be included and (2) a path may include not more than one bidirectional edge. Each edge in a path has a quantitative coefficient associated with it, as does any path. Table 3.3 shows some of the rules of path coefficient calculation.

Table 3.3 Calculating path coefficients

- The path coefficient for a bidirectional edge between exogenous variables is just the correlation or covariance.
- For a single causal directed edge, the path coefficient is the regression or correlation coefficient.
- For a path with several links: the product of the individual path coefficients.
- Variables may be connected by more than one causal path and the path coefficients are then the partial regression or partial correlation coefficients, e.g.:

$$r_{XY.Z} = \frac{r_{XY} - r_{XZ}r_{YZ}}{\sqrt{1 - r_{XZ}^2}\sqrt{1 - r_{YZ}^2}}$$

- The path coefficient associated with an edge for an error variable is the residual variance of the (sink) variable itself, e.g.:

$$1 - R_x^2 \text{ or } \sqrt{1 - R_x^2}.$$

- The path coefficient for a bidirectional edge (unanalysed or unresolved) between endogenous variables is the appropriate partial correlation or covariance coefficient. The total effect is the sum of direct and indirect causal effects.
- The sum of path coefficients for all paths between two variables is the bivariate correlation between them.

From Grace (2006).

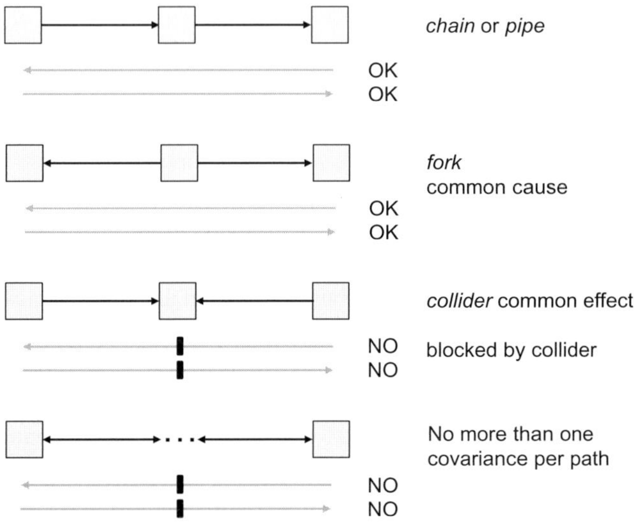

Figure 3.22 Permitted path tracing in a directed acyclic graph (DAG) model of a multivariate causal hypothesis (on 3-node motifs): chain or pipe, fork and collider.

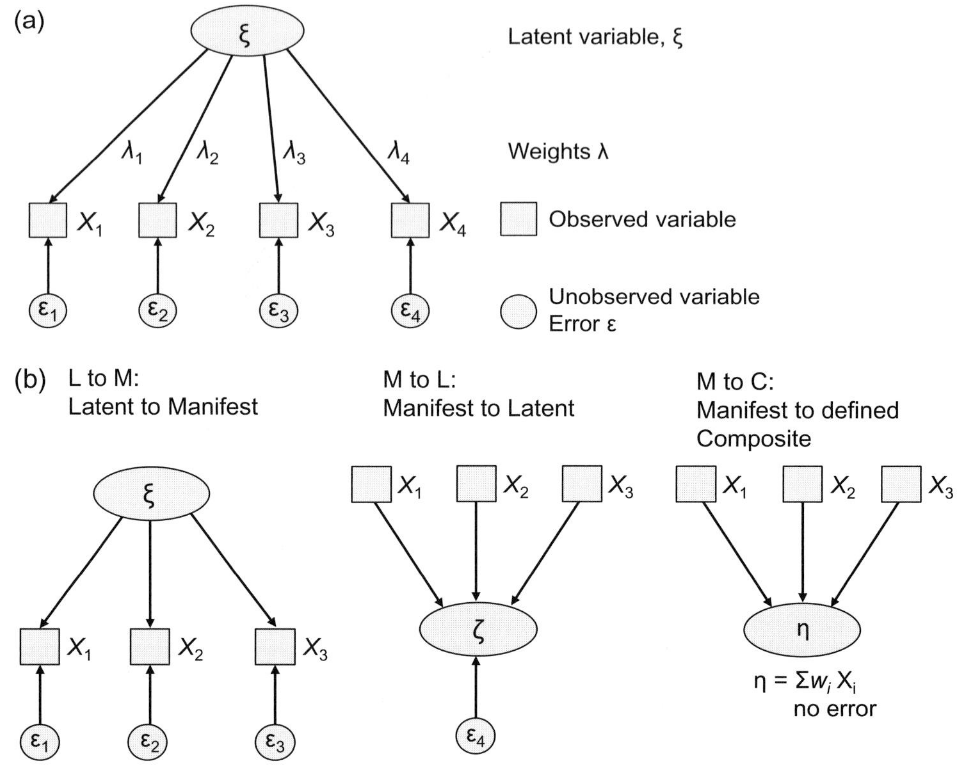

Figure 3.23 (a) Factor analysis examines the effects of latent (unobserved) variables on those that are measured. "Rule of thumb": three indicator variables (observed or manifest) is the minimum for one latent variable. (b) Structural equation modelling: types of composite variables.

Path analysis, as originally formulated or in its more evolved form, includes only observed variables, and as such, it is a particular case of the approach known as Structural Equation Modelling (SEM; Grace 2006) with only observed variables, while the full version of SEM can include unobserved or *latent* variables. Ecological studies are highly reliant on latent variables; consider a quantity like *photosynthetic rate*: we cannot actually measure it, but we can measure variables that act as indicators of it, such as gas exchange (see Ackerly & Stuart 2009; Shipley 2016). The inclusion and consideration of latent variables in data analysis may be familiar from the technique known as Factor Analysis (Figure 3.23a; Legendre & Legendre 2012). In addition to the inclusion of latent variables, another strength of the general SEM methodology is the potential to include composite variables, made from compounding two or several variables into a single one, whether the component variables are latent or observed (Figure 3.23b). A characteristic such as soil fertility is a composite variable when we create it by combining measured variables associated with fertility: N mineralisation, P, K, and so forth (Shipley 2016).

As an example of the effectiveness of the SEM approach, we will examine a study by Whalen et al. (2013) described in Grace and Irvine (2019). The dominant herbivores of epiphytes in marine seagrass beds (*Zostera maritima*) are microcrustaceans (Gammarids and Caprellids) which are important prey for small fish. The microcrustaceans may benefit the plants by grazing on the epiphytic algae on the leaves which might otherwise decrease photosynthesis and survival. A top-down chain begins with a decline in large fish numbers due to overfishing, allowing small fish to increase. More small fish could suppress the microcrustacean densities, increasing epiphytes and thus contributing to the decline of the seagrass beds. This top-down hypothesis was investigated based on the effects of a treatment of slow-release insecticide (a binary variable) on the microcrustaceans and the epiphytic algae they consume in a completely randomised experiment. In addition to the abundances of these organisms, the density of *Zostera* and of macroalgae in the seagrass bed were measured as covariates.

Figure 3.24a and b illustrates a number of ways that the data might be analysed in the ANOVA (analysis of variance) approach (including a multivariate version, MANOVA, and a version with covariates, ANCOVA). Figure 3.24c (left panel) shows a possible SEM for the study, in which any effect of the treatment on the epiphytes is completely mediated through the abundances of the two microcrustaceans. There is no direct effect in this model. However, the empirical evaluation of the conditional independences in the data found three non-independencies that deserved added links: the macroalgae are refuges for the microcrustaceans, thus having an indirect negative effect on the epiphytes, and greater *Zostera* density can also enhance the Gammarids (Figure 3.24c, right panel).

This example illustrates several points; we can start with two:

1. The inclusion or omission of any single directed edge in the model is the equivalent of a hypothesis to be evaluated.
2. A key step in evaluating a causal model is determining the conditional dependence or independence of any pair of variables in the context of the other variables included.

These are both important principles for the application of causal analysis to causal networks in ecological studies, as we will now describe in greater detail.

Causal analysis is the evaluation of the causal content of hypothesised models, making assumptions explicit to assess them. Causal analysis is based on a hypothetical causal model, and a causal model consists of

- A set of exogenous (background) variables, U, which are not explained by the model.
- A set of endogenous variables, V, that are the focus of the model and to be explained by it (U and V may contain latent variables).
- A set of functions, F, that form a mapping from U to V.

Every causal model has associated with it a directed acyclic graph (hereafter DAG) with the variables as nodes and each edge directed to a member of V from a member of U or from a parent variable in V (Figure 3.25).

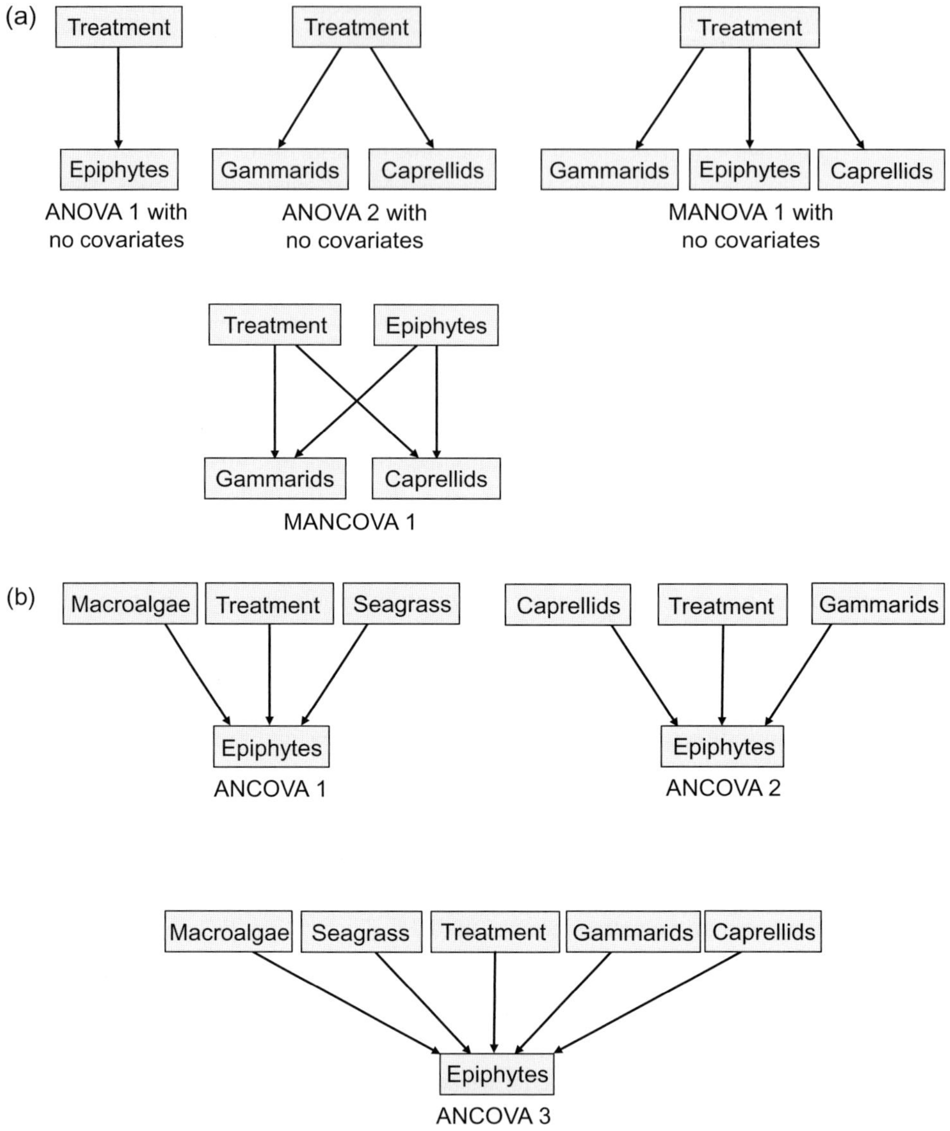

Figure 3.24 Marine food web analysis models: (a) Analysis of variance and multivariate analysis of variance. (b) Analysis of covariance.
[Redrawn from Grace, J. B., & K. M. Irvine. (2019). Scientist's guide to developing explanatory statistical models using causal analysis principles. Extended pre-publication draft of Grace & Irvine, *Ecology* (2020), **101**, e02962, with permission.] (c) SEMs with and without a direct effect of insecticide treatment on epiphytes.

A key part of this approach is to determine independence and dependence among the variables in the DAG that portrays their relationships. The notation used is $A \parallel B \mid C$ for "*A* and *B* are independent when conditioned on *C*," and $A \sim\parallel B \mid C$ for "*A* and *B* are not independent when conditioned on *C*" (other typographical conventions are

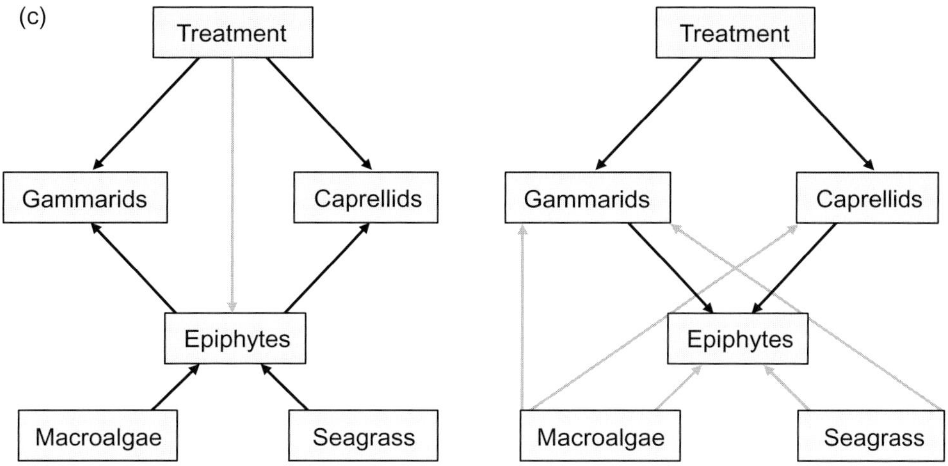

Figure 3.24 (cont.)

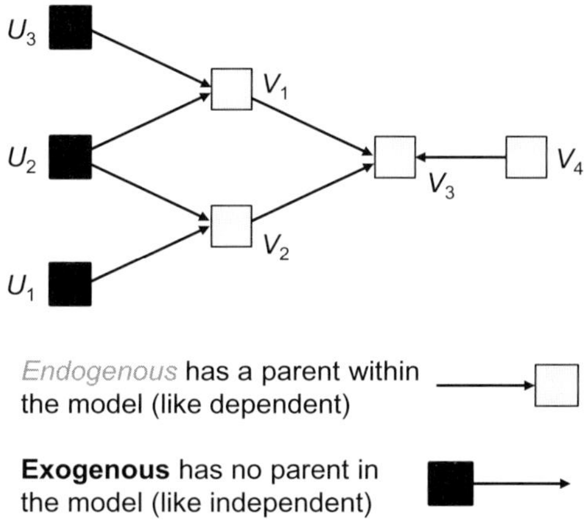

Figure 3.25 Causal model as a DAG. An endogenous variable has a parent variable within the model (like a dependent variable in regression). An exogenous variable has no parent in the model (like an independent variable). The variables V_1 and V_4 are not naturally independent, but they are independent when conditioned on V_3. That is, $V_1 \sim\| V_4$ V_4 but $V_1 \| V_4 \mid V_3$. Variable 3 is a causal mediator between variables 1 and 4.

possible). For example, in Figure 3.25, the variables V_1 and V_4 are not naturally independent but they are independent when conditioned on variable V_3. This can be expressed as $V_1 \sim\| V_4 \mid \emptyset$ or $V_1 \sim\| V_4$, but $V_1 \| V_4 \mid V_3$. Conditioning on a variable is like holding its value fixed, or like dividing the data into separate parts each with a different set value of the conditioning variable. For example, for a binary variable Z, conditioning on Z looks at partitioning the data into two subsets, one with $Z = 0$ and one with $Z = 1$.

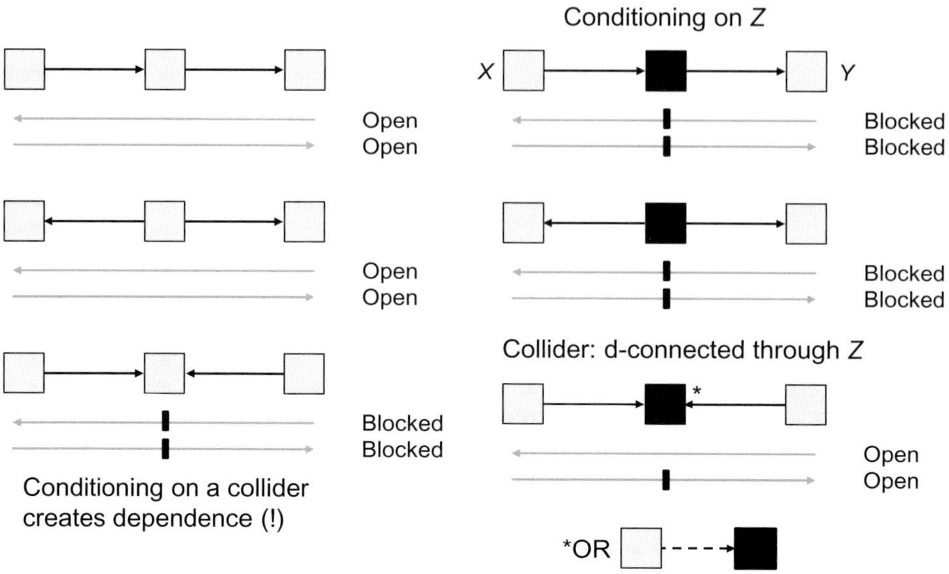

Figure 3.26 Directional separation in paths (3-node motifs). Conditioning on a collider creates dependence.

In general, the determination of dependence and independence uses the rules of *directional separation* (often called "d-separation" or "d-sep"). They resemble the rules for path analysis, with the added component that the relationship between two sets of variables is affected by conditioning on a third set of variables (Figure 3.26). That conditioning switches dependence to independence and vice versa, as shown. The most puzzling switch may seem to be the effect of conditioning on the collider variable, Z, which converts the source variables X and Y from independent to dependent. Consider a simple form of causal dependence in which the variable Z has value 0 or value 1 in the case that one or both of X and Y has value 1, themselves taking values 0 and 1 independently with some probability. If Z is unknown, X and Y remain independent, but if Z is known, that independence is lost. For example, if Z is known to be 1, then $X = 0$ forces $Y = 1$ (Figure 3.27). Conditioning on a collider, or on any of its descendants, induces an association between the two shared cause variables.

Determining dependence and independence is a critical component of the correct interpretation of causal models, and the d-separation approach allows us to do that. The basic rule is that

{X} and {Y} are d-separated by {Z} if for any *undirected* path U between X and Y:

- There is a non-collider in U that is in Z.

OR

- There is at least one collider in U that is not in Z and has no descendent in Z.

In the first case, Z blocks the path; in the second, the collider blocks the path so that in either case X and Y are separated.

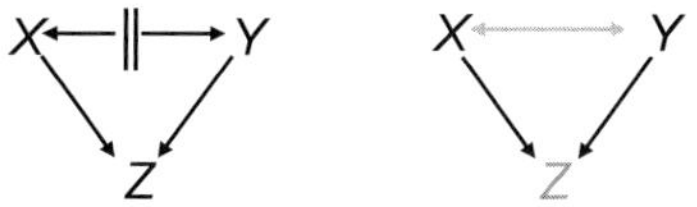

Figure 3.27 Collider conditioning. Conditioning on a collider or any descendent of a collider induces association. Consider: X and Y each take value 0 or 1 independently. Z takes value 1 whenever either X or Y is 1 and 0 otherwise. If you do not know Z, you can learn nothing about Y from the value of X and nothing about X from the value of Y. Knowing Z: if Z is 1, then $X = 0$ forces $Y = 1$. X and Y are no longer independent when conditioned on Z.

While complicated for large causal structures, the d-separation algorithm allows us to read off all conditional dependence and independence between any pair of variables, based on conditioning on any other variables, "directly" from the graphical model, because d-separation has been proven to correctly encode them all.

Readers may remember the statistical notation that looks like $P(Y = y, X = x)$, which denotes the probability that variable Y takes a particular value y, given that variable X takes a particular value x. A similar notation has been introduced to indicate the intervention that variable X is forced to take the particular value x. What is called the *do*-operator represents such an intervention: $P(Y = y \mid do(X = x))$ means the probability that variable Y takes a particular value y, given that variable X has been set at the particular value x (Pearl 2009).

The concept of *confounding* in statistical analysis refers to the situation where the effects of different variables cannot be distinguished because they are somehow mixed together. A simple observational study comparing weight loss in men and women might easily confound gender-linked factors such as body mass, metabolic characteristics, and preferred mode of exercise, which may differ systematically between men and women. More specifically, "confounding" often refers to the common cause structure in which the positive correlation of variables X and Y can be partially or wholly explained by their shared dependence on a third variable Z. This is referred to as spurious correlation, but is better called *induced correlation* because the observed association may be a combination of true interdependence and the shared cause effect.

One version of confounding is that it is the difference between $P(Y = y, X = x)$ and $P(Y = y, do(X = x))$ (Pearl 2009). That explanation may seem over-simplistic, but it is related to what is called the *backdoor* path between X and Y, with an edge directed from X to Y, through a shared cause, Z (with edges $Z \rightarrow X$ and $Z \rightarrow Y$; Figure 3.28a). If this backdoor path is not blocked, for example by conditioning on Z, an association between X and Y is induced, and they are not independent. More formally, for the ordered pair of variables, $X \rightarrow Y$, Z satisfies the backdoor criterion if no node in Z is a descendent of X, and Z blocks all paths between X and Y with an edge directed to X. This backdoor criterion helps to identify variables to control or to condition upon if confounding is to be avoided. Then,

$$P(Y = y, do(X - x)) = \Sigma_z P(Y = y, Z = z, X = x) P(Z = z) \tag{3.7}$$

Therefore, if there are no backdoor paths from X to Y, we then have

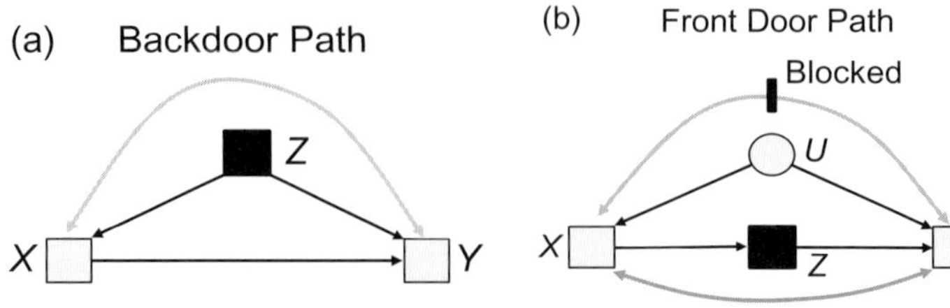

Figure 3.28 (a) Backdoor path. For the ordered pair of variables, X and Y, Z satisfies the backdoor criterion if no node in Z is a descendent of X, and Z blocks all paths between X and Y with an edge directed to X. Then, $\mathrm{P}(Y = y, do(X = x)) = \Sigma_z \mathrm{P}(Y = y, Z = z, X = x)\ \mathrm{P}(Z = z)$. The *do* operator represents an intervention that makes X take the value x. If there are no back door paths from X to Y, then $\mathrm{P}(Y = y, do(X = x)) = \mathrm{P}(Y = y, X = x)$. (b) Front door path. U is unobserved and cannot form a backdoor path. For X and Y, Z satisfies the front door criterion if (a) Z intercepts all directed paths from X to Y; (b) there is no unblocked back door path from X to Z; and (c) all backdoor paths from Z to Y are blocked by X. Then,
$\mathrm{P}(Y = y, do(\mathrm{X} = x)) = \Sigma_z \mathrm{P}(Z = z, X = x)\Sigma_{\mathrm{x}} \mathrm{P}(Y = y, X = x, Z = z)\mathrm{P}(X = x)$

$$P(Y = y, do(X = x)) = P(Y = y, X = x) \tag{3.8}$$

that is, there is no confounding, as suggested.

Consider a causal chain $X \rightarrow Z \rightarrow Y$, with an *unobserved* shared cause for X and Y, call it U, with edges $U \rightarrow X$ and $U \rightarrow Y$ (see Figure 3.28b). Note that U is unobserved and is therefore not eligible for a backdoor path. Under these conditions, for the ordered pair of variables, X and Y, Z satisfies the frontdoor criterion if (1) Z intercepts all directed paths from X to Y, (2) there is no unblocked backdoor path from X to Z; and (3) all backdoor paths from Z to Y are blocked by X.

Then,

$$P(Y = y, do(X = x)) = \Sigma_z P(Z = z, X = x)\Sigma_x P(Y = y, X = x, Z = z)P(X = x), \tag{3.9}$$

where Z can be seen as a mediator between X and Y, and there is a technique known as mediation analysis, commonly applied in epidemiology and other medical applications (Pearl 2009; Pearl et al. 2016). Here, the causal chain $X \rightarrow M \rightarrow Y$ is augmented by an edge from X to Y (Figure 3.29). The goal is to partition the total effect of X on Y into the direct effect and the indirect effect mediated by M. The "natural" indirect effect (i.e. no confounding) is, for example

$$\Sigma_M E[Y|X = x, M = m][P(M = m, X = 0) - P(M = m, X = 1)]. \tag{3.10}$$

This is the "mediation formula" for indirect effects (Pearl 2009; Pearl and MacKenzie 2018). Figure 3.30 gives an example in which photosynthetically active radiation (PAR) and soil temperature have direct causal effects on net ecosystem exchange and indirect effects through respiration (simplified from Fan et al. 2016).

In using and interpreting SEMs and causal models in ecology, one feature to be wary of is the presence of loops and autocorrelation in these network structures. For example, we know that phytoplankton and zooplankton in aquatic systems can affect

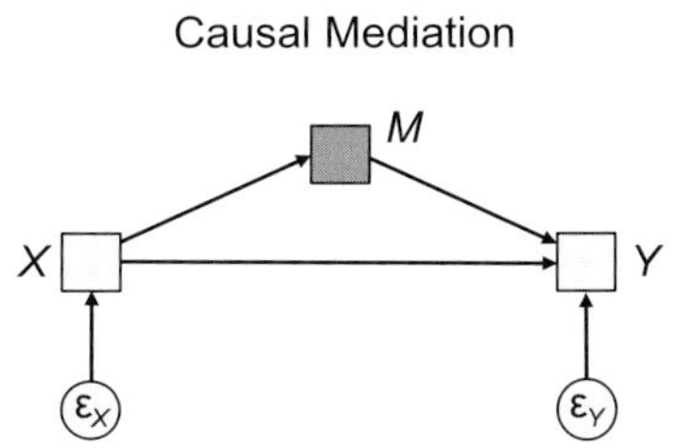

Figure 3.29 Causal mediation. The total effect of X on Y can be partitioned into the direct effect and the indirect effect mediated by M. The natural indirect effect (with no confounding) is $\Sigma_M \mathrm{E}[Y|X = x, M = m][\mathrm{P}(M = m, X = 0) - \mathrm{P}(M = m, X = 1)]$. This is the "mediation formula."

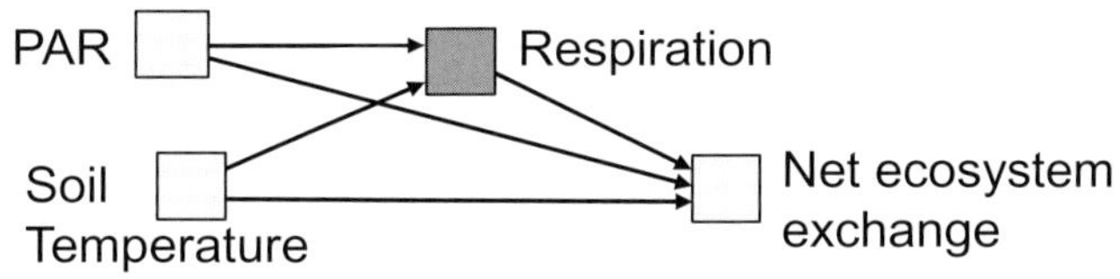

Figure 3.30 Causal mediation example. PAR (photosynthetically active radiation) and soil temperature have direct causal effects on net ecosystem exchange and indirect effects through respiration. [Simplified from Fan, Y., Chen, J., Shirkey, G., John, R., Wu, S. R., Park, H., & Shao, C. (2016). Applications of structural equation modeling (SEM) in ecological studies: an updated review. *Ecological* Processes, **5**, 19. Copyright 2016, SpringerNature; based on Shao et al. 2016.]

each other directly, but also indirectly through other factors. That gives two different kinds of loops (Figure 3.31), direct and indirect, but even the indirect effects may be reciprocal (the third loop structure in the upper panel of Figure 3.31); for a discussion of these features in multivariate experiments, see Grace (2006). Similarly, we have to be aware of the possible effects of autocorrelation on the behaviour of a variable or the whole model, which can be seen as a kind of self-loop (Figure 3.31, lower panel). Autocorrelation also arises through time for repeated measures; for example, if we are measuring the density of *Poa* in a field, the density in 2008 is probably not independent of the amount measured in 2007 (Figure 3.32, lower right).

The example involving *Poa* in Figure 3.32 is simplified from a study by White et al. (2013) on the invasion of *Poa pratensis* ("*Poa*") into a native community ("cover") which had explicitly serial data. Figure 3.32 shows the principal effects determined from the study, with nitrogen and summer defoliation as manipulations, and with *Poa*, native community cover, and species richness as measured variables. Both *Poa* and cover have obvious temporal autocorrelation, but species richness less so. This version of the SEM graphical model raises the question of how to include something like the possible interaction between defoliation and nitrogen treatments. Is there an implicit, rather than explicit, inclusion of interaction in the way the model is constructed? The concept of interaction needs some clarification.

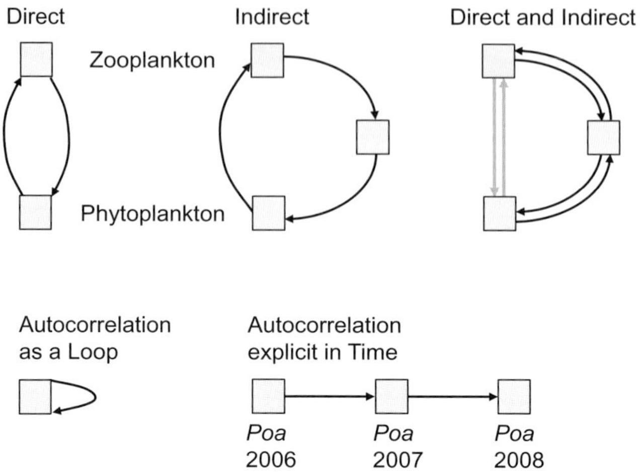

Figure 3.31 Loops and autocorrelation. Upper panel: feedback loops: direct, indirect, and both together. Lower panel: autocorrelation as a self-loop and autocorrelation explicit through time.

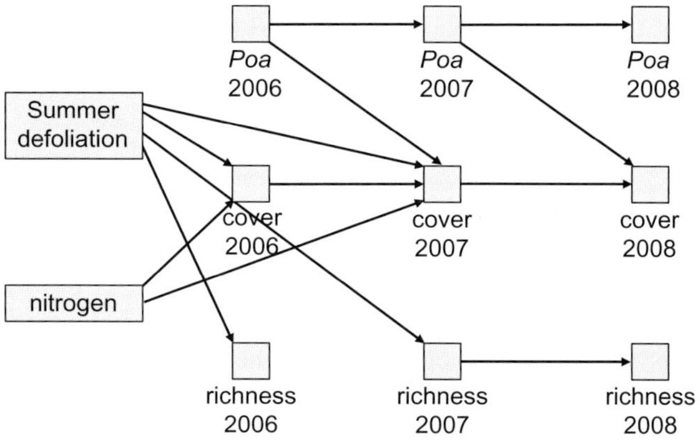

Figure 3.32 SEM example. Largest significant positive effects in a study of the invasion of *Poa pratensis* (variable "*Poa*") into a native community (variable "cover") with explicitly serial data. For details, see White et al. (2013). Is there an implicit interaction term in this diagram? Perhaps N × defoliation?

[Redrawn with permission from White, S., Tannas S., Boa, T., Bennett, J. A., Bork, E. W., & Cahill, J. F., Jr. (2012) Using structural equation modelling to test the passenger, driver and opportunist concepts in a *Poa pratensis* invasion. *Oikos* **122**(3), 377–384. © 2012, the authors.]

We have been using the term "interaction" to refer to a relationship between species, to cover both positive relationships such as mutualism and negative relationships such as predation. In the same context, we have also referred to indirect interactions, for example, when a third species modifies a two-species interaction (see Figure 1.4). In statistics, however, "interaction" refers to the combined or mutual

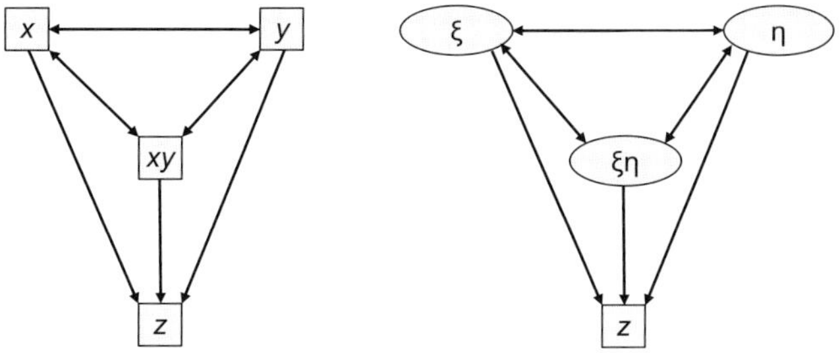

Figure 3.33 Interaction terms. Including an interaction term in a causal model or SEM: whether the predictors are observed, xy, or latent, $\xi\eta$.

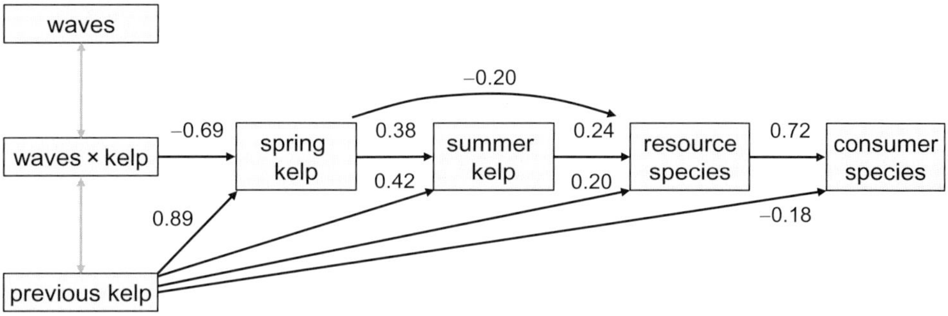

Figure 3.34 SEM with interaction term. Example: Kelp and storm frequency. [Simplified with permission from Byrnes, J. E., Reed, D. C., Cardinale, B. J., Cavanaugh, K. C., Holbrooks, S. J., & Schmitts, R. J. (2011). Climate-driven increases in storm frequency simplify kelp forest food webs. *Global Change Biology*, **17**, 2513–2524. © 2011 Blackwell Publishing Ltd.] Resource species are algae and sessile invertebrates; consumer species are the next trophic levels. This example has an interaction term, waves × kelp, and temporal autocorrelation in the three kelp variables.

effect of two variables. For example, a regression model of the effects of x and y on z may include the possibility that the influence of x may vary with y (or vice versa; it is symmetric) accounted for by an xy term:

$$z = \alpha + \beta_1 x + \beta_2 y + \beta_3 xy \tag{3.11}$$

Here, xy is called an interaction term. In causal models, graphical models, or SEMs such terms can also be included, as indicated by Figure 3.33, whether the predictor variables are observed or latent (Batista-Foguet et al. 2004).

The inclusion of interaction terms in causal models needs to be considered carefully, and may depend on whether the research is observational or experimental. Byrnes et al. (2011) provided a good example in their study of the effects of increasing storm frequency on kelp forest food webs. Figure 3.34 gives one version of their model with wave disturbance and the amount of kelp the previous summer, and their interaction, acting as predictors of spring kelp, summer kelp abundance, the richness

Table 3.4 Principles for causal analysis

- Graphical representation is an indispensable tool for causal analysis and explanatory modelling.
- Equations that can represent probabilistic causal network relations are also needed.
- Extracting scientific interpretations from a set of correlated variables requires a theory for why the variables are correlated.
- The inclusion of mediation pathways leads to unsaturated networks that represent hypothesised mechanisms with testable implications.
- Included links are testable given appropriate data.
- Omitted links are testable via evaluations of conditional independence.
- Backdoor (common cause) relationships are a leading concern for confounding, but they can be eliminated by backdoor adjustment.
- Confounding can also be eliminated via frontdoor identification.
- Mistaking collider/descendent variables for covariates can lead to inadvertent bias.
- Sequential learning is the ultimate route to confident interpretations.

From Grace (2019), unpublished draft, abbreviated to table 4 in Grace and Irvine (2019).

of resource species (algae and sessile invertebrates) and the richness of consumer species (the next trophic levels). Both the SEM and experiments suggest that increasing frequency of large storms will cause the trophic network to decrease in diversity and complexity, mainly due to species loss in higher trophic levels (Byrnes et al. 2011; see also Lefcheck 2016). As a second example, Buck et al. (2016) studied the effects of increased nitrogen and phosphorus on the dynamics of a chytrid fungal pathogen on larval amphibian hosts. Nutrient addition benefitted the tadpoles through increased algal growth for their nutrition and the infected tadpoles underwent metamorphosis sooner than the unexposed. Using an SEM approach, the authors expected and looked for an interactive effect of nutrients and the pathogen, but they found only additive effects.

We will end this subsection with a summary of principles for these kinds of analysis techniques (Table 3.4) from Grace & Irvine (2019). An imported concept is the incorporation of these methods into the general endeavor of advancing scientific knowledge, which can be called "sequential learning." Path analysis, SEMs, and causal analysis are intended to be confirmatory, not exploratory, but each analysis can and should be a step in the iterative process of learning. Understanding process and function in ecological systems requires background knowledge, and what we learn from any one analysis becomes background knowledge for next steps, even if we reject a hypothesis or discard a model as inadequate.

There is much more that could be said about this constellation of methods, and there are excellent books (Grace 2006; Pearl et al. 2016; Shipley 2016) that do so in more detail, but it is an important background for our study of ecological networks in general. It is especially important for the quantitative analysis of these networks, and for thinking clearly about the relationship between network structure and ecological process, and about the interaction between them.

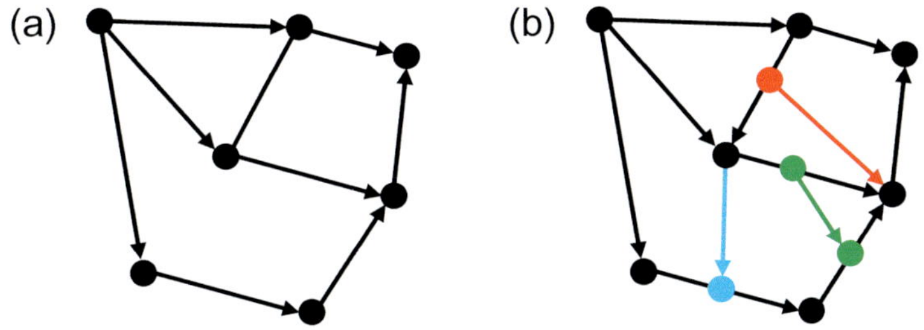

Figure 1.4 Interactions among interactions. Starting with 9 two-species directed interactions (edges) among 7 species (nodes) in (a), modifications can occur in which edges now act secondarily as nodes. The secondary interactions are shown by coloured edges. One modification is for a third node to affect another two-species interaction (black species node to blue interaction node). Another variant is for a two-species interaction to affect a third individual species (red interaction node to black species node). Last, one two-species interaction can affect another two-species interaction (green interaction node to green interaction node). (A black and white version of this figure will appear in some formats.)

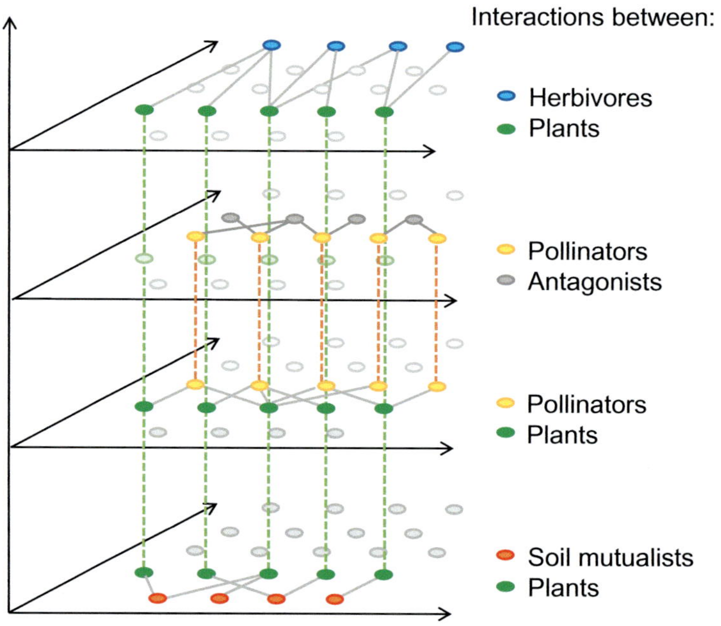

Figure 1.7 Multi-interaction network as layers. Layers show species nodes and edges for particular interactions: herbivory, pollinators and their antagonists, pollination mutualism, and soil-based mutualisms (e.g. mycorrhizae). Identity edges (dashed lines) join instances of the same node. (A black and white version of this figure will appear in some formats.)

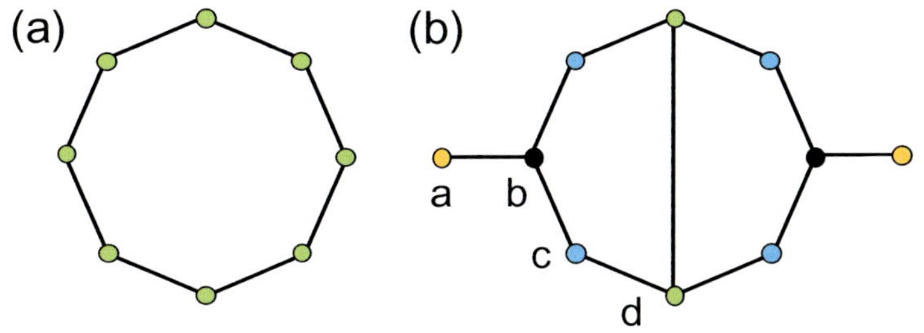

Figure 2.9 Graph symmetries and automorphism orbits. (a) The eight nodes are indistinguishable; all are in a single automorphism orbit. The graph has many symmetries. (b) The ten nodes occur in four distinct topological positions: two in each of *a*, *b*, and *d*, and four in *c*: the four orbits are indicated by shading. Nodes of the same shade can be mapped onto each other by automorphisms. (c) The eight nodes are all distinct, each in their own automorphism orbit. The graph has no symmetries. (A black and white version of this figure will appear in some formats.)

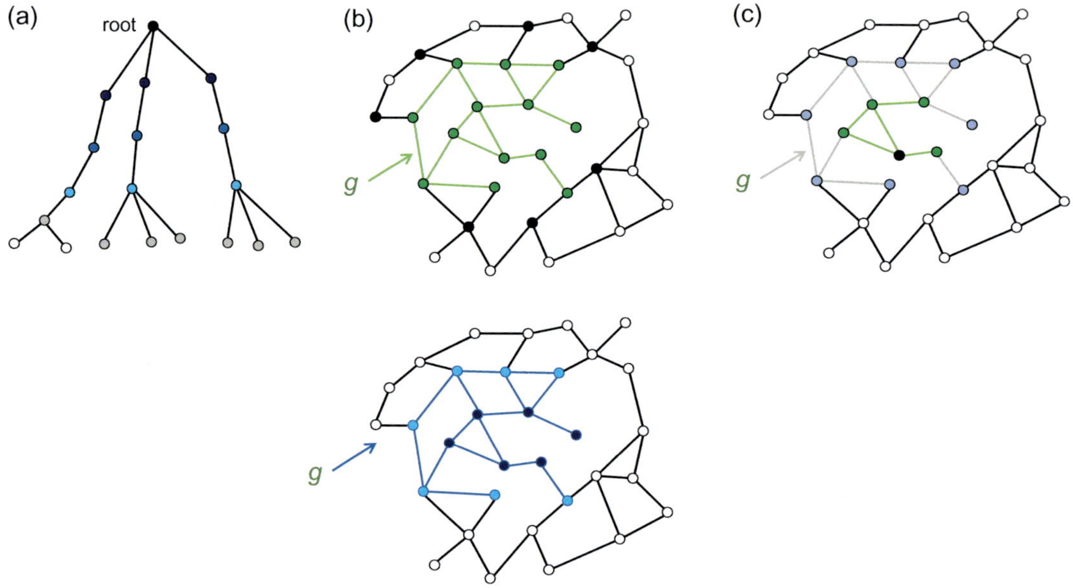

Figure 2.15 Hierarchies in networks. (a) Rooted tree. A rooted tree creates a hierarchy of nodes based on distance from the root. It is a partial order. Nodes in grey are all distance 2 from the root node (black). (b) Subgraph dilation and erosion. The dilation of the subgraph, g (green nodes), includes the black nodes. The erosion of the subgraph, g (all green nodes), includes only the darker blue. (c) Iterated dilation of a single node. The dilation of a single node (black) gives 1° neighbours (dark blue) nodes, 2° neighbours (light blue), and 3° neighbours (white). (A black and white version of this figure will appear in some formats.)

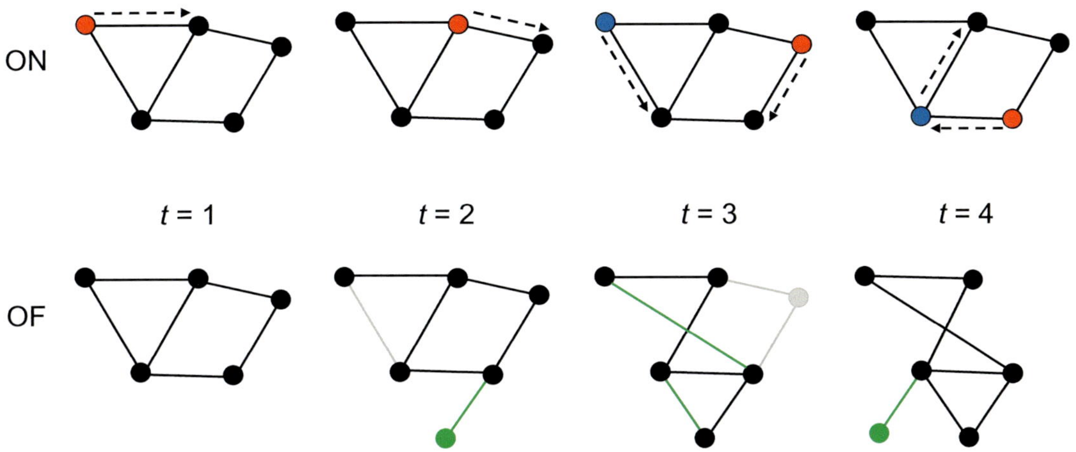

Figure 3.1 Dynamics **on** a network and dynamics **of** a network. Dynamics on a network: node state affects neighbour states (or not) as indicated; no structural change. Dynamics of a network: nodes and edges are lost (grey and then absent) or gained (green and then black) between time intervals. (A black and white version of this figure will appear in some formats.)

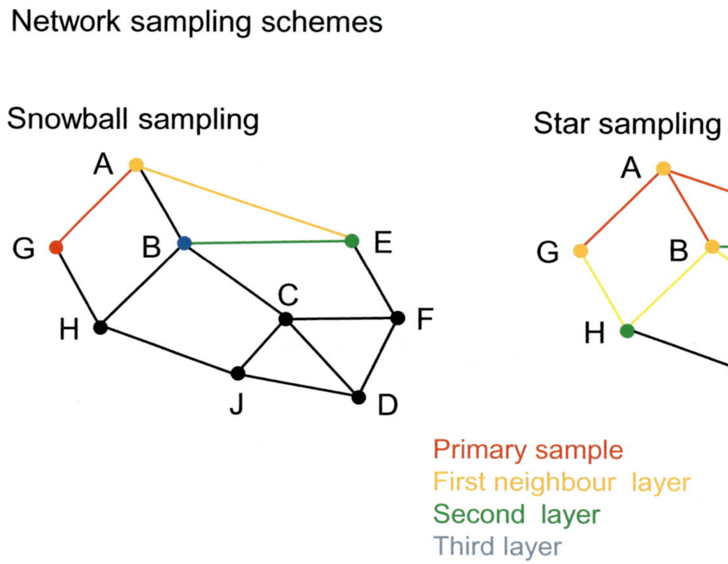

Figure 3.6 Network sampling schemes: snowball sampling and star sampling. Nodes coded by shading to show primary sample and subsequent neighbours. (A black and white version of this figure will appear in some formats.)

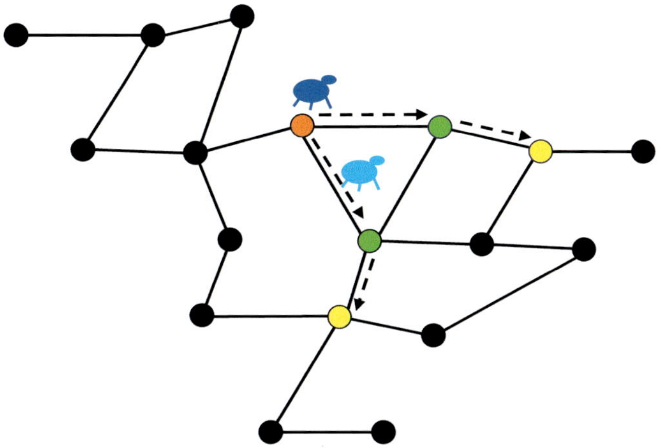

Figure 3.7 Crawler methods of sampling: a crawler starts at a random node and explores the network following its own rules. The crawler may replicate itself and follow many routes at once. (A black and white version of this figure will appear in some formats.)

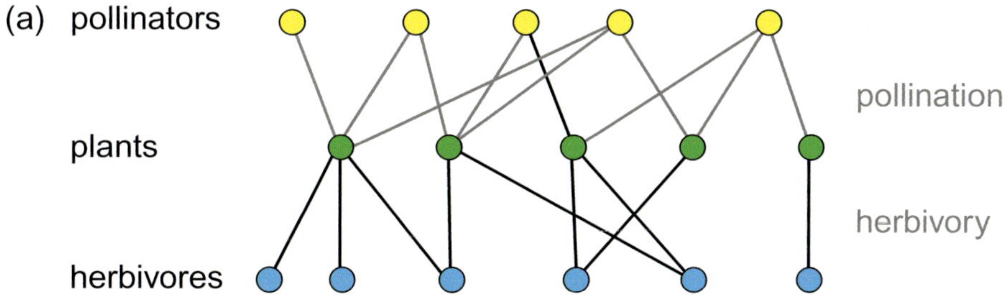

Figure 4.1 Merging two bipartite networks. (a) Pollination and herbivory: one layer. (b) Two layers, one for each interaction. (c) Merging interaction networks: layers representing interactions, *not* trophic levels. (A black and white version of this figure will appear in some formats.)

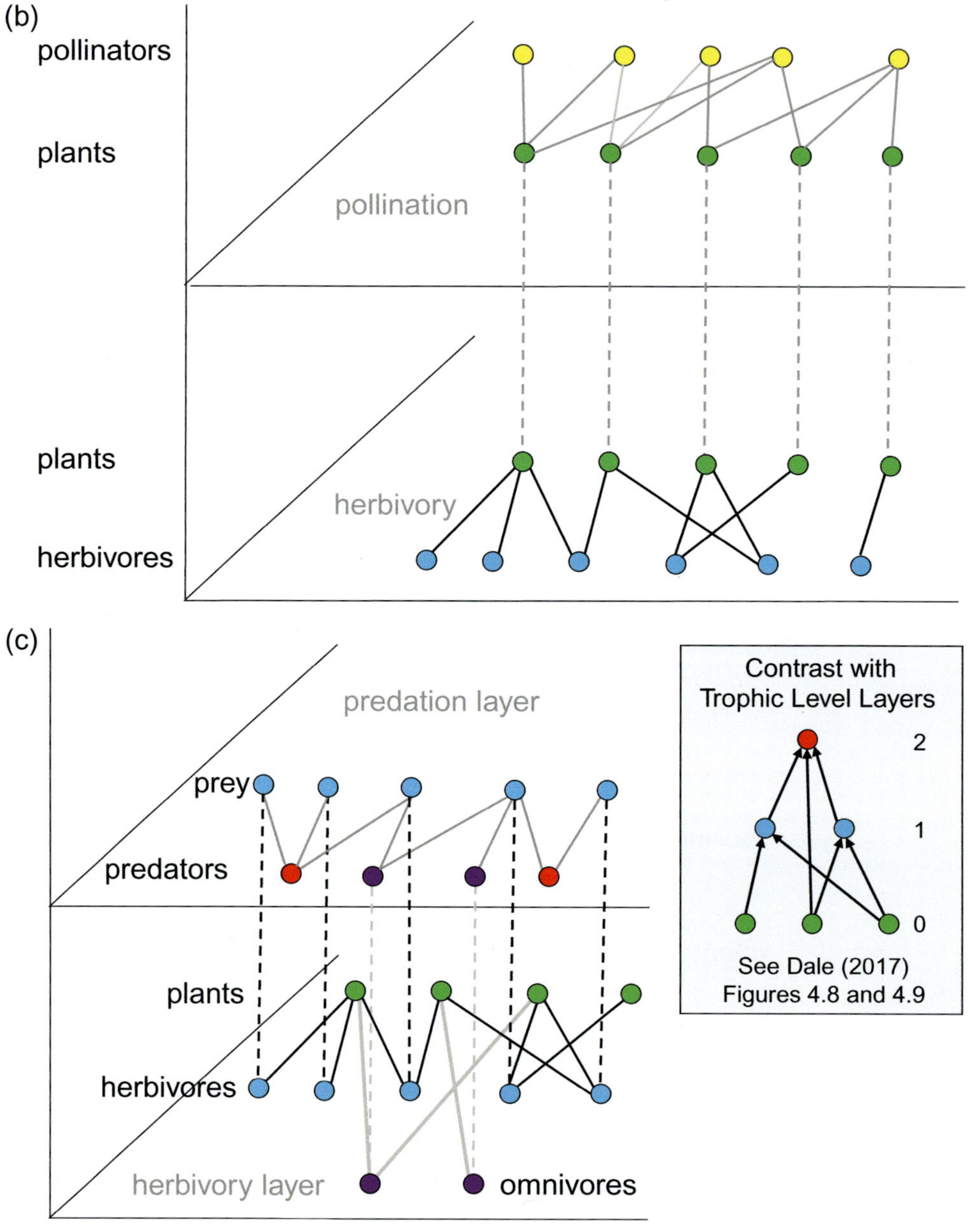

Figure 4.1 (cont.)

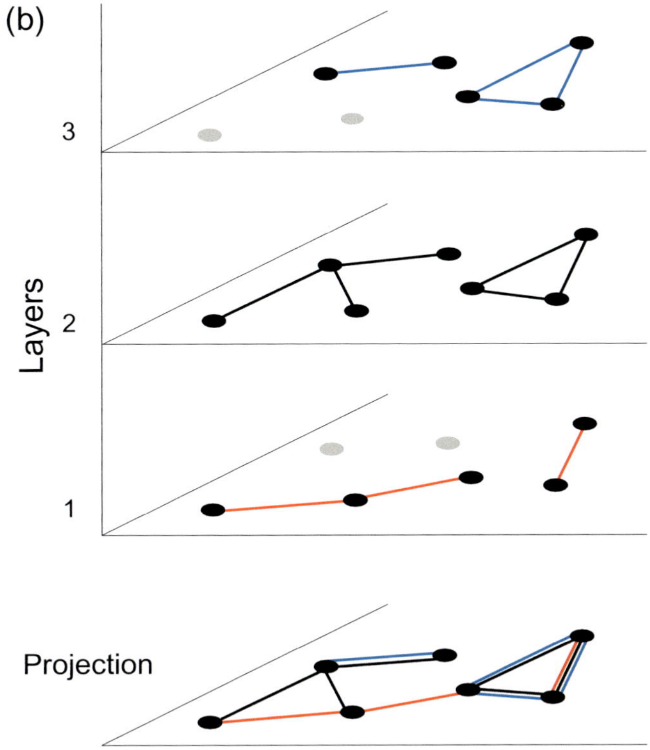

Figure 4.3b Projection of a multilayer network to form a multigraph with coloured edges (grey nodes are absent from that layer). (A black and white version of this figure will appear in some formats.)

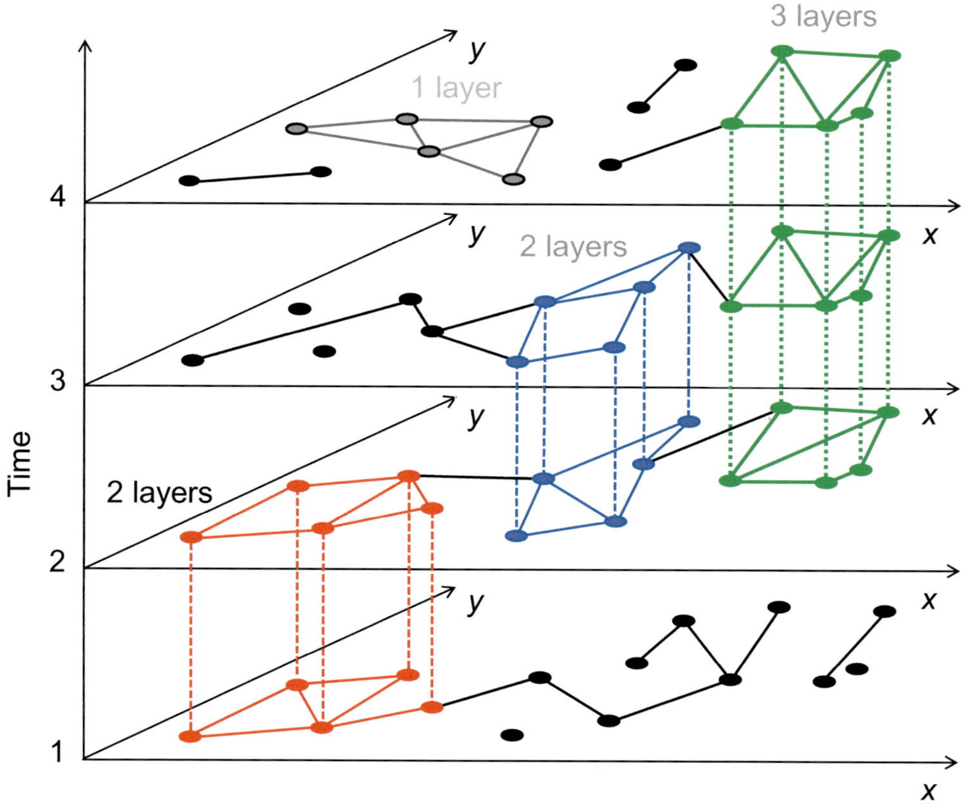

Figure 4.6 Multilayer clusters: space and time. (A black and white version of this figure will appear in some formats.)

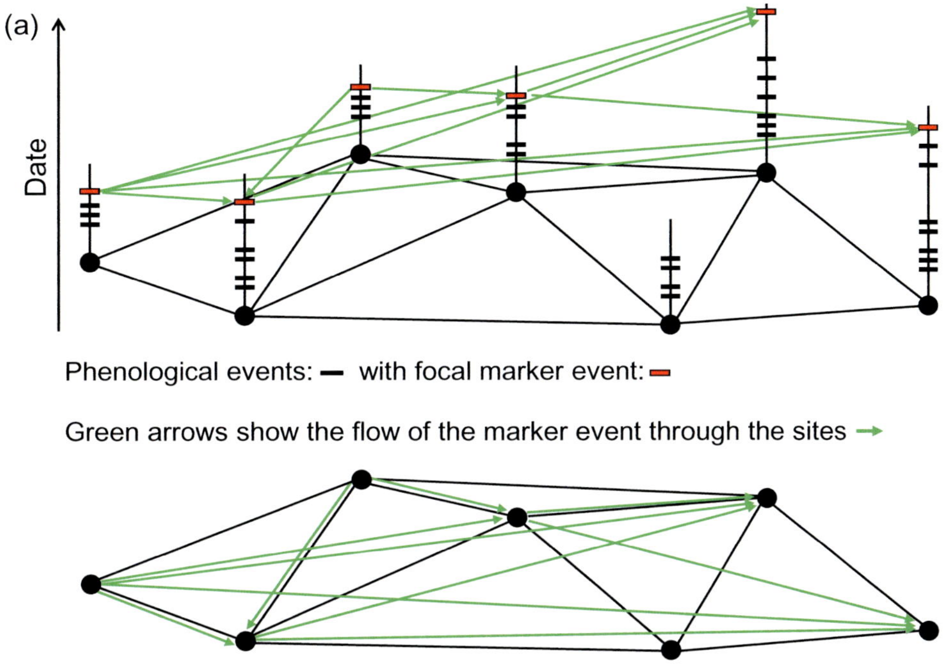

Figure 4.11a (a) Phenology at permanent sites; each node has a temporal sequence of events. Topological definition of spatial edges (Delaunay triangulation). One event type: look for burstiness and synchrony, perhaps as a function of absolute or relative locations. Phenology at regional sites: leaf production is indicated by the short horizontal bar. Position on vertical timeline indicates date. Partial order of sites by event date. (A black and white version of this figure will appear in some formats.)

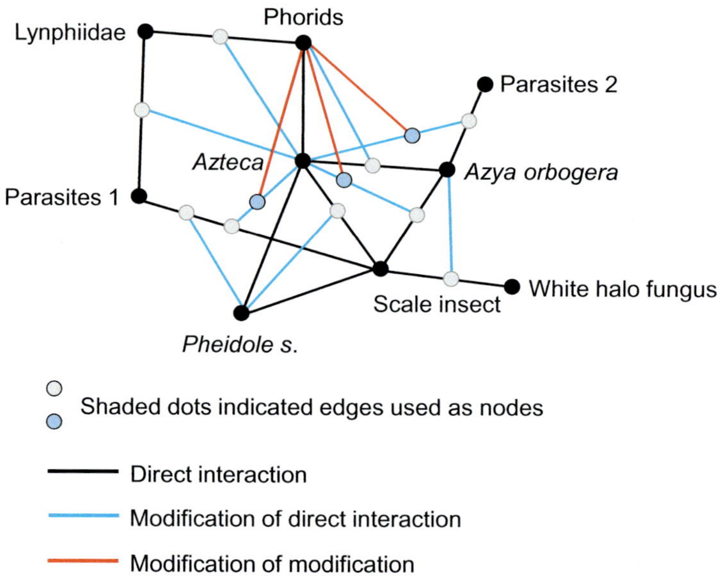

Figure 4.13 Indirect effects: two levels of modifications of ecological interactions (see Hsieh et al. 2012; Perfecto et al. 2014; Golubski et al. 2016). Part of a coffee agro-ecosystem, showing a hierarchy of interaction modifications: edges at the lower level of the hierarchy become nodes for edges at the next. *Azteca* and *Pheidole* are ants. [Redrawn from Hsieh, H.-Y., Liere. H., Soto, E. J., & Perfecto, J. (2012). Cascading trait-mediated interactions induced by ant pheromones. *Ecology & Evolution*, **2**, 2181–2191 © 2012 the authors; Perfecto, I., Vandermeer, J., & Philpott, S. M. (2014). Complex ecological interactions in the coffee agroecosystem. *Annual Review of Ecology, Evolution, and Systematics*, **45**, 137–158; Golubski, A. J., Westlund, E. E., Vandermeer, J., & Pascual, M. (2016). Ecological networks over the edge: hypergraph trait-mediated indirect interaction (TMII) structure. *Trends in Ecology and Evolution*, **31**, 344–354. © 2016 Elsevier Ltd. All rights reserved.] (A black and white version of this figure will appear in some formats.)

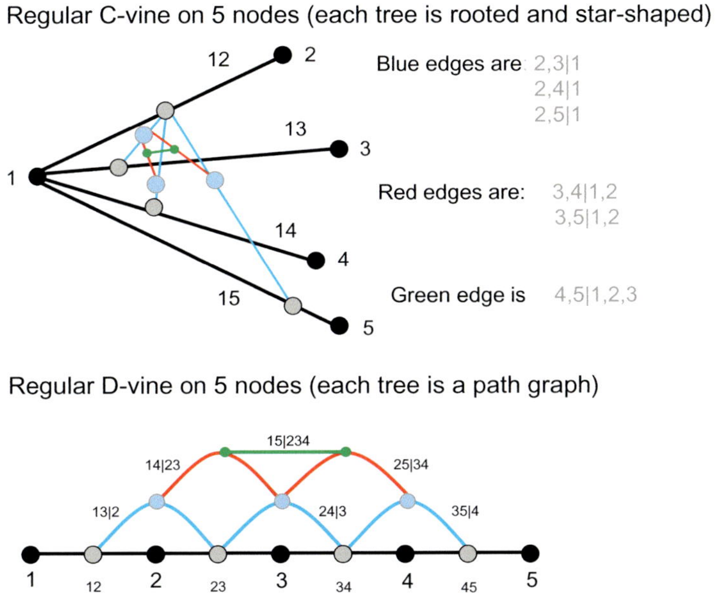

Figure 4.14 Vines are hierarchically structured graphs in which the edges at a lower level are used as nodes for further edges at the next level. A regular C-vine on five nodes: each tree is rooted and star-shaped. A regular D-vine on five nodes: each tree is a path graph. (A black and white version of this figure will appear in some formats.)

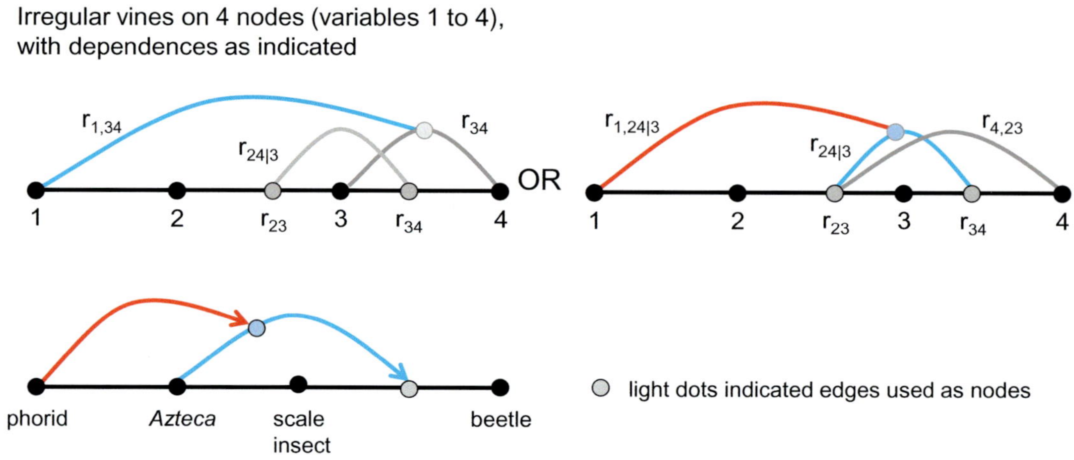

Figure 4.15 An *irregular vine* [new term] allows the mixing of hierarchical levels joined by pairwise edges, for example, a level-1 node joined to level-1 edge by level-2 edge, thus allowing either a taxon or a taxon–taxon interaction to modify other interactions. Irregular vines on 4 nodes (variables 1–4), with dependences as indicated. Irregular vine version of Golubski et al. (2016). *Azteca* affects the interaction of scale and beetle; phorids modify that indirect interaction. [Reprinted with permission from Golubski, A. J., Westlund, E. E., Vandermeer, J., & Pascual, M. (2016). Ecological networks over the edge: hypergraph trait-mediated indirect interaction (TMII) structure. *Trends in Ecology and Evolution*, **31**, 344–354. © 2016 Elsevier Ltd. All rights reserved.] (A black and white version of this figure will appear in some formats.)

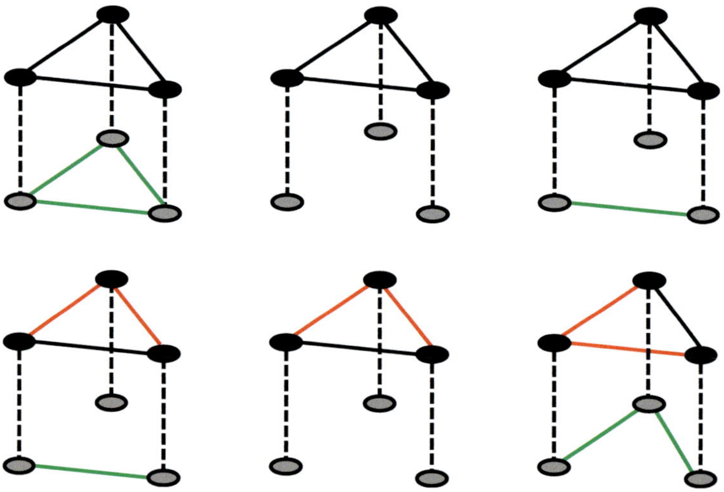

Figure 4.25 Frequent bilayer motifs in brain networks: Upper: function layer with positive (black or solid) and negative (red or dashed) edges. Lower: structural layer, edges grey or green. Vertical edges are interlayer identity edges. [Based on findings of Battiston, F., Nicosia,V., Chavez, M., & Latora, V. (2016). Multilayer motif analysis of brain networks. *Chaos*, **27**, 047404. With permission from AIP Publishing.] Note edge coding difference. (A black and white version of this figure will appear in some formats.)

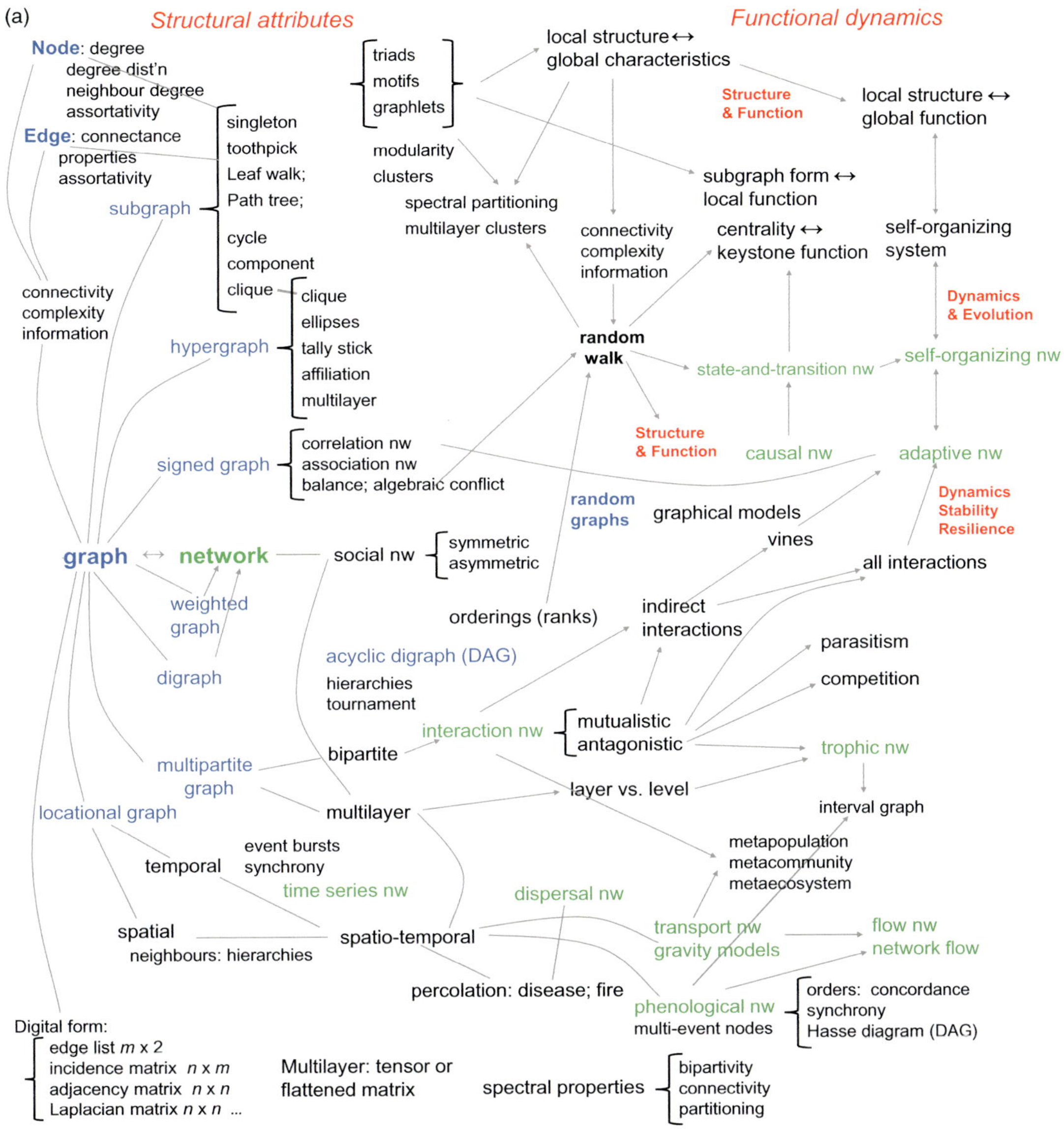

Figure 5.3 Two versions of the atlas of concepts, structures, and methods. Only the most important connections are included. (a) A version without miniature figures. (b) A version with miniature structural figures, which help the memory but make the diagram more cluttered. Choose the one that works best for you. (A black and white version of this figure will appear in some formats.)

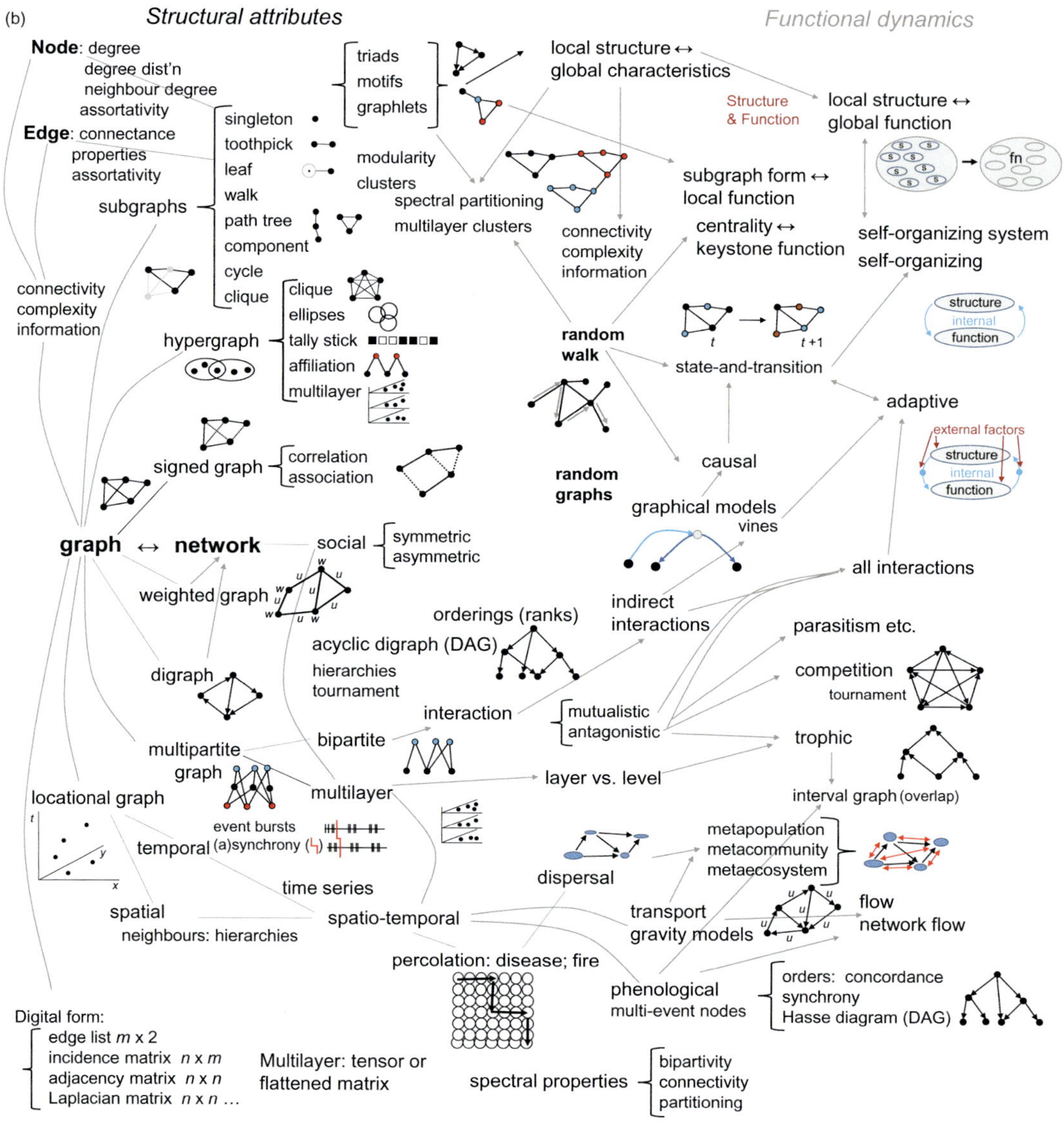

Figure 5.3 (cont.)

3.6 Concluding Comments

The quantitative analysis of ecological networks and the statistics for this analysis are often computationally complex, even for static networks of fairly simple structure. If the network is dynamic and observed through time, the analysis can be expected to be challenging, the data for static or one-observation networks can still be very large. These large data sets provide a number of challenges to their analysis. The most obvious is how the computational resources required scale with "size," as measured by the number of nodes or edges (see Newman 2010). The computational complexity of an algorithm is expressed as its order, for example order n if it increases linearly with n, written $O(n)$. The simplest forms of analysis are often problems that can be solved in polynomial time, whether that is $O(n)$, $O(n^3)$, or $O(n \log n)$, and thus referred to as solutions in class P (solution in *polynomial* time; Goldreich 2010; Moore & Mertens 2011). More complex problems may require exponential time (class *EXP*), such as $O(2^n)$, which quickly become impractical for increasing n (see the video by 3Blue1Brown on what 2^{256} implies: *youtube.com/watch?v=S9JGmA5_unY*). Between classes P and *EXP*, there is a whole set of famous problems that may not have polynomial solutions but can have their solutions verified in polynomial time; think of a Sudoku puzzle with n rows and n columns: hard to solve but easier to check (Goldreich 2010; Moore & Mertens 2011). These fall into class *NP* (*non-deterministic polynomial*). That computational classification may seem unimportant for ecological network analysis, but consider the calculations required for a measure of centrality (it is $O(n(m + n) \log n)$) or finding all complete subnetworks as the number of nodes in the network grows. While shortest path algorithms in two dimensions are in P, shortest path algorithms in three dimensions are in *NP*, as are problems such as finding the largest complete subgraph in a network. Obviously, the question of computational complexity for algorithms used for the statistics and analysis of networks is critical for our understanding of ecological networks as the information available and the size of data sets continues to grow (see Pržulj & Malod-Dognin 2016).

A final comment on computational complexity for ecological networks refers back to the beginning of Section 1.2, where we mentioned some of the differences between ecological networks and the standard social and information networks that tend to focus on hubs and power laws and small worlds. Often discussions of network complexity will assume that the graph is "sparse" (Newman 2010) in that the number of edges grows only linearly with the number of nodes, giving $O(m) = O(n) \ll O(n^2)$ and an edge density, ρ, that decreases with increasing n. This is certainly true of some ecological networks, think of a spatial lattice like a triangulation, in which node degree is constant with n, but it will fail as a general assumption. In particular, food webs tend to have constant edge *density*, with increasing average node degree with increasing n (Martinez 1992; Dunne et al. 2002). Clearly the network's sparseness or edge density can have a large effect on the computational complexity of particular characteristics of interest, and may affect our choice of edge density when there is one. For correlation networks, for example, there may be a computational advantage to limiting the number of edges per node, but that limit may also represent a loss of important information. The weighted correlation analysis (or WGCNA), mentioned earlier, retains all edges of the correlation network and their weights, and so $O(m) = O(n^2)$ in that case; at least it is still polynomial!

Another form of complexity in ecological network analysis that we see arising more and more is the elaboration of basic network structure. For example, starting with an observational study of flowering plants and their pollinators in our favorite meadow in mid-July, we might extend the study with equivalent observations at the same meadow in two-week periods from early May to late August, or to equivalent short-term observations at several meadows over a range of elevations. An alternative is to expand the range of interactions in the study by including data on the plants' herbivores and mycorrhizal associations and the pollinators' predators and parasites. In any of these examples, we end up with a network that consists of several layers, allowing different kinds of analysis within layers, between layers, and over the whole network. The treatment of such multilayer networks deserves a full chapter of its own, and that is what follows.

4 Multilayer, -type, and -level Networks

4.1 Introduction

Ecological networks include a great range of complex phenomena and there are many ways in which basic network topology can be translated into structures that include and take advantage of those complexities. As with other areas of network science, which has evolved from or through many different disciplines, there is a diversity of terminology associated with these diverse topological structures which may appear to lack systematic or consistent application. Our endeavour here is to present that diversity of structures and clarify their application in the ecological context.

In the past, studies of ecological networks have tended to consider one kind of interaction at a time: a food web, pollination mutualism, or host–parasite relations. Recently, there has been an increasing tendency to consider two kinds of interaction at once (e.g. trophic and non-trophic or antagonistic and mutualistic; Bastolla et al. 2009; Fontaine et al. 2011; Kéfi et al. 2015, 2018; Thébault et al. 2018) or even several kinds in the same study (Ings 2009; Pocock et al. 2012; Emer et al. 2018; Kéfi et al. 2018). In considering more interactions, the goal is to understand the complexity of structure and dynamics of ecological assemblages or communities through investigating the ways that the various interactions affect each other, both generally and specifically (Fontaine et al. 2011; Kéfi et al. 2018).

As an introductory example, consider an assemblage of flowering plants together with both their herbivores and their pollinators. The resulting network can be seen as the union of two bipartite graphs, one for pollination and one for herbivory (Figure 4.1). To be thorough, the edges in the two parts could be coloured or shaded to indicate that they represent different kinds of interactions (Figure 4.1a). The two interactions, with the plants as participants in both, can then be compared for characteristics such as nestedness and modularity, and how these topological characteristics do or do not align (Fontaine et al. 2011). An alternative presentation is to separate the two interactions into network layers, perhaps with tie lines to emphasise the shared plant species as nodes in the two layers (Figure 4.1b). This two-layered network is a simple instance of a *multilayer network* (Boccaletti et al. 2014; Kivelä et al. 2014; Pilosof et al. 2017), which is one modification of a basic network structure. Networks with layers representing different interactions that are connected by interlayer edges only for the shared species are said to be "diagonally coupled" (Pilosof et al. 2017) based on their matrix representation, as we will explain. There can be no meaning for an edge between the

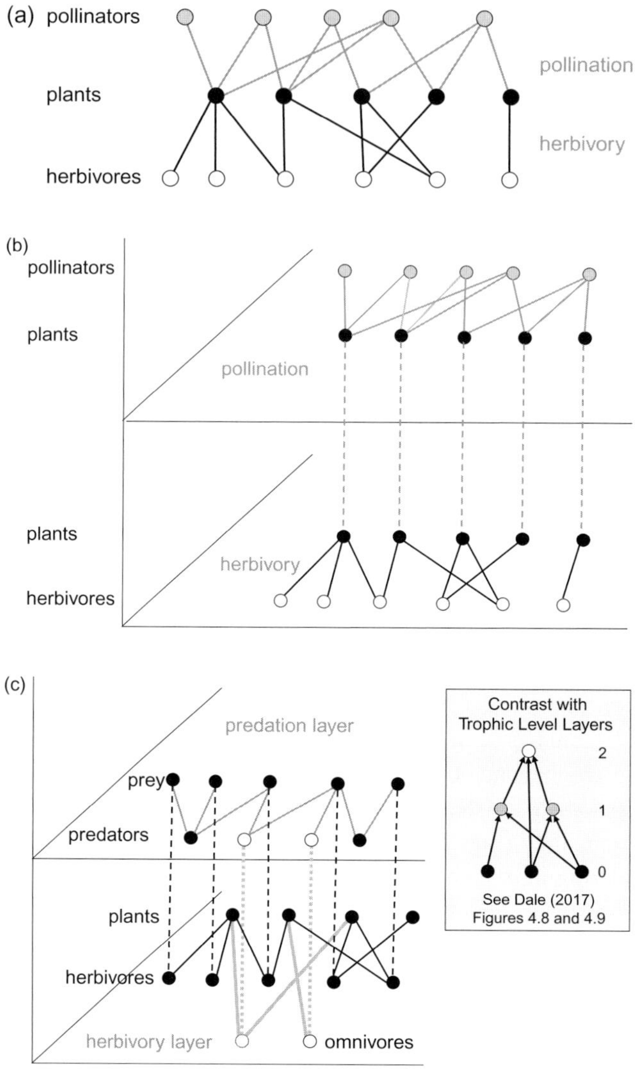

Figure 4.1 Merging two bipartite networks. (a) Pollination and herbivory: one layer. (b) Two layers, one for each interaction. (c) Merging interaction networks: layers representing interactions, *not* trophic levels. (A black and white version of this figure will appear in some formats. For the colour version, please refer to the plate section.)

layers of this structure except for those indicating identity between instances of the same plant species, sometimes called *coupling edges* or *edges of identity*. Food webs are often presented as if with layers representing trophic levels (0 for primary producers, 1 for their consumers, and so on; see inset of Figure 4.1c) with edges between the trophic layers representing consumption; the alternative is to have separate layers for the different interaction types, as in the main panel of Figure 4.1c. Because the layers represent interactions, not trophic levels, omnivores appear in both layers, once for herbivory and once for predation on herbivores.

As the title of this chapter indicates, we will describe several elaborations of the basic network structure (even if they represent complex networks), which are typified by a single type of object as nodes, a single kind of interaction between pairs of nodes as edges and representing a single point in time or a single stage of development. The elaborated structures are often described by terms with the prefix "multi-" to include more than one type of node, more than one type of edge, more than just pairwise interactions, or more than one time or stage of development. The first of these is the multilayer network.

4.2 Multilayer Networks

Many of the elaborations of network structure to include ecological complexities appropriately fall into the general category of *multilayer* (Boccaletti et al. 2014; Kivelä et al. 2014; Pilosof et al. 2017), and the availability of ecological data and the tools to analyse multilayer networks now provides the opportunity to explore "the multilayer nature of ecological networks" Pilosof et al. (2017). This form of network consists of several layers, each of which holds a network graph of functions or interactions within it and, possibly, sets of edges between layers (Figures 4.1 and 4.2; Boccaletti et al. 2014). The sets of nodes in the layers may or may not be the same. Each layer is essentially its own network but when the layers are considered together it creates a multilayer network (picture the layers stacked like a deck of cards, as in Figure 4.2). Two general examples are human social multilayer networks in which different layers represent the different contexts for contacts between individuals, for example, work versus sports versus former school, and transportation networks in which each airline has a layer of its own, but the layers combine to form the entire air transport network (Boccaletti et al. 2014). In ecological applications, the layers can be determined by locations such as landscape patches, by observation times, by habitat types (Timóteo et al. 2018), or by separate food webs arising from these locations or other factors. The layers can also be determined by categories of ecological interactions (plant–pollinator, competition, predator–prey, host–parasite...) or by levels of organisation (population, community, metacommunity). Essentially, layers may be defined by almost anything: space, time, interaction type, level, and so on. For many ecological networks, although it seems to be an elaboration to divide the network into layers, there is often the advantage of greater clarity of structure in the division into layers, both for visualising a complex network and for its analysis and interpretation; similar comments apply generally to envisioning information (Tufte 1990). Figure 4.2 shows examples of multilayer networks in which each layer may be spatial with its own intralayer edges and the layers represent different times. Identity edges run between the layers, joining instances of the same node, whether stationary (Figure 4.2a and b) or moving (Figure 4.2c), and there may be interlayer edges representing other relationships among different nodes at different times (Figure 4.2a).

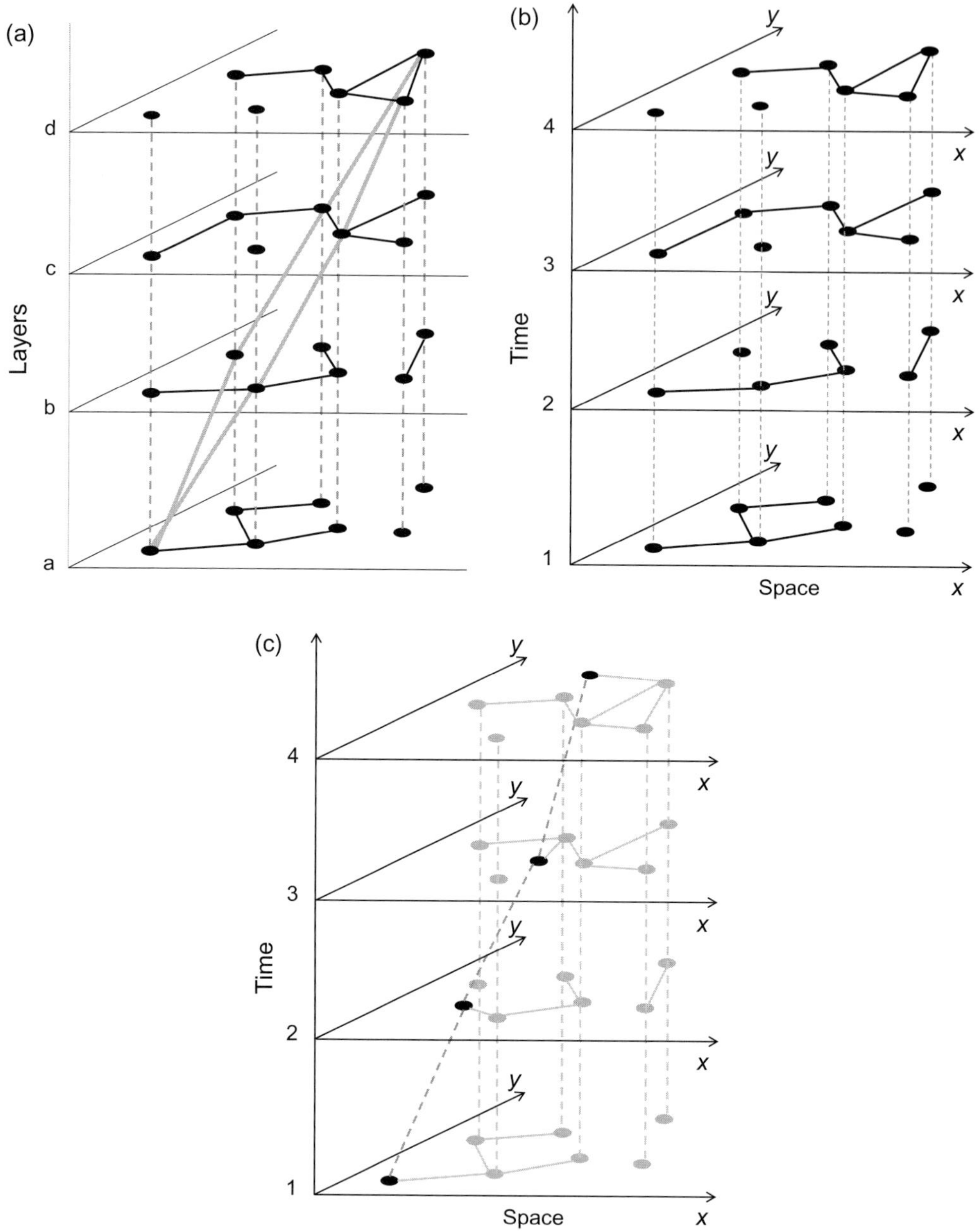

Figure 4.2 Multilayer network. (a) Like a deck of cards, the whole network consists of a stack of layers; each layer has its own edges, creating a network layer graph. Here, the nodes are the same in all layers. Identity edges can join instances of the same node in different layers, shown by dashed lines. There may also be edges between nodes in different layers (dark grey). (b) Multilayer network in space and time. Time slices show a series of edges through time. Identity edges join instances of the same node in different times, shown by dashed lines. (c) Time slices show a series of edges through time, with the shared nodes having spatial locations in the x–y plane. Identity edges (dashed lines) join instances of the same node even when it moves in space.

The term "multilayer" is very general as applied in the context described by Pilosof et al. (2017) and the other references cited, although no list will exhaust the possibilities. "Multilayer" is the best established of the terminology available as possibilities, but other similar terms have been suggested, including "multiplex," "multislice," "multirelational" and "multilevel"; and in the diversity and confusion that has resulted, some explication of the classification of such terms is a pre-requisite to further discussion of these network structures. In fact, Kivelä et al. (2014) have already provided a very impressive list of the many (many!) terms that have been used for the various elaborations of the basic networks and their graphs (Kivelä et al. 2014). We will not repeat all that material, but instead we introduce terms as needed, as consistently as possible, and with some formality to help restrict their application, because one source of confusion is the informality with which terms are sometimes added for consideration and use. We will resist the temptation to provide our own rationalised classification scheme.

Following Boccaletti et al. (2014), a multilayer network is a family of M network graphs, call it $M = (G, C)$, where G is the set consisting of layer graphs indexed by $\alpha = 1$ to M:

$$G = \{G_\alpha \,|\, G_\alpha = (V_\alpha, E_\alpha)\}, \tag{4.1}$$

G_α consisting of the nodes, V_α, and edges, E_α, of layer α. C is the set of interlayer edges:

$$C = \{E_{\alpha\beta}\varepsilon(V_\alpha \times V_\beta) \,|\, \alpha, \beta = 1, \ldots M, \alpha \neq \beta\}. \tag{4.2}$$

The set of all edges is therefore the union of inter- and intralayer edges:

$$E(M) = C \cup \{E_\alpha | 1, \ldots M\}. \tag{4.3}$$

For a monolayer network, the edges are coded by the $n \times n$ adjacency matrix, $\mathbf{A}$, with entries $a_{ij}$for the edges between nodes i and j; for a multilayer network, the connections are coded by an $n \times M \times n \times M$ array with entries $A_{j\beta}{}^{i\alpha}$ for edges between node i in layer α and node j in layer β. The basic entries are 1 for an edge and 0 for no edge, but weights and signs and directions can be accommodated by replacing the 1's with the appropriate values, just as in the familiar adjacency matrix. The multilayer network can also be represented by a "supra-adjacency" matrix. This is a square $(n \times M) \times (n \times M)$ matrix, which is an $M \times M$ array of M^2 $n \times n$ adjacency matrices with the intralayer edges coded in the small matrices on the main diagonal of the large array. The off-diagonal matrices in the large array give the interlayer edges. If the only such edges are the nodes' identity edges, those off-diagonal matrices are themselves diagonal. The matrix entries are $a(i, \alpha, j, \beta) = 1$ if there is an edge between node i in layer α and node j in layer β, and 0 if there is none. The need for weights and so forth can be accommodated in the usual way. An intralayer edge is given by $a(i, \alpha, j, \alpha) = 1$; and interlayer edge is $a(i, \alpha, j, \beta) = 1$ with $\alpha \neq \beta$; and in particular, $a\ (i, \alpha, i, \beta) = 1$ is an edge of node identity or a coupling edge for node i in layer α and node i in layer β.

Multilayer networks have given rise to their own vocabulary, reflecting emergent properties not found in monolayer structures (following Kivelä et al. 2014). If all layers contain the same set of nodes, this is *node-aligned*. If each node occurs in only one layer, this is *layer-disjoint*. If the only interlayer edges are coupling edges indicating identity, this is *diagonally coupled* (already mentioned). A *node-coloured network* is one in which the nodes belong to several categories and each category has its own layer, so this is also node-disjoint. This could also be called a network with *multitype nodes*. For example, if each trophic level has a separate layer as labelled in Figure 4.1c-inset, it is node-coloured and node-disjoint. An *edge-coloured network* joins nodes in pairs using different categories of relationships, one "colour" per layer, and the layers may all have the same set of nodes (thus node-aligned). This could also be called a network with *multitype edges*. Figure 4.1b is edge-coloured.

The term *multiplex* is usually reserved for multilayer networks in which the sets of nodes are the same for all layers ($V_1 = V_2 = \ldots = V_M$), so that only the edges differ between layers; and edges between layers, if any, are restricted to joining instances of the same node (Boccaletti et al. 2014; Figure 4.2). The term "multiplex" has also been used to indicate networks in which the edges belong to more than a single category (e.g. trophic and non-trophic interactions; see Kéfi et al. 2018; also Kao & Porter 2017), so that each layer represents a different kind of interaction, as in Figure 4.1b. The study of multiplex networks is currently a "hot" topic in the statistical physics of complex networks (but not just in physics), and there is much to be learned about and from them (Mondragon et al. 2018); Lee et al. (2015) have provided a good summary.

The *projection* of any multilayer network graph, P(*M*) as in Figure 4.3a, is a single layer with node set that is the union of all *M* layer node sets, and edge set that is the union of all edges between different nodes (whether intra- or interlayer). For a multiplex network, the projection graph has the same node set as the whole network ($V_1 = V_2 = \ldots = V_M$), and includes all intralayer edges: $\text{U}_\alpha\{E_\alpha | \alpha = 1, \ldots M\}$ (Figure 4.3a). The projection graph can be considered as an aggregation of all layers, using a union-of-edges rule to determine where edges occur (Boccaletti et al. 2014). Aggregation can occur for any two or more layers and any of several rules for the edges in the aggregation can be applied:

1. Union-of-edges, so that one edge results where there was one or more in any of the layers being aggregated (and none otherwise).
2. Summation of edge weights, so that the edge in the aggregation of the layers is the sum of the weights of the corresponding edges, where they exist (or just a count of edges in that position).
3. Average of edge weights, where they exist.
4. Multiple edges in a multigraph: all instances of an edge are maintained so that some node pairs may be joined by many edges, one for each layer in which the edge was present before aggregation.

In the last case, the layers that are the sources of the projected edges in a multiplex network can be accounted for using a multigraph with coloured edges, each colour representing a separate layer. A *multigraph* is a structure with nodes and edges like an ordinary graph, but any pair of nodes can be joined by more than one edge. A multilayer graph becomes a projection multigraph with coloured edges by assigning

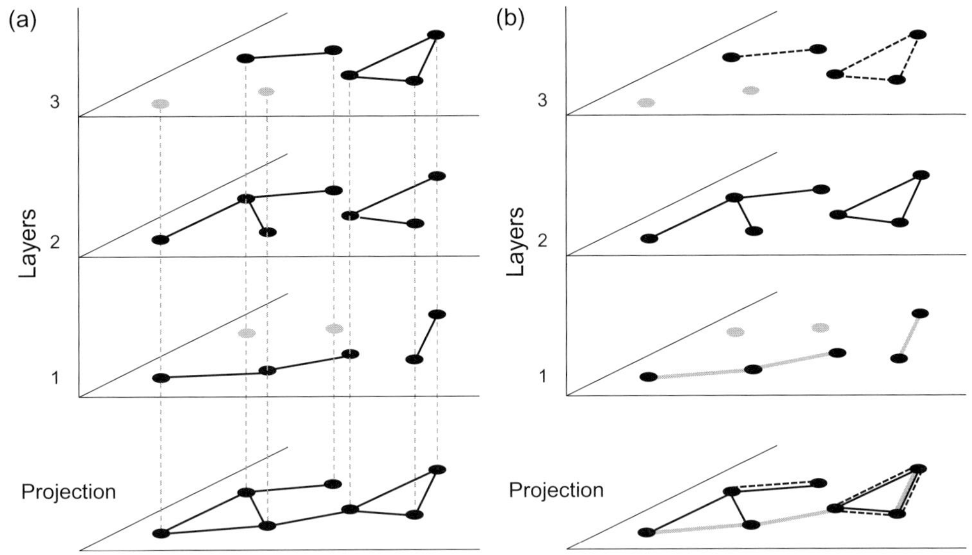

Figure 4.3 (a) Projection of a multilayer network. Identity edges join instances of the same node in different layers. Grey nodes are absent from that layer. (b) Projection of a multilayer network to form a multigraph with coloured edges (grey nodes are absent from that layer). (A black and white version of panel b will appear in some formats. For the colour version, please refer to the plate section.)

a different colour to each layer and including all coloured edges in the projection graph (Figure 4.3b; see Kivelä et al. 2014).

The choice of aggregation rule for network layers, just like the choices for other forms of aggregation in ecological data (e.g. taxonomic aggregation) can have profound effects on the outcomes and their interpretation, especially when it comes to the evaluation of robustness and resilience (Pilosof et al. 2017).

Given a layered network, one of the key questions is how the statistics and measures of monolayer networks translate to accommodate the multilayer structure. For example, how does modularity include nodes from more than one layer, and how does centrality account for the layered structure? In some layered networks, "edges of identity" join instances of the same node in different layers, as used in temporal or spatiotemporal graphs to track identifiable individuals through time and space (Dale 2017). Many of the challenges and solutions for multilayered network statistics and analysis are relevant to, or already solved for temporal graphs, spatially referenced networks (networks-of-networks, graphs-of-graphs, or metanetwork) or spatiotemporal graphs for complex networks (Dale & Fortin 2014; Dale 2017; Martensen et al. 2017). Therefore, one approach to multilayer network statistics and other metrics is to use a combination of local measures within the layers (like the time slices) and complete network measures for all layers together. This is not a universally applicable solution, and many of the familiar graph theoretical measures will require modification to be truly useful for multilayer networks (Kivelä et al. 2014), especially if the dimensions within and between layers are not commensurable, but for some concepts it is an excellent first approach.

When nodes are missing from some layers of a multilayer network (Figure 4.3a), the temptation is to fill in those absences with dummy nodes with no edges, so that the topological structure is fully node-aligned. For many purposes, this stratagem will work well, but there is the danger it will affect summary statistics such as edge density and average degree, unless properly accounted for.

4.2.1 Node Degrees in Multilayer Networks

We should examine the concept of a node's degree in multilayer networks. Within a multiplex layer, a node's degree can be based on the number of intralayer edges that attach to it: $k_\alpha(i)$. The total intralayer degree for node i is just the sum of these over all M layers (whether or not the nodes are constant through all layers):

$$k_{\text{layers}}(i) = \Sigma_\alpha k_\alpha(i) \tag{4.4}$$

That sum does not include the interlayer edges of node i, $k_C(i)$, and so the total (sometimes called "global") degree of node i is

$$k_{\text{tot}}(i) = k_{\text{layers}}(i) + k_C(i) \tag{4.5}$$

Similarly, when assessing triad completion (the clustering coefficient), which is the proportion of triplets of nodes that have two edges that have all three, it is easy to distinguish between the ("local") intralayer measure and the ("global") all-layers-included measure (cf. Boccaletti et al. 2014). (We do not recommend the use of "local" versus "global" as substitutes for terms that are explicit about which layers are included.)

An obvious characteristic of interest for a node-aligned multiplex network (the same nodes in every layer) is the correlation of node degrees in different layers: the correlation of $k_\alpha(i)$ and $k_\beta(i)$ for all α and β, using any one of several measures of correlation (Pearson, Spearman, Kendall, etc.; see Boccaletti et al. 2014). Under the same circumstances and focusing on the edges instead of the nodes, the characteristic of *edge overlap* may be of interest (Bianconi 2018); this is the number or proportion of edges in common between layers (i.e., considered two at a time). In ecological contexts, 'overlap' is used for other properties and a better choice of term might be *edge matching coefficient*. The edge matching coefficient for layers α and β localised at node i is $o_{\alpha\beta}(i) = \Sigma_j\, a_\alpha(i,j)\, a_\beta(i,j)$. Then the total edge matching for the two layers is $O_{\alpha\beta} = \Sigma_i\, o_{\alpha\beta}(i)$ (Boccaletti et al. 2014). Edge matching obviously contributes to any measure of the overall similarity between layers in a multiplex network.

4.2.2 Walks and Paths

For any multilayer network, the familiar concepts of walks and paths can be extended from simpler monolayer graphs, based on the set of all edges, both inter- and intralayer:

$$E(\mathcal{M}) = \mathcal{C} \cup \{E_\alpha | \alpha = 1, \ldots M\} \tag{4.6}$$

A *walk* is an alternating sequence of nodes and edges that connect them (a *closed walk* if it begins and ends at the same node); Figure 4.4 shows a multilayer random walk. A *path* is a walk that uses no element twice. A *cycle* is a path that begins and ends at the same node. If there is a path between any two nodes, then the network is *connected*. The *length* of a walk or a path is usually and simply defined as the number of edges, or, if the edges are weighted, the sum of the weights or of the weights' inverses. However, this becomes an interesting matter in a multilayer network, because there are choices to be made on how interlayer edges are included: Are they equivalent to intralayer edges and counted in the same way? Do they belong to a separate category to be counted as a separate kind of distance for calculating the total length of a path? Again, this is a familiar set of questions from multilayer temporal or spatiotemporal graphs in which the dimensions of space and time are not directly commensurate (Figure 4.2b and c). In spatiotemporal networks, the spatial distance is included in *topological length*, the number of steps from source to sink, and the temporal length is the time elapsed from source to sink, sometimes called *latency* (Williams & Musolesi 2016). Then, the spatiotemporally shortest path is the temporally shortest path with the minimum spatial length (Williams & Musolesi 2016). This definition can be generalised for multilayer networks: the shortest multilayer path between two nodes is a path that has both the shortest interlayer length and the fewest intralayer steps or the shortest intralayer distance.

Simple counts of edges in either category (geodesic distance) or sums of inverse weights if the weights have similar meanings are obviously the most straightforward measures of interlayer paths (Boccaletti et al. 2014; Kivelä et al. 2014). Cozzo et al. (2015) used the term "supra-walk" to signify a walk in a multiplex network in which intralayer edges can have an interlayer edge inserted before or after, but there cannot be consecutive interlayer edges. In some algorithms, the steps of a random walk from node to adjacent node can be supplemented by rare "teleport" jumps to any node in any layer (Figure 4.4b), for example for PageRank centrality measures (De Domenico et al. 2015).

Another emergent measure for multilayer networks is *interdependence*: this is the proportion of all shortest paths that include nodes in more than one layer, and thus at least one interlayer edge (Cozzo et al. 2015). This concept is illustrated in Figure 4.4b.

4.2.3 Centrality and Node Ranking

Having defined walks and paths and path lengths, an obvious property to consider next is centrality. Layers can be compared for any of the intralayer measures that are familiar from simple (monolayer) network graphs including various measures of centrality. Node degree is one measure of centrality already discussed, but there are many choices for approaches to measuring this property (e.g. De Domenico et al. 2013). Another obvious measure of node centrality is the proportion of all shortest paths between nodes in which the focal node occurs, irrespective of how length is defined for the system being studied.

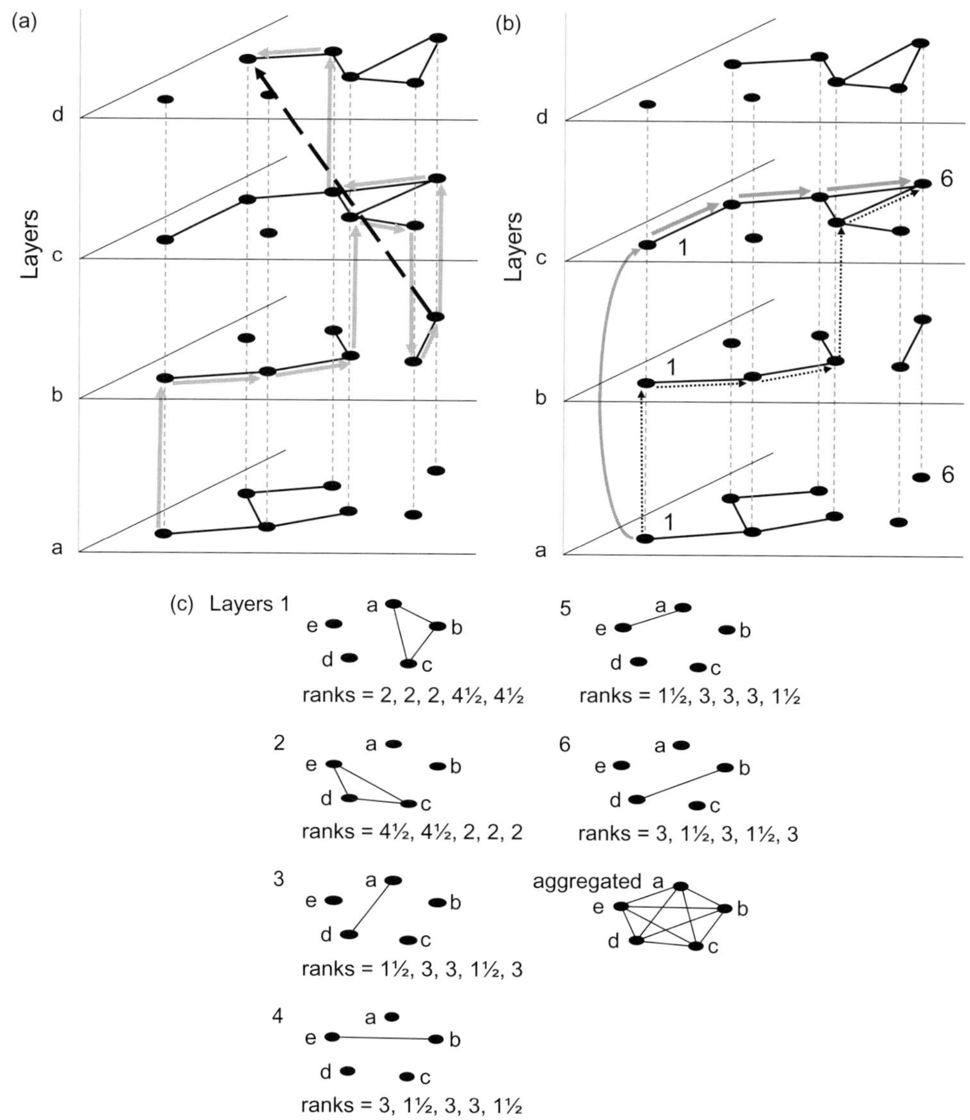

Figure 4.4 (a) Random walk in a multilayer network; the nodes are the same in all layers. Identity edges join instances of the same node, shown by dashed lines, with a multilayer teleport jump in long-dashed edge. Walk + teleport is basic to the PageRank approach to node ranking or centrality measure. (b) Interdependence: some shortest paths need more than one layer, for example, any shortest path from node 1 to node 6 must use at least two layers. The two different paths depend on whether interlayer edges occur only between adjacent layers. (c) Versatility, centrality, and ranking. Centrality ranks by layer; versatility is an equivalent measure for multilayer networks.

Based on almost any of the monolayer centrality measures, a multilayer value can always be created with a weighted sum or an average of the values from the individual layers. See the material in table 3.1 and figure 3.8 of Dale (2017) for a summary of centrality measures for monolayer graphs. This aggregated measure may be easy to calculate but it may not embody the hoped-for concept. De Domenico et al. (2015) introduced an alternative measure, similar to centrality, which they call a node's *versatility*. This value summarises the centrality values of each node *calculated over the entire multilayer structure*, that is, not by a single layer at a time. They explain the logic of this approach with an authorship example (laid out in their figure 2) and provide two versions: *eigenvector versatility* and *PageRank versatility*. Figure 4.4c provides an alternate exposition of a similar example. Here, the six layers of Figure 4.4c represent six sites and the edges represent significant positive correlations between pairs of the five species (a, b, c, d, and e) present at all sites. We rank the species by centrality in each layer, with nodes of equal centrality receiving tied ranks. In Layer 1, nodes a, b, and c have equal centrality, which is greater than that of nodes d and e (also tied). De Domenico et al. (2015) assigned these ranks of 1 for nodes a, b, and c, with ranks of 4 for nodes d and e. Layer 2 resembles Layer 1 with three edges forming a triangle, but the remaining layers have only one edge each, so the ranks are 2 1's and 3 3's in De Domenico's assignment. To ensure that the rank sum is constant (it should be 15), we suggest assigning ranks that are averaged for tied nodes, so the ranks are now (2, 2, 2, 4½, 4½) rather than (1, 1, 1, 4, 4) for Layers 1 and 2 and consisting of 2 1½'s and 3 3's in the remaining layers (Figure 4.4c). In either rank assignment scheme, when the layer ranks are summed for each node, the consensus places node c as the least central. When the layers are aggregated, the aggregated network is a complete graph in which all nodes have equal ranks. Using the entire multilayer network, even in a simple way, recognising the different weights for the layers based on the number of edges, so that Layers 1 and 2 have larger weights (3 versus 1), node c is then recognised as the highest rank of versatility (multilayer centrality). As stated, this result is consistent with that of De Domenico et al. (2015) for both eigenvector versatility and its PageRank version.

4.2.3.1 Eigenvector Versatility

Eigenvector versatility is based on the solution of the equation (using Einstein notation for summation; in this case the sum is over all values of i and α):

$$A_{j\beta}{}^{i\alpha}\Theta_{i\alpha} = \lambda_1 \Theta_{j\beta}. \tag{4.7}$$

Here, λ_1 is the largest eigenvalue and $\Theta_{j\beta}$ gives the versatility of node j in layer β (see De Domenico et al. 2015). The "global" versatility of node j is just the sum of these values over all layers. In addition to the PageRank version, described in terms of random walk steps to adjacent nodes and random "teleport" jumps to nodes in any layer (Figure 4.4a), De Domenico et al. (2015) also have explained "hub & authority" versatility (HITS algorithm), Katz versatility, and betweenness versatility in some detail. They have also noted that the aggregation of layer-by-layer centrality scores can be misleading and provide examples to illustrate the point (De Domenico et al. 2015).

Multilayer connectivity is based on walks and paths, as is any evaluation of multilayer modularity. Timóteo et al. (2018) used a multilayer network approach to investigate the spatial structure of seed dispersal interactions, with habitats in the Great Rift as the layers. They found that the importance of dispersers to this system was captured by multilayer versatility but not by the more standard metrics based in monolayer structures.

4.2.4 Clusters and Clustering

Clustering and modularity in multilayer networks are of obvious interest in many ecological applications; clusters of nodes, whether the nodes are sites, species, or individuals are important features in any network. Before turning to multilayer clusters in general, it is useful to begin with examining *triadic closure* (also known as *triad completion* or the *clustering coefficient*) in the multilayer context. This measures the proportion of triplets of nodes with two edges that are closed by the presence of the third edge, thus forming a triangle. Cozzo et al. (2015) used the concept of a supra-walk (in which intralayer edges may have one, but not more than one, interlayer edge inserted in the walk) to examine triad completion in multiplex networks. Figure 4.5 illustrates the five versions of triadic closure assessment in a multiplex network. Examining all possible configurations, the final measure is just the proportion of such node triplets that are actually closed by the third edge.

There are many approaches to multilayer clustering, and Figure 4.6 illustrates just one. This is a multiplex network with layers possibly representing times with nodes arranged in space. In each layer, the maximal highly connected subgraphs are identified in an iterative process, and then those that are most similar to an identified module in an adjacent layer. Three multilayer clusters are indicated by the three shading

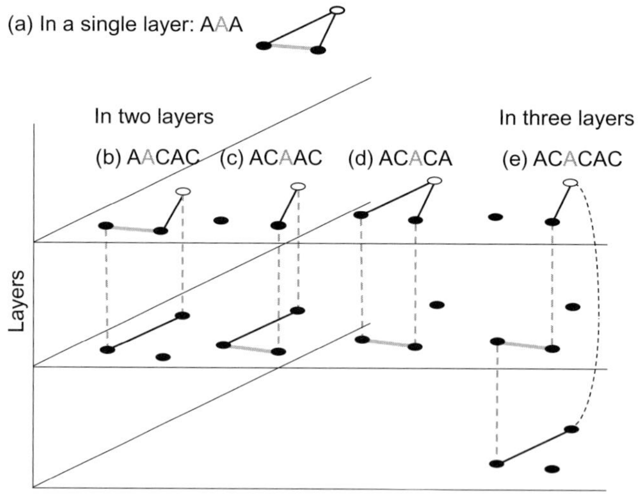

Figure 4.5 Triad closure. Identity edges join instances of the same node in different layers. The grey edge and corresponding letter is the closing edge; A indicates within layer and C between. (a) Single layer. (b) Two layers. (c) Three layers.

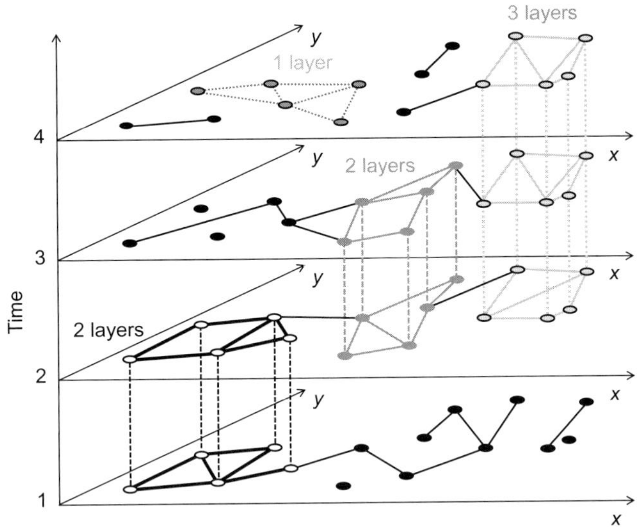

Figure 4.6 Multilayer clusters: space and time. (A black and white version of this figure will appear in some formats. For the colour version, please refer to the plate section.)

intensities in Figure 4.6, including nodes in two, two, and three layers. The cluster that is in only the top layer may or may not be counted, depending on the application, because it occurs in only one layer.

Another form of cluster analysis for multiplex networks is to cluster complete layers based on structural similarity between pairs of layers (Kao & Porter 2017). Kao & Porter (2017) referred to the concept as "connection similarity" and illustrated the technique with data from airlines and from research areas in physics. As with other characteristics of multilayer networks, there is a wide range of choices for how structural similarity is quantified.

One of the most useful guides to multilayer cluster analysis is the survey by Kim & Lee (2015). They surveyed a number of proposed methods for cluster detection in layered networks of two and of more than two layers, including methods based on matrix factorisation (spectral analysis; Brouwer & Haemers 2012) and on pattern mining. They considered 18 publicly available data sets and evaluated the methods using seven properties. For ecological networks, we suggest that the following three properties given by Kim & Lee (2015) are the most important ones:

- Property 2: Application to more than two layers. Most ecological networks that are multilayer will have more than two layers, and so this is a necessary property. Methods that apply only to bilayer networks can be ruled out.
- Property 4: Algorithm insensitivity. Some multilayer approaches are tightly linked to particular graph clustering algorithms and many algorithms perform well or poorly depending on the characteristics of the graphs. It is desirable to have the ability to apply different algorithms that can improve cluster detection.

- Property 6: Independence from the order of layers. For some kinds of ecological networks, such as those with time slices, sensitivity to the order in which layers are processed may be an acceptable property, but in multilayer networks which have no natural order, sensitivity to the order of processing should be avoided.

Based on this set of filters and the analysis in Kim and Lee's (2015), the methods most to be recommended are those using spectral analysis, whether based on the adjacency matrix (Tang et al. 2009) or on the Laplacian (Dong et al. 2011). DeFord and Pauls (2017) also have provided an evaluation of spectral clustering, specifically for multiplex networks, and this is obviously an area of current and ongoing research, which will undoubtedly find useful application in ecological network analysis.

4.2.5 Spectral Properties

As for unlayered or monolayer networks, the spectral properties of the graphs of layered networks contain much information about their structure, whether based on the adjacency (or other) matrices of the layers, the adjacency matrix of the projection, or on other matrices unique to layered networks such as the supra-adjacency matrix (Figure 4.7a; Boccaletti et al. 2014). This matrix reduces the multidimensional adjacency structure (tensor) to a two-dimensional matrix of matrices, showing the intralayer adjacencies in the sub-matrices along the main diagonal and the interlayer adjacencies in the blocks off the main diagonal (Figure 4.7a–d; Boccaletti et al. 2014). If the supra-adjacency matrix is $\mathbf{A_M}$, and $\mathbf{D_M}$ is the diagonal supra-matrix of the nodes' intralayer degrees, a "supra-Laplacian" can be derived as $\mathbf{L_M} = \mathbf{D_M} - \mathbf{A_M}$. The dynamics of layered networks can be closely related to the spectral properties of their Laplacian or supra-Laplacian matrices (Boccaletti et al. 2014; Kivelä et al. 2014). For example, the second smallest eigenvalue, the *algebraic connectivity*, and its eigenvector give important information about the network topology and structure (Kivelä et al. 2014). Other spectral properties, such as spectral radius, provide information on connectedness: for example, the spectral radius of a strongly connected graph or digraph decreases when nodes or edges are removed (Brouwer & Haemers 2012), which has implications for the spread of an epidemic through the multilayer system (Kivelä et al. 2014). As an example, Gomez et al. (2013) studied diffusion processes in a two-layered network and found that the spectral analysis based on the supra-Laplacian was highly informative. Solé-Ribalta et al. (2013) more generally investigated the Laplacian spectral properties of multiplex networks and the insights this method can provide, and further research is undoubtedly ongoing.

Given the complexity inherent in the layered structure of these networks, the pursuit of spectral properties becomes equally complicated, and we will refer the reader to Boccaletti et al. (2014), Kivelä et al. (2014), and references therein for further details. As a simple summary of results and expected outcomes, we suggest that the properties and dynamics of multilayer networks, *even when there are only two layers*, will turn out to be quite different from the dynamics of otherwise similar monolayer networks. Getting out of the Flatland of monolayer networks can radically

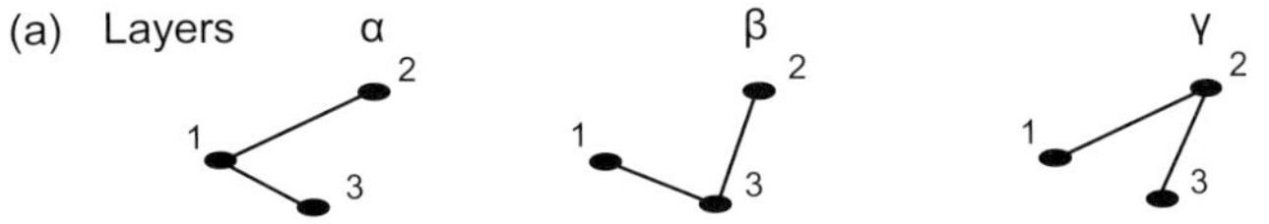

0	1	1	0	0	0	0	0	0
1	0	0	0	0	0	0	0	0
1	0	0	0	0	0	0	0	0
0	0	0	0	0	1	0	0	0
0	0	0	0	0	1	0	0	0
0	0	0	1	1	0	0	0	0
0	0	0	0	0	0	0	1	0
0	0	0	0	0	0	1	0	1
0	0	0	0	0	0	0	1	0

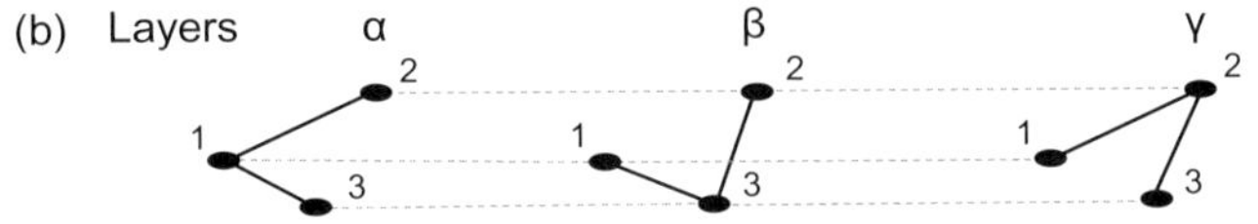

0	1	1	1	0	0	0	0	0
1	0	0	0	1	0	0	0	0
1	0	0	0	0	1	0	0	0
1	0	0	0	0	1	1	0	0
0	1	0	0	0	1	0	1	0
0	0	1	1	1	0	0	0	1
1	0	0	0	0	0	0	1	0
0	1	0	0	0	0	1	0	1
0	0	1	0	0	0	0	1	0

Figure 4.7 (a) Supra-adjacency matrix: no inter-layer edges. (b) Supra-adjacency: only "diagonal" interlayer edges (identity ties) between neighbour layers. (c) Supra-adjacency: with non-identity interlayer edges. (d) Supra-adjacency matrix: no inter-layer edges. For a multiplex network in which each layer represents a different interaction (edge shaded), only adjacencies are required.

"change the game." Appropriately, "Escaping Flatland" is the title of one chapter in the book by Tufte (1990) on effective methods for envisioning information; it is no coincidence that another chapter describes the importance of layering information as a way of decluttering and clarifying the material presented.

4.2.6 Resilience, Robustness, and Fragility

The comment about Flatland applies with equal strength to properties related to resilience, robustness, and fragility; getting out of Flatland really changes the game,

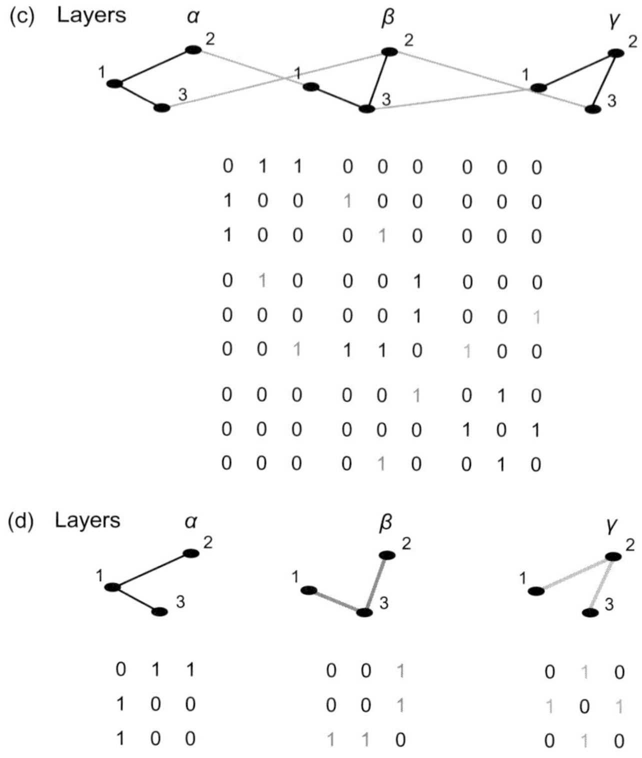

Figure 4.7 (cont.)

even if there are only two layers. (Think of the change of dynamics and tactics for tic-tac-toe in three dimensions rather than just two.) For example, Pilosof et al. (2017) studied two-layer networks of plants and their flower visitors, and plants and their leaf-miner parasitoids and found that the multilayer structure changes the *qualitative* conclusions about robustness to extinctions. The effects of topology on network resilience or robustness can be unexpected (Boccaletti et al. 2014). For example, a failure or attack in one layer can give rise to a cascade of failures in other layers. Yet, work by Kéfi et al. (2018) has suggested that in ecological multilayer networks, specific combinations of interactions can improve species persistence, productivity, and community resilience. The dynamics of these networks are usually discussed in terms related to percolation or the spread of disease (Kivelä et al. 2014). Depending on the model investigated, the addition of layers to a multiplex network of percolation can make it more fragile (such as additional species in a random food web model), but under some circumstances, the addition of layers increases robustness by creating redundant interdependencies among layers (Radicchi & Bianconi 2017). The overall effect of multilayer structures compared with monolayer networks is that the layering introduces new degrees of freedom which have important consequences for network dynamics, but these are not currently well understood (Kivelä et al. 2014). This topic is an area of very active research and new insights will result from the ongoing efforts to understand their dynamics, both those focused on ecological networks and more generally.

4.2.7 Comments on Multilayer Networks

As suggested by the examples given, many characteristics of multilayer networks can be initially approached using now-familiar graph theoretical properties for monolayer networks, but Kivelä et al. (2014) have pointed out that it is necessary to define emergent properties or concepts that are intrinsic to multilayer structures and to develop the right measures for them. Returning to spatiotemporal networks for an example, consider the spread of disease through a sessile population with the individuals as the nodes and time-labelled edges for date of the nodes' contemporaneous disease appearance; the orderliness of disease spread can be measured by the shortness of these edges and the rarity of cross-temporal edge crossings (Dale 2017). For any network in which the layers represent time slices, whether the data are spatiotemporal or aspatial and temporal, differences between layers can be interpreted in the context of change or ecological memory within the system, with emergent properties to match. Given the complexities and sophistication available with layered networks, as well as the great variety of contexts in which they occur, there are undoubtedly many more emergent properties to be described and discovered.

Clearly there are many (many!) potential applications for multilayer networks in ecology (e.g. Boccaletti et al. 2014 cite animal social networks as well as aquifer modelling; Finn et al. 2017 on animal sociality), and well beyond what we may currently imagine. Applying a multilayer approach can help to clarify the complexities of network data when an appropriate division of information into layers can be found. Pilosof et al. (2017) have pointed out also that the strength of multilayer network analysis is that it allows us to formulate and evaluate complex systems while explicitly incorporating processes that operate both within and between layers. Think about the complexities available by adding directions, signs, or weights to the edges, both intra- and interlayer. Currently in the literature, applications to social structure and to the influence of network layers on the spread of disease may seem to be the closest to true ecological research, but many that may seem more distant may prove to be closely analogous to studies we might perform (e.g. economic or transportation networks). There is much that can and will be done with multilayer networks for ecological applications. "Stay tuned!"

4.3 Beyond Multilayer

In addition to elaborations that result in multilayer networks, we should also consider

- Multitype nodes and multitype edges: the nodes in the network include different kinds of entities within the same network, for example, species, sites, variables, and so on. For example, the classic bipartite graph of species versus sites to examine nestedness uses two types of nodes. Multitype edges can arise even when nodes are of a single type, such as edges between species in the same network may be mutualism or parasitism or predation. If the nodes are multitype, there may be different categories of edges within each node type and between node types. For example, if the nodes are sites and species, the edges may indicate species presence at any site, pairs of sites that are near spatial neighbours, and species that are closely related or share a habitat.

- Multiple orders of interaction modification: the interaction between any two species, A and B, can be modified by other factors; for example, by the presence/absence of species C, by the interaction of species A and C, by the interaction of species C with its parasite P, and so on. This network elaboration includes both the familiar indirect interactions (Dale 2017) and the hierarchical modification or interaction of interactions: possibly of first order, second order, and so forth. (Golubski et al. 2016).
- Multinodal edges: multiway interactions portrayed by edges that join more than two nodes at a time; these structures are *hypergraphs*.
- Multistem structures: the complexities brought on by some of these elaborations may be best captured by graphs which allow branching and anastomosing of basic tree-like graphs; these are known as *vines*.

We will work through this list with more detail on each, but all these elaborations allow us to test hypotheses that we would not be able even to formulate without them.

4.4 Multitype Nodes

So far, our treatment of ecological networks has mainly dealt with networks that are somewhat homogeneous, with the nodes in any one network representing the same kinds of objects, and the edges representing the same kinds of relationships, even if they differ in quality (e.g. positive versus negative species associations) or quantity (e.g. intersite stream flow rates). In some of these networks, it makes sense to consider alternate routes for materials, energy, or information to move from one node to another, and that is where the concept of the "sum of all paths" (as in circuit theory) comes in: alternate routes may be used simultaneously, rather than only the most efficient (Figure 4.8; see Doyle & Snell 1984; McRae et al. 2008). Real ecological networks typically include more than one type of object and more than one type of interaction, even when we do not realise it. This is known to be typical of microbial communities, which may consist of taxa that act collectively with emergent properties and community metabolisms, (e.g. Paerl & Pinckney 1996; Bell 2009; Faust et al. 2012; Layeghifard et al. 2017), but it is also true of other ecological networks of mutually interacting taxa (see Ings et al. 2009; Kivelä et al. 2014; Kéfi et al. 2015; García-Callejas et al. 2018). For example, Pilosof et al. (2017) have discussed the mutual influence between food webs and host–parasite

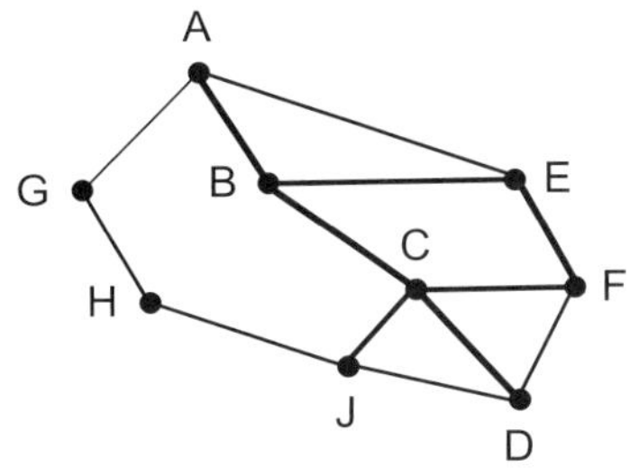

Figure 4.8 Sum of all paths in a physical context (like circuit theory). Thicker lines indicate less resistance or more capacity. The flow from A to D may have the least resistance along paths A, B, C, D, but the total flow includes all possible paths, for example, A, E, F, C, J, D.

relationships to illustrate the application of multilayer networks in ecological studies; these are sometimes treated as networks-of-networks, matching our ongoing discussion of graphs-of-graphs (Gao et al. 2012, 2014; among many). With increasing diversity of taxa and their interactions in a network, the concept of the sum-of-all-paths for the entire networks becomes much more complicated (Figure 4.9) but also more important. Pilosof et al. (2017) have argued that incorporating more interaction types will help identify keystone species in the multilayer networks that result. Similarly, to evaluate the robustness of particular function-based networks (e.g. trophic, pollination, parasitism, seed dispersal) within the "network-of-networks" at the whole farm scale, Pocock et al. (2012) examined 1,501 interactions among 560 taxa in the Norwood Farm study. They found that the subnetworks varied considerably in their robustness (pollination being especially fragile), and the covariance of robustness was not strong overall. Many of the subnetworks were bipartite, consisting of two layers, and some were three-layer networks, such as plant–aphid–aphid-parasitoid. Interestingly, their correlation network of the subnetworks' robustness has no substantial negative values! Lee et al. (2015) cited this work as a good example of multiplex networks on the ecological context. More research of this scale, detail and diversity is urgently needed to provide information on the restoration of damaged ecological networks.

The sum of *all* paths for species interaction networks must include indirect interactions and interactions modified by other actors or interactions (Figure 4.9): How does pathogen guild D affect the competition between A and C? How does mycorrhizal guild M affect the mutualism of A and P? There are also many interactions of interaction paths: competition with predation, mutualism with omnivory, and so on; but there are also interactions within guilds, for example, among parasites in the same host (Pedersen & Fenton 2006), but that happens in many communities, not just in microbial ones. These complications may be another source of perceived disconnection between network analysis and community ecology in addition to sampling limitations and bias and confounding or confusing niche-based and impact-based interpretations (cf. Blüthgen 2010).

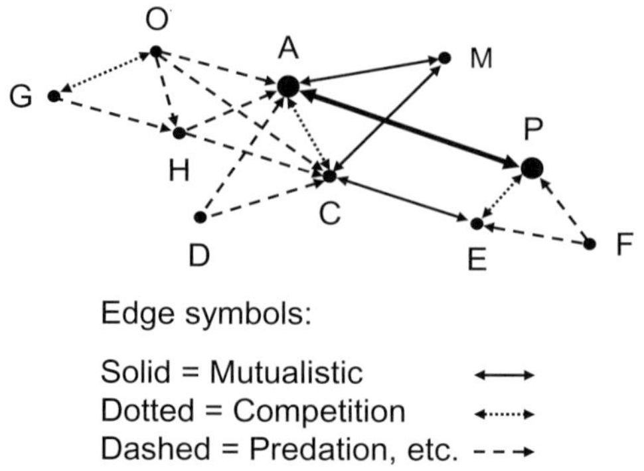

Figure 4.9 The sum of *all* paths for species interaction networks must include indirect interactions and interactions modified by other actors or interactions: How does D affect the competition between A and C? How does M affect the mutualism of A and P? Nodes A and P represent single species; the other nodes are sets of species. Plant A and pollinator P have competitor sets C and E. M are the mycorrhizal fungi of A and C; D are their pathogens. F are predators and parasitoids of P and E. H are the herbivores of A and C. G are predators of H. O are omnivores. Different edge types code different interactions. Identity edges (dashed lines) join instances of the same node. Competition within functional groups is not included, but it could be.

4.4.1 Phenology Networks

Phenology is the sequential appearance of seasonally determined characteristics of plants, such as leaf-out in spring or anthesis in summer, or of other organisms, such as amphibians' annual mating or egg hatching in insects. Large-scale patterns and changes in climatic conditions can be studied by examining the phenology of chosen species and their key events in a geographic network of sites through the years. Such research is currently of special importance in the context of climate change and its effects on the ecology of everything on the planet. The study of phenological networks of marker stages of development gives rise to several more variants of multitype node networks (or multilayer networks) for consideration:

1. Multievent nodes, in which each node is a site that has a series of events, possibly of several different types, associated with it in a temporal sequence, or equivalently a time sequence of distinct events at each site, and
2. Multitype node and edge networks, in which the nodes are the events, not the sites, and different node types give rise to two or more kinds of edges associated with different relationships among the nodes.

This network topological structure is most easily explained with reference to a detailed, if artificial, example. At a number of sites in a broad but defined geographic area, researchers record each year the dates at which distinct events occur in the sequences of plant phenology, such as the first flower opening (anthesis) of *Corylus avellana* or *Syringa vulgaris* or the first unfolding of leaves (leaf out) of *Aesculus hippocastanum*. For example, Schröder et al. (2014) described a study of 6,500 sites in

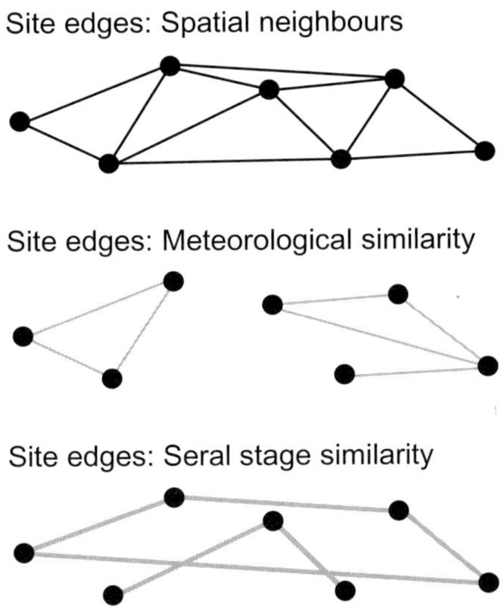

Figure 4.10 Choices for determination of edges between sites: spatial neighbours, meteorological similarity, or seral stage similarity.

Germany using 35 different phenological events. The phenological events can be treated as nodes in a spatiotemporal context with the sites as nodes of constant location persisting through time. This structure can also be treated as a multilayer network, with one layer per day, but most of the layers would be empty of event nodes, and so other network treatments make more sense.

There are many choices for edges in this phenological network. For example, the sites have a spatial relationship based on location and so a spatially defined network of neighbours makes sense (Figure 4.10), but we might also use correlation edges between sites based on similarity of elevation or meteorological records, or subnetworks of the sites based on categories such as vegetation type or seral stage. If there is only a single type of event that occurs multiple times at each site, characteristics such as "burstiness" (temporal clustering; Nicosia et al. 2013) or synchrony of equivalent events between sites may give intersite edges indicating greatest similarity for that character (Figure 4.11a). In Figure 4.11a, the vertical line representing time can be considered as a multinode edge that joins all events to their own site, but there can also be edges between events at different sites. For each event type, directed edges can be constructed between subsequent instances of the same marker event (ties omitted); thus, each event type imposes a partial ordering on the sites (Figure 4.11a). These also can be compared to evaluate similarities and differences in the (spatiotemporal) flow of the various events through the study region, rather like the flow of disease through a population or the percolation of fire through a landscape. Of course, the flow of phenological markers we observe is the result of the interaction between the flow of environmental conditions, such as day-length, maximum and minimum daily

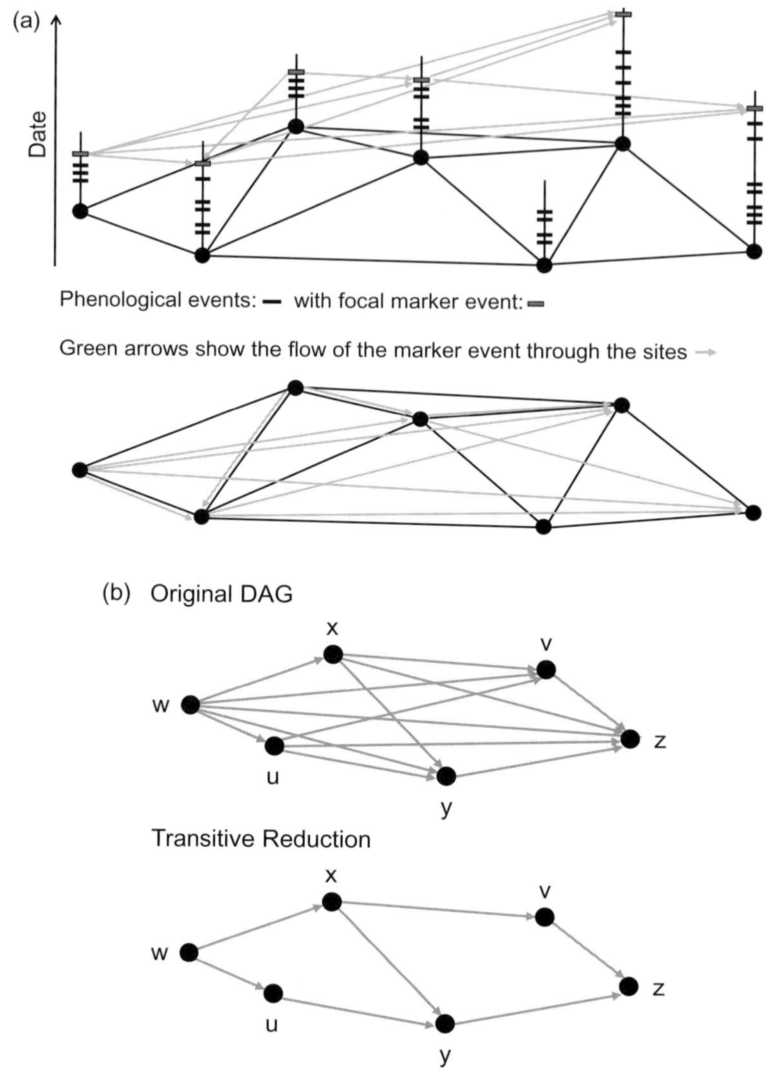

Figure 4.11 (a) Phenology at permanent sites; each node has a temporal sequence of events. Topological definition of spatial edges (Delaunay triangulation). One event type: look for burstiness and synchrony, perhaps as a function of absolute or relative locations. Phenology at regional sites: leaf production is indicated by the short horizontal bar. Position on vertical timeline indicates date. Partial order of sites by event date. (b) Directed acyclic graph (DAG) and its transitive reduction (also called Hasse diagram). Rule: include edge $x \to y$, when $x \to y$ and there is no z such that $x \to z \to y$. (c) For a bipartite or bilayer network with spatial and nonspatial components, the spatial relationships among sites can be compared with nonspatial relationships among the species, such as taxonomic relatedness. (d) Phenological studies often include site-specific time-series of climatic or meteorological conditions, for example, temperatures postdicted from pollen records at study sites. (A black and white version of panel a will appear in some formats. For the colour version, please refer to the plate section.)

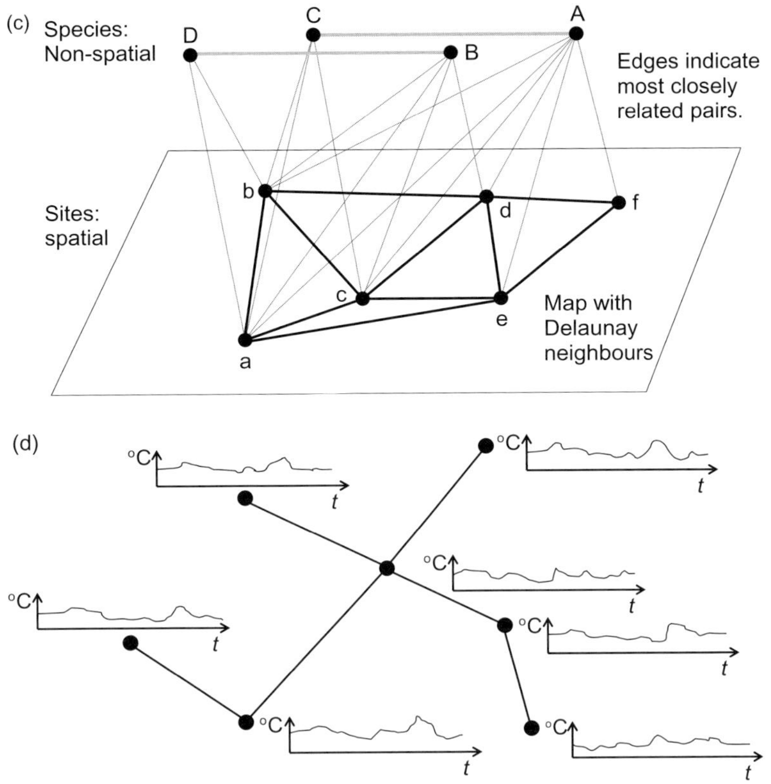

Figure 4.11 (cont.)

temperatures, and the physiological responses of the plants, which may vary with their ecotypes as adapted to particular locations (Chiune & Régnière 2017).

Given several marker events and their locations in time at each site (and thus their locations in space), an obvious approach to analysis is to consider the concordance of the (partial) orders of these events between all pairs of sites in the network. [See Conover (1980) for a discussion of measures of ranking concordance, with or without ties.] Then, consider the average concordance between pairs of sites as a function of physical distance using appropriate bins, or, using a spatial topological network like Delaunay triangulation (Dale & Fortin 2010, 2014), the average concordance between sites that are first-order neighbours, second-order, third-order, and so on.

A second visualisation of such data is to focus on one marker event at a time and use directed edges to show temporal precedence between pairs of sites. Looking at Figure 4.11b, it is obvious that including all edges for all cases of "event x at site a significantly precedes event x at site b" makes a very cluttered figure. One way to improve this presentation and its analysis is to take this *directed acyclic graph* (DAG) of all these edges and abstract its *transitive reduction*. This reduction is created using the edge rule that the edge $x \rightarrow y$ is included only if $x < y$ there is no node z such that $x < z < y$; an example is given in Figure 4.11b. These transitive reductions of DAGs are sometimes referred to as *Hasse diagrams* (Brüggemann & Carlsen 2006), which themselves have many applications in environmental studies (see Helm 2006; Sørensen et al. 2006). If there are no ties between the events' dates, the original data creates a tournament for each marker event with a directed edge in one direction or the other for each pair of sites. In that case, the transitive reduction process is like converting the results of a tournament to a win-loss score hierarchy with an ordering of the sites based on that score (see Dale 2017). With such a hierarchy or the Hasse diagram, a key characteristic to be investigated is the variability or consistency of the orders of events.

What is special about the data in phenological networks is that not only does each phenological marker event provide a partial ordering of the sites, but also over all the sites, the observations provide a partial ordering of those marker events. If the aim is to provide a general order for the sites, earliest to latest, and to compare these orders year-to-year, the marker events can provide this, just as a the one-time observations of several chemical components at each site can provide an order of the sites from most polluted to least polluted (Helm 2006). The result, therefore, is a two-way structure of partial orders: temporal partial orders of sites, one for each marker event, and a consensus partial order of the marker events based on their order at each site.

Phenological networks also provide the opportunity for investigating a range of ecological and evolutionary questions. What are the benefits and risks for greater or less synchrony of equivalent events (e.g. flowering or fruiting) in different sympatric species? How close or loose does that synchrony have to be for the benefit or risk to be effective? At what spatial distances do the forces for or against synchrony act? (See Johansson et al. 2015; Chuine & Régnière 2017.)

Phenological networks may be most useful, especially in relation to questions about the effects of changing climate, when there are networks of recorded environmental data in matching time series for comparison. A time series network consists of several located sites, each with an associated time series of environmental data such as air temperature or relative humidity (Figure 4.11d). These are analysed for spatiotemporal characteristics such as trends, cycles, and synchrony, which can then be compared with similar patterns in the phenological data. Often the time series network is sparser than the phenological network (e.g. Schröder et al. 2014), and the environmental data must be interpolated for the more numerous phenology sites, often by Kriging (Gerstmann et al. 2016). Equally informative, the phenological data may itself be the basis for interpolation to sites not monitored (Ruiz et al. 2012; Gerstmann et al. 2016), and the relationship between environment and phenology can be investigated using a technique such as co-Kriging, although it has assumptions that may not fully hold (Dale & Fortin 2014).

The basic form of these phenological networks is similar to the bipartite networks of sites and species where there are three or more kinds of edges: spatial neighbours or similarity measure edges among sites (intralayer); phylogeny or relatedness (or interactions like competition) among species (intralayer); and occupancy or affinity edges linking species with the sites where they occur (interlayer). This form is illustrated in Figure 4.11c. Another related spatiotemporal network is the spatial network of site-specific time series, which can be used in conjunction with phenological data to interpret or predict changes in conditions over longer periods of time. Another interpretation of multisite phenological observation data sets is that they form a network of sites, each with an associated time-series, not of continuously valued weights, but of a series of time-stamped 1's among a constant background of 0's, each labelled with the kind of phenological stage it represents.

Then, the timestamps themselves form a data series for analysis. At each site, the question may be whether there is evidence of consistent and significant change of dates for distinct events. For example, Schröder et al. (2014) found a significant shift of plant phenological phases of a week or more in Hesse and other parts of Germany, with an overall prolongation of vegetation of about three weeks. Most of these changes were correlated with air temperature, with for example, the flowering of apple trees (*Malus domestica*) starting earlier in the year as the mean air temperature (March–June) increased (Schröder et al. 2014). In analysing such data, we must account for the fact that they are swimming in a sea of autocorrelation (spatial, temporal, and spatiotemporal)! Schröder et al. (2014) corrected for the presence of autocorrelation in calculating the correlation between phenology and air temperature using Dutilleul's (1993) correction (see also Dale & Fortin 2009). It is also important, however, to take full advantage of the properties of the network itself in which the data reside.

4.5 Multiple Interactions and Multiorder Interactions

We have already mentioned the common phenomenon of three-way effects, described in correlation networks (Chapter 3, Section 3.2.2) as "3-feature interactions" (Weiss et al. 2016). For example, the observed relationship between species A and B can change markedly depending on the presence or absence of species C, due to the interactions of species A and C or B and C, or the indirect effects due to C on the shared environment or through other organisms. Three-way interactions are both common and significant in co-occurrence microbial networks with important implications for the functioning of soil microbiomes (Connor et al. 2017). These multiway effects are not limited to only three features or species or factors, but can obviously include four, five, or more. Dale (1999) has described an approach to studying multiway (k taxa) associations based on 2^k tables of presence and absence, allowing some insights into multitaxa effects and the interactions of pairwise interactions. The complexities of the interaction of interactions may suggest using hypergraphs, in which edges can connect more than two nodes, for ecological networks (e.g. Estrada & Rodriguez-Velazquez 2006; Golubski et al. 2016). Hypergraphs would allow several taxa to be brought together in a single multispecies interaction, although the details of the interactions might be obscured. Cascading tree-like graphs called *vines* are an alternative; a vine is a "braided" structure (like a braided stream) in that its branches both diverge and fuse (anastomose) where the original first-order edges act as nodes for more edges to attach in subsequent levels. We will describe both these structures and compare their suitability for multiway or multiorder interactions in ecological networks.

4.5.1 Multinode Edges

A *hypergraph*, like a graph, consists of a set of nodes, *V*, and a set of edges, *E*. The difference is that the edges can join the nodes into unordered sets that include more than two nodes, e.g. $V = \{1, 2, 3, 4, 5, 6, 7\}$; $E = \{(1, 2); (2, 3, 4, 5); (5, 6, 7)\}$, as in the first panel of Figure 4.12. Any hypergraph can be converted to a multilayer graph by assigning one layer to each hyperedge (Figure 4.12, lower left; see Boccaletti et al. 2014). Other approaches to displaying hypergraphs are available, as shown in Figure 4.12: colour-coded cliques in a standard pairwise graph (upper right panel), bipartite graph of an affiliation network (middle right, the two parts are edge-sets versus nodes; see Newman 2010), or presence-absence tally-sticks, one for each hyperedge (lower right, Dale & Fortin 2014). Notice that the incidence matrix of the hypergraph has more than two nodes recorded per edge, whereas the adjacency matrix, which must deal with the nodes in pairs even if the edges themselves contain more than two nodes, is the same as the adjacency matrix for the "clique" version (Figure 4.12). The bold zeroes on the diagonal of the adjacency matrix represent self-adjacency; with some logical justification, these could be replaced by 1's, thus making it different from the matrix for the "clique" version.

In the last version, the tally sticks, which represent species combinations like a 2^k table of presence and absence, could themselves become the nodes of a "correlation network" with edges between the combination-nodes based on the (dis)similarity of pairs of these species combinations (as in Dale & Fortin 2014). A related comment is that lumping together groups of species as if they were one (e.g. competitor set C or set E in Figure 4.9) has a similar effect as that of using a hypergraph for those several species and any other species with which they interact.

Hypergraphs do not generally capture the analytical hierarchy required to analyse correctly the hierarchical structure of interactions of interactions, where the edges between nodes that represent first-order interactions themselves interact, so that they essentially become like nodes with second-order edges representing the second-order interactions between them. All nodes included in one hyperedge are equivalent within that edge; a hierarchy would require a hierarchy of nested hyperedges. Indirect effects must be included because the same spectrum of structures will have different functional implications depending on the interactions coded by the edges and the interactions between them. Figure 4.13 illustrates the direct, indirect, and modified indirect interactions among some of the taxa in a coffee agroecosystem (Golubski et al. 2016): *Azteca* ants, scale insects, a beetle (*Azya*), other ants (e.g. *Pheidole*), spiders (Lynphiidae), and Phorid flies. Some of the direct interactions are modified by the presence or actions of other taxa (dark grey edges), and some modifications are

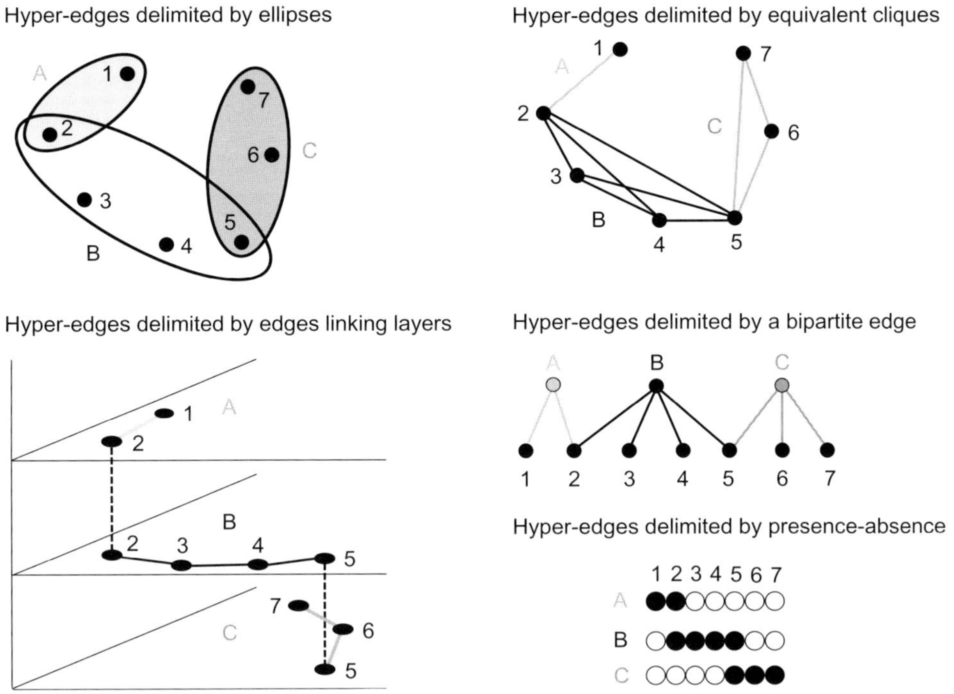

Hypergraph matrices

Incidence matrix:

	Nodes			
Edge 1	1	2		
Edge 2	2	3	4	5
Edge 3	5	6	7	

Adjacency matrix:

Nodes →

Nodes ↓

0	1	0	0	0	0	0
1	**0**	1	1	1	0	0
0	1	**0**	1	1	0	0
0	1	1	**0**	1	0	0
0	1	1	1	**0**	1	1
0	0	0	1	1	**0**	1
0	0	0	0	1	1	**0**

Figure 4.12 Hypergraph. A set of nodes with edges each of which can join more than two nodes at once. Hyperedges can delimited by ellipses, by a multilayer network with one layer per hyperedge, by equivalent cliques of pair edges (colour-coded) within each hyperedge, by a bipartite edge-membership graph (an affiliation network; see Newman 2010), or by presence-absence tally-sticks (Dale & Fortin 2014), one per hyperedge. The incidence and adjacency matrices for a hypergraph, as illustrated. The bold zeroes on the diagonal represent self-adjacency; with some logical justification these could be replaced by 1's, thus making it different from the matrix for the "clique" version.

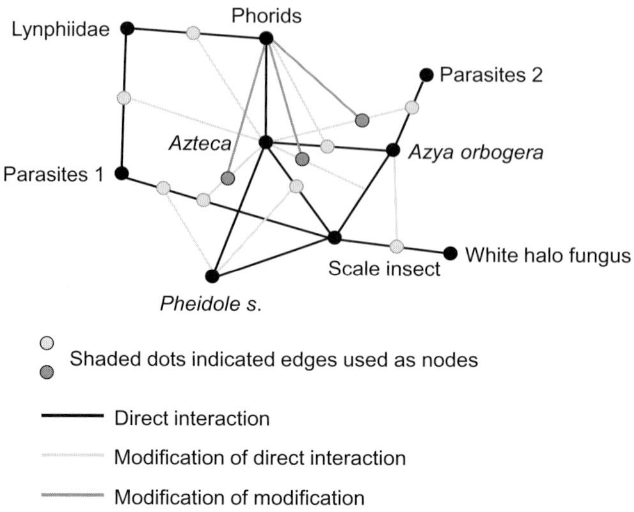

Figure 4.13 Indirect effects: two levels of modifications of ecological interactions (see Hsieh et al. 2012; Perfecto et al. 2014; Golubski et al. 2016). Part of a coffee agro-ecosystem, showing a hierarchy of interaction modifications: edges at the lower level of the hierarchy become nodes for edges at the next. *Azteca* and *Pheidole* are ants. [Redrawn from Hsieh, H.-Y., Liere. H., Soto, E. J., & Perfecto, J. (2012). Cascading trait-mediated interactions induced by ant pheromones. *Ecology & Evolution*, **2**, 2181–2191 © 2012 the authors; Perfecto, I., Vandermeer, J., & Philpott, S. M. (2014). Complex ecological interactions in the coffee agroecosystem. *Annual Review of Ecology, Evolution, and Systematics*, **45**, 137–158; Golubski, A. J., Westlund, E. E., Vandermeer, J., & Pascual, M. (2016). Ecological networks over the edge: hypergraph trait-mediated indirect interaction (TMII) structure. *Trends in Ecology and Evolution*, **31**, 344–354. © 2016 Elsevier Ltd. All rights reserved.] (A black and white version of this figure will appear in some formats. For the colour version, please refer to the plate section.)

themselves modified by other species (light grey edges) or possibly other interactions. For example, while the *Azya* beetle interacts with the scale insect directly by predation, *Azteca* modifies the interaction between the beetle and the scale insect by protecting the herbivore, and that effect is also modified by the presence of Phorids which parasitise the ants, and whose mere presence reduces that ants' protective activity (Figure 4.13; see Perfecto et al. 2014).

4.5.2 Multistem Structures

Rather than hypergraphs, the best alternative may be particular braided tree-like graphs, with structures that branch and anastomose, known as "vines" (Joe 1997; Bedford & Cooke 2004). A *regular vine* for n variables is a set of nested trees in which the edges of tree j are used as the nodes of tree $j + 1$ and two edges in tree j are joined by an edge in tree $j + 1$ only if they share a node in common (Figure 4.14). The edges at each iteration are weighted by the partial correlation appropriate to their end

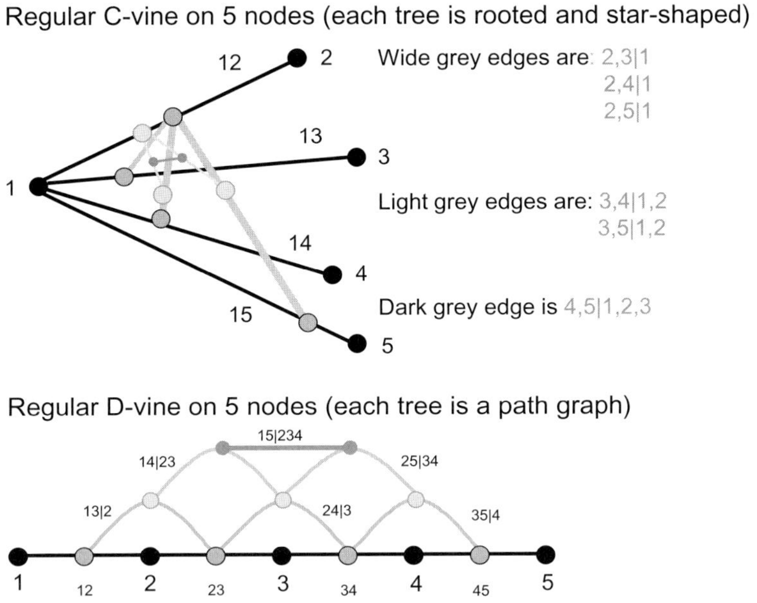

Figure 4.14 Vines are hierarchically structured graphs in which the edges at a lower level are used as nodes for further edges at the next level. A regular C-vine on five nodes: each tree is rooted and star-shaped. A regular D-vine on five nodes: each tree is a path graph. (A black and white version of this figure will appear in some formats. For the colour version, please refer to the plate section.)

"nodes." In a regular *C-vine* ("C" for canonical), each tree is a rooted tree and star-shaped (Figure 4.14, upper panel). A regular *D-vine* consists of trees that are path graphs as in Figure 4.14 (lower panel).

This graphical model is a statistical model of the structure of dependence among the *n* variables once a copula function has been attached to each edge (Kurowicka & Cooke 2002). We can modify this very useful concept to be even more flexible and even more in line with the structures encountered in studies of how ecological interactions interact with and modify other ecological interactions. An *irregular vine* (a new term) allows the mixing of the hierarchical levels of interacting interactions, such as a species (taxon) modifying the interaction between two others (Figure 4.15) as well as the modification of interactions by interactions (see also Golubski et al. 2016). In the particular case of ants and scale insects and coccinellid beetles and phorid flies, the ant pheromones may be the key factor in these trait-mediated cascades of interactions (Hsieh et al. 2012). Figure 4.16 gives an example from the rocky intertidal with the crab *Carcinus* as the top predator, preying on the carnivore *Nucellus* and the herbivore *Littorina*, which consume the "resources" of barnacles (*Semibalanus*) and fucoids (*Ascophyllum*); see Trussell et al. (2017) for details. The interactions have only two levels: direct and indirect, as illustrated, rather than the three in the coffee agrosystem example (Figure 4.13). In a related study of an intertidal

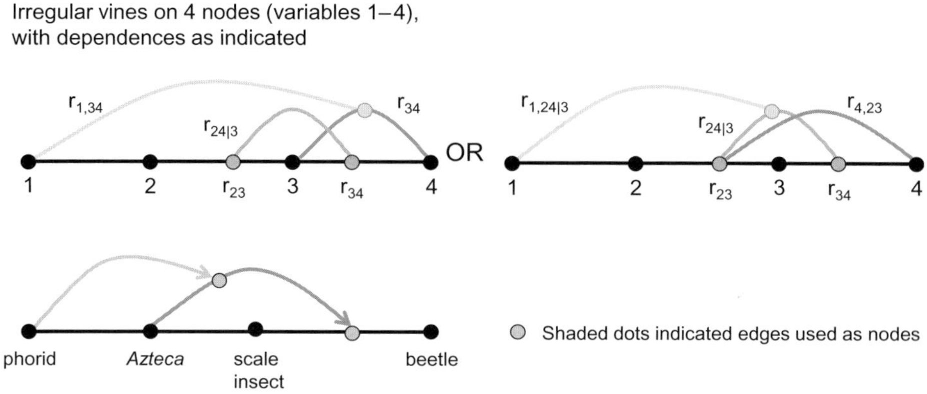

Figure 4.15 An *irregular vine* [new term] allows the mixing of hierarchical levels joined by pairwise edges, for example, a level-1 node joined to level-1 edge by level-2 edge, thus allowing either a taxon or a taxon–taxon interaction to modify other interactions. Irregular vines on 4 nodes (variables 1 to 4), with dependences as indicated. Irregular vine version of Golubski et al. (2016). *Azteca* affects the interaction of scale and beetle; phorids modify that indirect interaction. [Reprinted with permission from Golubski, A. J., Westlund, E. E., Vandermeer, J., & Pascual, M. (2016). Ecological networks over the edge: hypergraph trait-mediated indirect interaction (TMII) structure. *Trends in Ecology and Evolution*, **31**, 344–354. © 2016 Elsevier Ltd. All rights reserved.] (A black and white version of this figure will appear in some formats. For the colour version, please refer to the plate section.)

system that included the shore birds, Wootton (1994) combined path analysis and experiments to predict the direct and indirect effects (the use of path analysis to study causal networks is summarised in Section 3.5, and further details and guidance can be found in Shipley (2016).

The hierarchical depiction of indirect interactions seems to be more compatible with the concepts described by Golubski and Abrams (2011) and Delmas et al. (2019) or Golubski et al. (2016) as in Figure 4.16. Similarly, two pathogens may compete for the resources of the same host or they may weaken the host facilitating a combined attack; an opportunistic pathogen may be required for an endemic pathogen to become established in a host, or an opportunistic pathogen may be able to invade only after a preceding pathogen infection. These irregular vines as graphical models or as structures for further network analysis seem to be more appropriate than hypergraphs, the multinode edges of which may confound the effects that we want to distinguish and characterise.

The flexibility of this approach can be tested further by examining the distinction between density-mediated indirect interactions (DMII) and trait-mediated indirect interactions (TMII) and how they can be quantified (Abrams 1995; Okuyama & Bolker 2007). Following Okuyama & Bolker (2007), consider a three-level system consisting of a resource (e.g. green plants), a forager (e.g. grazing herbivore), and a predator (e.g. quadruped carnivore), as shown in Figure 4.17. The presence of predators can affect either the number of herbivores (say h when predators are present and H when they are absent, $h < H$) or the per-herbivore foraging resource uptake or

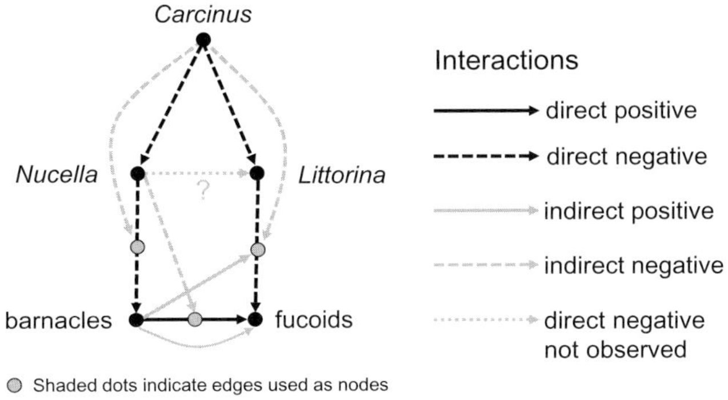

Figure 4.16 Irregular vine for indirect interactions: rocky intertidal food web, restructured from network in Trussell et al. (2017). Positive and negative interactions, direct and indirect as indicated. *Carcinus* is a crab; *Nucella* (whelk) and *Littorina* (snail) are molluscs. [Redrawn with permission from Trussell, G. C., Matassa, C. M. & Ewanchuk, P. J. (2017). Moving beyond linear food chains: trait-mediated indirect interactions in a rocky intertidal food web. *Proceedings of the Royal Society B*, **284**, 20162590. © 2017 the authors. Published by the Royal Society. All rights reserved]

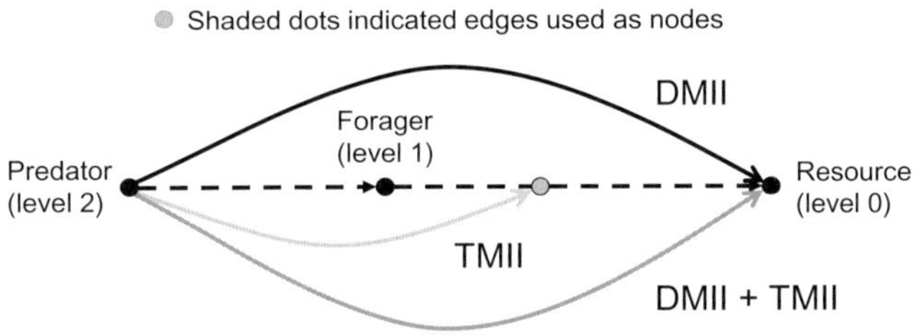

Figure 4.17 Irregular vine for indirect interactions. Irregular vine on three nodes (predators, foragers, and resources), with dependences as indicated. Density-mediated indirect interaction (DMII) is the effect of predation on the resource abundance and removal through the effect of predators on forager density. Trait-mediated indirect interaction (TMII) is the effect of predation on the resource abundance and removal through the effect of predators on forager traits or behaviour.

efficiency, say f when predators are present and F when they are absent. Because antipredator behaviour reduces foraging effort or efficiency, $f < F$. If the predators are both present and actively taking prey (the "true" predation treatment), the resource uptake is fh; if there are no predators, it is FH. If the predators are present and affecting forager behaviour but are prevented from removing prey (the "threat" predator treatment), resource uptake becomes fH. Last, if predators are absent but the herbivores are experimentally removed at the same rate as true predation (the "culling" treatment), uptake is Fh. The predator threat treatment creates TMII only (no numerical effect), culling creates only DMII, and the true predation treatment creates both. Therefore, these indirect effects can be measured by the differences [e.g. "cull" – "true" = $h(F - f)$]

or the ratios (e.g. "cull" $\div$ "true" $= F \div f$) of the amounts observed in the four treatments, with some advantages in the use of ratios or log ratios (Okuyama & Bolker 2007). For example, $F \div f$ estimates the effect of TMII; but there are two versions of this ratio: (no predators) $\div$ (predator threat) estimates TMII in the absence of DMII, but (cull) $\div$ (full predation) estimates TMII in the presence of the DMII numerical effects (Figure 4.17).

The helpful distinction between indirect interactions that affect population density and those that affect the rates of direct interactions leads to a reconsideration of the generally accepted scheme for indirect interactions originally developed by Menge (1995), as shown in Figure 4.18. These indirect interactions are clarified when we can distinguish between effects on populations, as in keystone predation and trophic cascade, and effects on direct interactions. Indirect inhibition, apparent predation, indirect facilitation, a form of indirect parasitism are all easier to understand when the indirect effects through changing the direct interactions can be made explicit as in Figure 4.18c–f. The distinction may not always clarify or predict the outcome, as for example when a new exotic prey species is introduced into a system with native prey and an originally exotic predator, leading to hyperpredation or hypopredation on the native prey (increased or decreased predation rates as in the "rabbit, bird, cat" example, Figure 4.19; see Bate & Hilker 2012). Depending on the circumstances, hyper- or hypo-predation can look like apparent competition, apparent mutualism, or apparent predation in either direction (Bate & Hilker 2012).

Extending this simple scheme (TMII and DMII) to include several or many interacting participants will lead to irregular vines that are complicated but that, nevertheless, will help clarify the network and its indirect interactions. For example, networks of direct and indirect interactions in the microbial world of viruses, bacteria, and protozoans are both complex and critical to the functioning of aquatic microbial systems (Miki & Jacquet 2010). Simmons et al. (2018) have investigated the use of motifs to find indirect interactions in bipartite ecological networks. Given the importance and frequency of indirect effects, they argued that reliance on single simplified metrics, which lose large amounts of information, must miss substantial insights, and they present a method based on *bipartite motifs*. Their approach is very similar to the use of graphlets orbit frequency analysis because they count the occurrences of nodes in identifiable locations on all possible bipartite motifs (Figure 4.20 shows these motifs for $s = 2$, 3, and 4). The distinctions among nodes are based on the motif in which they occur and their position within the motif. The distinct positions are the automorphism orbits (Kashtan et al. 2004; Pržulj 2007) and so these motifs are essentially *oriented bipartite graphlets*. These graphlets are "oriented" because they have direction; for example, one prey and two predators is different from one predator and two prey (motif 2 versus motif 3 in Figure 4.20). It will be interesting to pursue this approach to include more layers and thus more kinds of interactions than just those considered so far. Working through this fascinating material emphasises the need for indirect interactions to be better and explicitly quantified, as well as better investigated across more tropic levels and in less studied systems (Sotomayer & Lortie 2015). Both specific surprises and more general insights undoubtedly await.

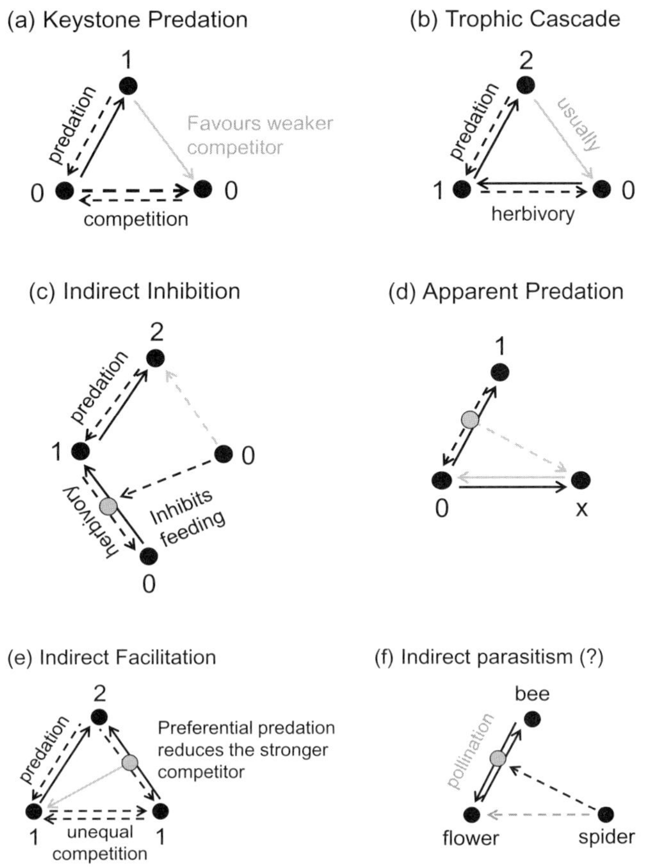

Figure 4.18 Indirect effects (see Menge 1995). Black nodes are species. Arrows toward affected species. Solid edge: positive effect. Dashed edge: negative effect. Grey edge: indirect effect. 0, 1, and 2 are nominal trophic levels. Grey node: a direct effect edge that is a 2° node. x is undefined trophic level. (a) Keystone predation (direct effect). (b) Trophic cascade (direct effect). (c) Indirect inhibition. (d) Apparent predation. (e) Indirect facilitation. (f) Indirect parasitism (spiders on flowers discourage bees, reducing seed set and plant fitness). [Redrawn with permission from Menge, B. A. (1995). Indirect effects in a marine rocky intertidal interaction webs: Patterns and importance. *Ecological Monographs*, **65**, 21–74. © 1995 John Wiley & Sons]

Including this magnitude of diversity in network analysis challenges the concepts in previous discussions of disentangling the relationships between structure and functional dynamics of ecological networks. We introduced, without details, the concept of graphlets for graphs with edges that are both directional and signed (Figure 3.19); perhaps that is what we need to extend the applications of graphlets beyond the basics we described in detail.

While the inclusion of more kinds of interactions in a network increase network complexity (always? usually?), multilayer inclusion is not necessarily a cause of

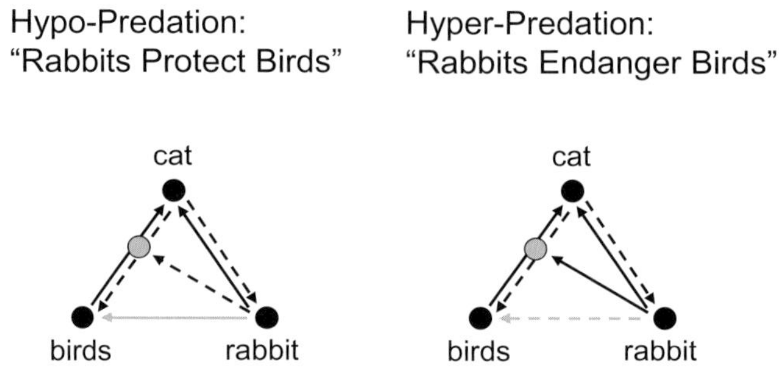

Figure 4.19 Hypo- and hyper-predation. Hypo-predation: "rabbits protect birds." As the preferred prey, rabbits benefit birds because their presence decreases the rate of cat predation on birds. Hyper-predation: "rabbits endanger birds." As the less preferred prey, rabbits endanger birds because their presence increases cat density and the rate of cat predation on birds. See Bate and Hilker (2012). [Redrawn with permission from Bate, A. M. & Hilker, F. M. (2012). Rabbits protecting birds: hypopredation and limitations of hyperpredation. *Journal of Theoretical Biology*, **297**, 103–115. Copyright 2012 Elsevier Ltd. All rights reserved]

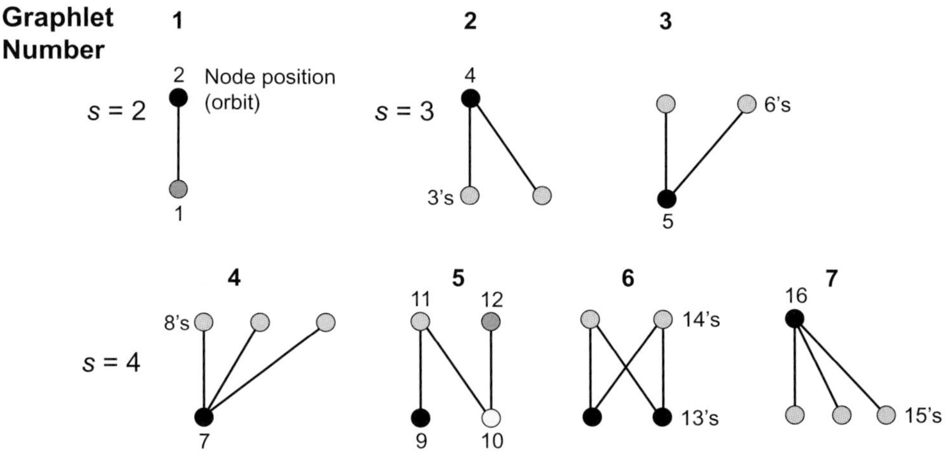

Figure 4.20 Bipartite motifs: oriented bipartite graphlets ($s = 2$–4). Within a single graphlet, nodes of the same shade are in the same orbit. [Following Simmons, B. I., Cirtwill, A. R., Baker, N. J., Dicks, L. V., Stouffer, D. B., & Sutherland, W.I. (2018). Uncovering indirect interactions in bipartite ecological networks. bioRxiv, DOI: 10.1101/315010]. The distinctions among nodes are based on the motif in which they occur (numbers in bold) and their position within the motif. The distinct positions are automorphism orbits (in regular font).

technical complexity, as discussed in Section 2.3. Greater diversity and broader inclusion can also be generated by using more than one site or more than one realisation of an ecological network, giving rise to a metanetwork related to metacommunities and their metawebs (Poisot et al. 2012).

4.6 Multilevel Networks

The last elaboration (well, almost the last) to consider in the list is the *multilevel network.*

A multilevel network consists of several levels, each with its own network graph, but the levels form a hierarchy or a nested structure and the nodes at different levels represent different kinds of objects. Another difference is that while multilayer networks may have many layers (think of time slices of the spatial locations of individuals in a herd of wildebeest or of their social interactions), the number of levels in a multilevel network tend to be small. The way we are using the term "multilevel" follows Kivelä et al. (2014), but not Boccaletti et al. (2014); it precludes a multilevel network from being multiplex, because the nodes in higher levels are aggregations, not individual instances of nodes in lower levels (e.g. populations as nodes are groups of individual organisms). Therefore, the nodes cannot be the same from one level to the next in the way they remain constant through the slices of a multiplex network. This use of "multilevel" follows the practice in statistical modelling and analysis, where it refers to a nested or hierarchical data structure (Snijders & Bosker 2012; Wang et al. 2013).

As a simple illustrative example, consider a multilevel correlation network based on the compositional similarity (or measured comparison of species diversity) among the epiphyte communities growing on tree branches. The levels to consider are trees within stands, stands within sites, and sites within regions (Figure 4.21a), just as in a hierarchical analysis of species diversity (Dale & Fortin 2014). (Remember the famous figure of *Clematis fremontii* with individuals in aggregations, in glades, in regions and then its complete geographic range; Erickson 1943 and often reprinted, e.g. Silvertown & Lovett-Doust 1993). This multilevel network is also *multiscale*, examining several spatial scales that match the organisational hierarchy of trees, stands, sites, and so on.

An obvious approach is to examine the nestedness of the "vertical" structure from one level to another, but another approach is first to examine the network in each level separately and then to compare or combine them with a "linked design" (Lazega et al. 2008). The linked design approach not only records the flow of resources between nodes within a level, but also indicates the matching of nodes in different levels and the flow of resources between levels. For example, one could compare the centrality of individual cancer researchers in a network of individuals with the centrality of their institution in a network of research institutions (Lazega et al. 2008) and then determine whether the flow of resources (information, money, research personnel recruitment) can explain observed similarities or differences. Figure 4.21b offers an elaboration of this concept (Lazega et al. 2008) with three levels: research institutions, research lab clusters within institutions, and individual researchers. In this artificial example, the network of collaborating lab clusters is quite different from those of institutions or research leaders, being obviously more local than the other two. The networks of institutions and research leaders may be based on reputation, research impact, and funding, whereas lab clusters' relationships

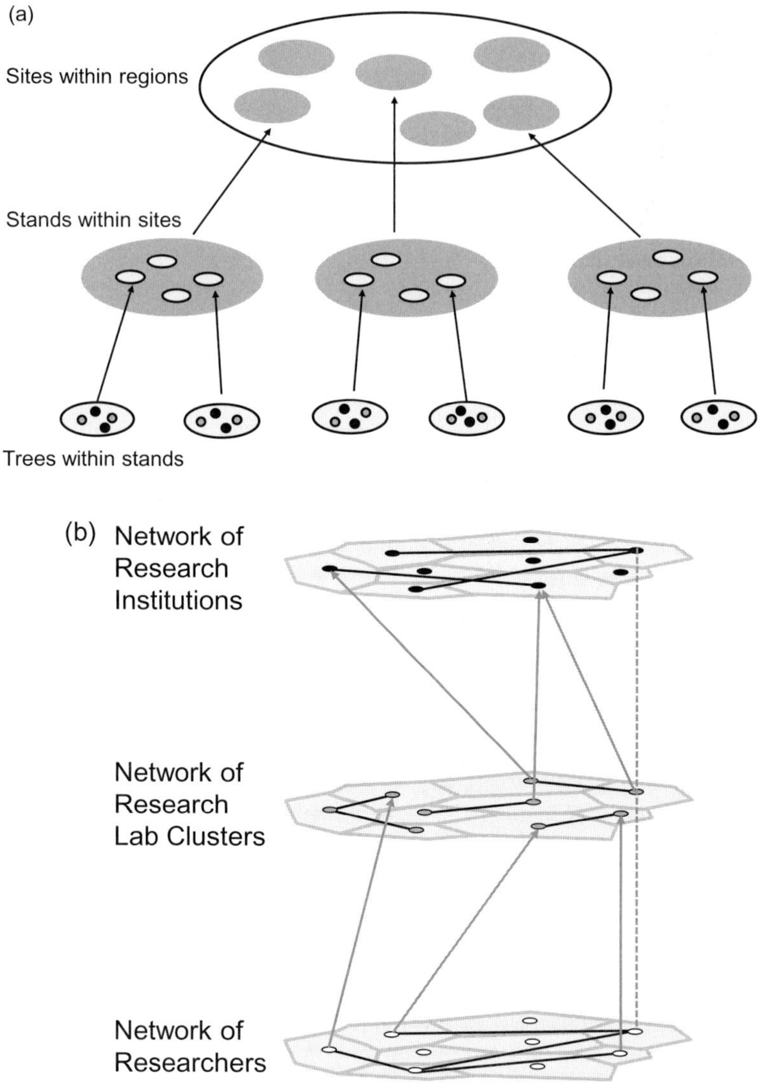

Figure 4.21 (a) Multilevel network: sites within regions, stands within sites, trees within stands. Edges omitted to avoid clutter. (b) Multilevel research network in its spatial context of provincial centres (each polygon is a province with one centre indicated). Dashed edge indicates matching nodes in different layers, not node identity. Edges within layers represent exchanges between peers; directed edges between layers indicate resource flow. The levels are networks of research institutes, of research clusters, and of researchers.

may be more practical, based on shared resources or shared facilities. On this example, the nodes in all three levels match, although the nodes do not actually represent the same entities but coincide by spatial location. A dashed edge joining the three nodes for each location could be used to indicate that matching, but not node identity as would be found in a multiplex network.

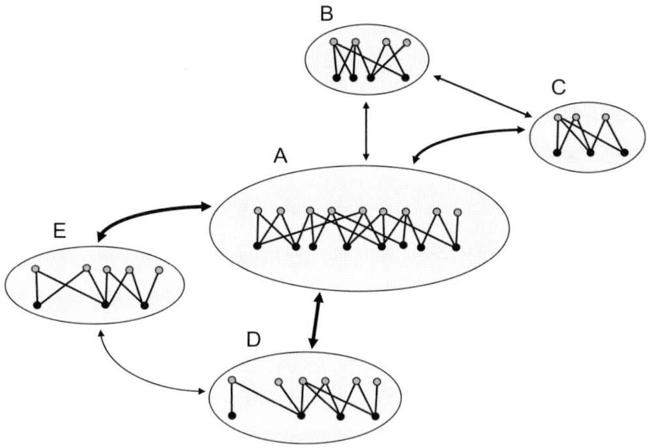

Figure 4.22 Locations as layers in a multilayer network. Graph-of-bipartite-graphs: pollination mutualisms in patches. Patches as nodes in a spatial network. The oval's size represents patch area; edge thickness indicates movement rates. The bipartite graphs show plants' and pollinators' mutualistic interactions.

A special case of multilevel networks is represented by a "graph-of-graphs" in which the nodes of a graph at one level each have an associated graph at another next level (Dale & Fortin 2014). For example, a spatial graph of study locations may have a pollination network graph for each site (Figure 4.22; from Dale 2017). The parallel concept is, of course, a network-of-networks; Pocock et al. (2012) have presented a correlation network based on the similarity of robustness values for subnetwork components in the Norwood Farm study such as flower visitors and seed feeding insects. The best approaches to the analysis of networks-of-networks (NoN) have been the object of much research since 2010, particularly in application to physics, and undoubtedly these investigations will continue apace (Gao et al. 2012, 2014). Some, but not all, of the most important applications involve networks embedded in two-dimensional space (Gao et al. 2014), as in our pollination example (Figure 4.22).

While many social applications of multilevel networks involve only two levels, ecological examples can be expected to have more, opening up the possible ways for analysis and modelling. This is another area of study with great potential for development and valuable application.

4.7 Multiscale Networks

The term "multiscale" means that the network somehow exists or at least can be studied at more than one spatial or temporal scale. Some multilayer and multilevel networks can also be seen as multiscale, where scale is interpreted as applying only in explicit spatial or temporal dimensions, having separate subnetworks available for each of the scales of interest. We noted that the "tree → stand → site → region" hierarchy for correlation networks of epiphyte communities was multiscale spatially as well as multilevel in its hierarchical structure. Spatial scale can also be based on a species' dispersal distance or spatial range, for example in comparing landscape connectivity for different organisms (Minor & Lookingbill 2010). One multiscale alternative is to create a multilayer network with each layer representing a different spatial scale of an ecological process, for example correlations of phenological marker events at distances of: <100 m, 100 m–1000 m, 1000 m–5000 m, and so on. Similarly, the multiple scales determined for the network may be temporal, for example, creating networks for interactions at the scales of days, weeks, months, and years. Given the many terms already in use, we suggest that "multiscale" be limited in its application only to explicitly spatial or temporal phenomena.

Given the long list of similar-sounding "multi-" terms, at this point, we will provide a brief summary of those we have described in Table 4.1.

4.8 Generalising Motifs (Multinode Motifs) and Graphlets

The last "multi-something" topic is not about an elaboration of the basic network structure, but an expansion of some of the subgraphs we can use to analyse networks, whether basic or themselves somehow elaborated. Yletyinen et al. (2015) expanded the familiar 3- and 4-node motifs for food webs to 5- and 7-node structures (Figure 4.23). These represent extended interactions such as apparent competition and omnivory (5 nodes), or exaggerations of the standard trophic roles of consumer and prey, to keystone species, generalist predator and highly predated prey (7 nodes). Here, then, familiar motifs with few nodes are generalised in structure by adding nodes.

Given any specific motif, in which particular nodes play particular roles, the idea is to generalise the motif adding nodes while acknowledging the roles of the additions. For example, in the feed-forward triad, we can name the roles as source, sink and intermediary as in Figure 4.24a (Kashtan et al. 2004 labelled them X, Y, and Z). The motif can be generalised by doubling any one of the roles to produce a 4-node motif (lower row of Figure 4.24a), or doubling all three. Each role can also be increased by using multiples of more than two (Figure 4.24a, lower row; Figure 4.24b). Further expansions are possible by having, for example, both multiple source nodes and multiple sink nodes (not shown). Kashtan et al. (2004) used the edge complement to distinguish between strong generalisations (all possible edges for the multimotif) and weak generalisations (only some of the possible edges), as illustrated in Figure 4.24b.

The reason for developing these "multimotifs" is to distinguish among structural, and by consequence, functional characteristics of different networks. For example, the transcription networks of bacteria and yeast and the neuronal networks of *C. elegans* both have the feedforward triad motif as a common constituent. The transcription networks have only the multiple sink generalisation and the neuronal network has only the multiple source generalisation, and the multiple-intermediary generalisation is rare in both (Kashtan et al. 2004). Similar analysis may be useful in studying ecological

Table 4.1 Summary of "multi-something" elaborations of networks

Multilayer network

Network of two or more layers, each having a network graph with intralayer edges; there may also be edges between the layers (Figure 4.2a).

Examples: social networks with layers defined by social context of interaction; pollination networks with layers defined by sites (space) or year (time); interaction networks with layers defined by interaction type (Figure 4.1b).

Multiplex network

Multilayer network with nodes the same in all layers, with different interactions in separate layers. Interlayer edges are "edges of identity" (coupling edges) joining instances of the same node (Figure 4.2b).

Multigraph format

A multigraph can have more than one edge between any two nodes; the multigraph derived from a multilayer network has all edges from all layers, with layers possibly distinguished by different labels or colours (Figure 4.3b). This is a projection graph of the original multilayer network.

Multitype nodes

The nodes belong to two or more categories, for example sites and species (Figure 4.11c). A variant is that each node is a location with an associated sequence of events (*multievent nodes*; Figure 4.11a).

Multitype edges

The edges belong to two or more categories, for example pollination versus frugivory or competition versus amensalism (Figure 4.9).

Multiorder interactions

Interactions between pairs of species are modified by the presence or absence of a third species. For more species, this can create a 2^k interaction as recorded in 2^k contingency tables. The multiorder phenomenon is closely related to indirect interactions between species (Figure 4.18).

Multistem structures (vines)

Where one species modifies the interaction between two others, or one interspecific interaction modifies another, these can be described by multistem structures called *vines*. Vines are similar to line graphs because edges in the original graph are used as nodes for a subsequent level of interactions (Figure 4.14).

Multilevel network

Several levels, each with its own network graph, but the levels form a hierarchy or a nested structure (Figure 4.21). The nodes at different levels (layers) represent different kinds of objects. A vine is a multilevel network because the nodes at each level are different kinds of objects: the edges on the first level become the nodes on the second level, and the edges on the second level become the nodes for the third level, and so on.

Multiscale

Almost any of the multi-something networks can also be multiscale when there are different subnetworks associated with their own spatial or temporal scales.

networks of different kinds. The important methodological contribution in the Kashtan et al. (2004) paper, however, is the concept of distinguishing the positions and thus the roles of nodes in subgraphs (here motifs) based on their automorphisms. This can be seen as an important conceptual step in the development of graphlet approaches to relating network structure and function through the orbits of structurally defined subgraphs.

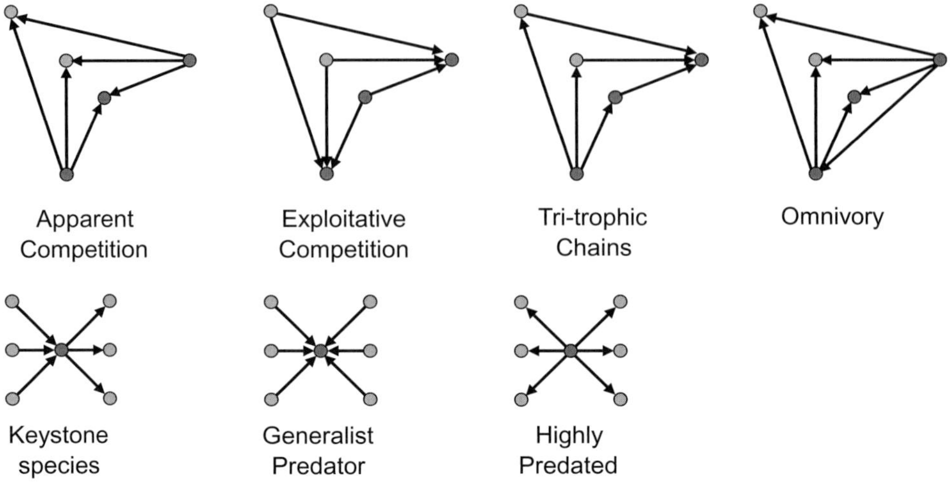

Figure 4.23 Expanded motifs for ERGM (Exponential Random Graph Model) configurations. [Based on Yletyinen, J., Bodin, Ö., Weigel, B., Nordström, M. C., Bonsdorff, E., & Blenker, T. (2015). Regime shifts in marine communities: a complex systems perspective on food web dynamics. *Proceedings of the Royal Society B*, **283**, 20152569]

In an unusual application of motifs, Battiston et al. (2016) were able to compare the structural layer of a bilayer network of the brain (the anatomical brain network) with its functional layer (the functional brain network) using multilayer motif analysis. Part of the challenge in this analysis was the fact that pairs of sites in the structural layer had either an edge or no edge (1 versus 0), whereas in the functional layer, the edges were positive, negative (in grey) or absent (+1 versus −1 versus 0). Under these conditions, there are 35 bilayer motifs of three nodes (Battiston et al. 2016), and of these, the six shown in Figure 4.25 were statistically more frequent than expected (based on their *Z*-score). Battiston et al. (2016) noted that in all six, the triangle in the signed layer was balanced with an even number of negative edges (remember that balance may not be an important property in all signed networks). The conclusion is that the functional coordination of brain functions is clearly constrained by the underlying structural network, but that structural connections may not be sufficient to determine functional correlation. As with most of the non-ecological examples we cite, this application of bilayer motifs can obviously provide inspiration for their application in ecological systems that parallel this situation.

Having considered the generalisation of motifs by doubling or multiplying nodes within them or by expanding them to versions in more than one layer, an obvious next step is to consider how the same modifications might work for methods based on graphlets. The answer is that this kind of generalisation of graphlets may not be so helpful. Generalising any motif creates a family of motifs each of which consists of two-to-many copies of the original motif that are attached to each other in different ways. The similar process of generalisation for a graphlet creates new

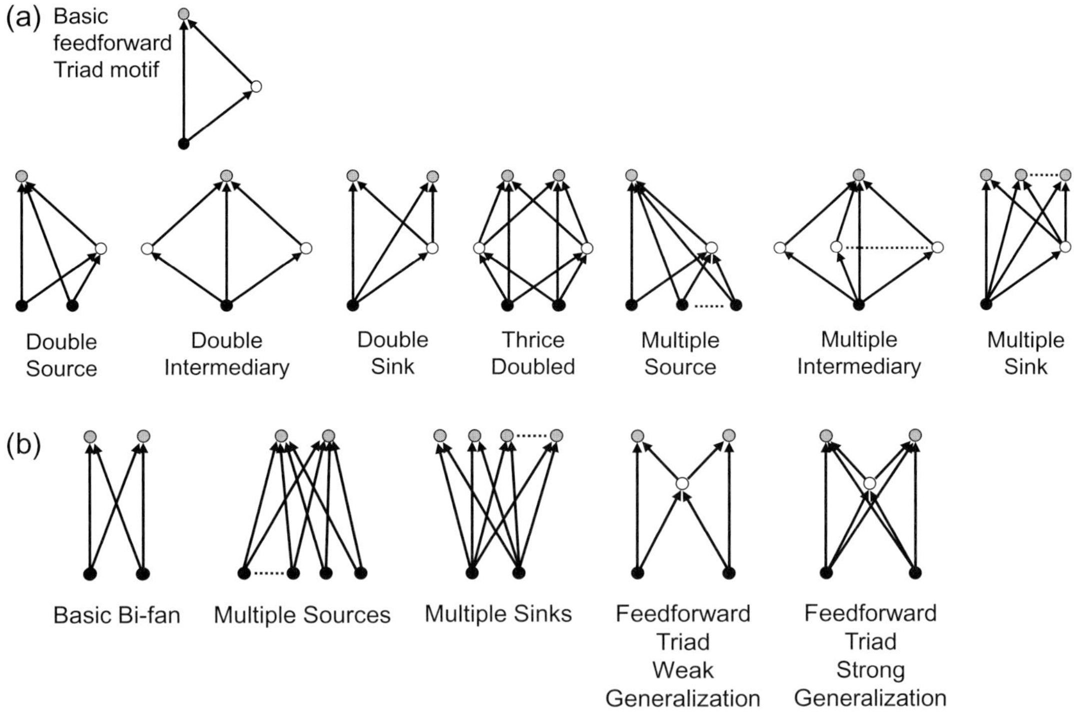

Figure 4.24 (a) Generalised motifs: based on feedforward. (b) Generalised motifs: based on bi-fan and strong vs. weak feedforward triad with double source and double sink. [Redrawn with permission from Kashtan, N., Itzkovitz, S., Milo, R., & Alon, U. (2004). Topological generalizations of network motifs. *Physical Review E*, **70**, 031909. © 2004 American Physical Society]

and larger graphlets with nodes in new orbits. Remember that the focus of graphlet analysis is counts of nodes in the defined orbits, not just counts of the subgraphs themselves. As an example, consider the feedforward motif; it is also is digraphlet 5, with nodes in orbits δ10, δ11, and δ12 (Sarajlić et al. 2016). The generalisation of this graphlet by doubling the intermediary node gives graphlet d27, with orbits δ85, δ86, and δ87; the "double source" generalisation produces d29, orbits δ91, δ92, δ93; and the "double sink" gives d30 with orbits δ94, δ95, and δ96 (Figure 4.26; cf. Sarajlić et al. 2016). Larger generalisations with node multiples will produce larger graphlets that may not have been catalogued. However, it is not clear that larger graphlets tell us more; in fact, Sarajlić et al. (2016) suggested that the smaller digraphlets are more effective for analysing these directed networks, in addition avoiding the challenges associated with counting graphlets and orbits with more nodes in the subgraphs, for large and complex networks. Similarly, Yaveroğlu et al. (2014) decreased the size of the largest undirected graphlets they used in their analysis from $n = 5$ to $n = 4$ with good results. Unless there is strong evidence to the contrary, we suggest that generalising graphlets by augmenting the set of nodes will not repay the effort.

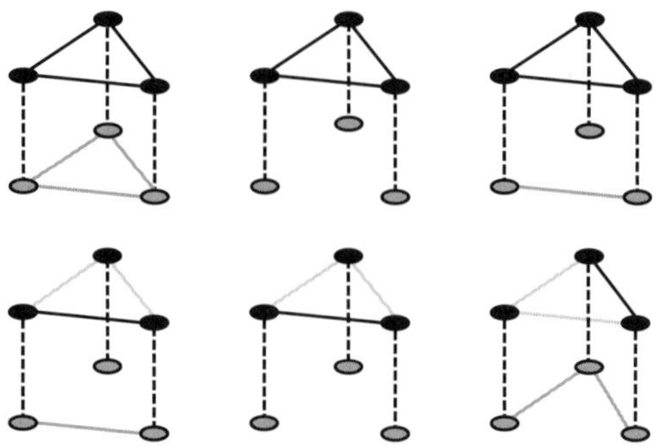

Figure 4.25 Frequent bilayer motifs in brain networks: Upper: function layer with positive (black or solid) and negative (red or dashed) edges. Lower: structural layer, edges grey or green. Vertical edges are interlayer identity edges. [Based on findings of Battiston, F., Nicosia, V., Chavez, M., & Latora, V. (2016). Multilayer motif analysis of brain networks. *Chaos*, **27**, 047404. With permission from AIP Publishing.] Note edge coding difference. (A black and white version of this figure will appear in some formats. For the colour version, please refer to the plate section.)

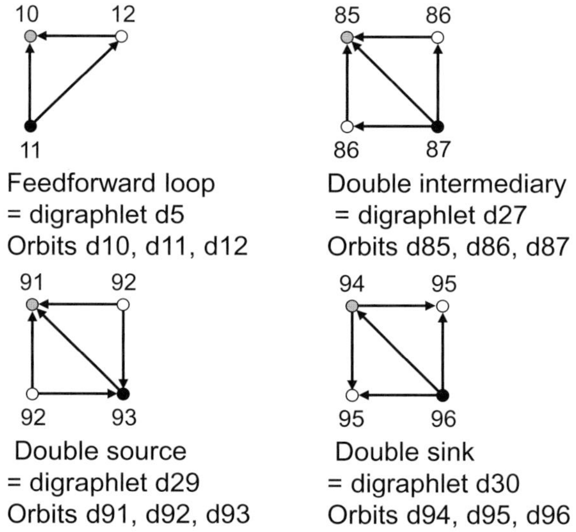

Figure 4.26 Generalisations of the feedforward loop motif, and relationships with digraphlets.

4.9 Concluding Comments

It is clearly a strength of the network approach to studying ecological systems that so many variations and complications of structure and function can be accommodated by elaborations of the basic network model of nodes and edges. Multilayer and multiplex network analysis is currently a hot topic in many branches of science, and it is easy to see why. There are many ecological, biological, and environmental phenomena that can only be portrayed, analysed, or interpreted in one of the "multi-something" network formats. The tie between multilevel statistical analysis and application of multilevel networks may be helpful in our understanding of both approaches. In some cases, while a multilayer approach is not absolutely necessary for the subject material, greater clarity and insight can be achieved by using this network format.

On the other hand, one very specialised kind of network that must be based on layers with different functions is the neural network approach to spatial analysis, image analysis and machine learning (Fausett 1994; Fischer 2009). Neural networks are based on the search for mathematical representations of information processing in neural systems, thus providing efficient nonlinear models for data analysis. A neural network is multilayer: one layer of nodes each for input and for output, with one or several processing layers between them, and weighted directed edges between the layers representing the flow of information: both feed-forward from input toward output, and back for iterative correction (Fausett 1994; Fischer 2009). Neural networks have not been much applied in ecological research, but it is another area of network studies for ecologists to pursue.

Most of the elaborated network structures are *conceptually* fairly simple, but converting that conceptual simplicity into precise mathematics and practical algorithms for analysis is more challenging. Fortunately for ecologists, much of the difficult detail has already been studied and both examples (mainly non-ecological), and software, either as dedicated packages or as modules in R or Python, are available for inspiration, assistance, and implementation. What is most exciting is that this is a very rapidly developing field and much more is to be expected as possibilities and options for multilayer/multiplex/multilevel/"multi-something" network analysis.

5 Tying It All Together
Summary and Synthesis

5.1 Network Thinking

This book has focused on the conceptual and quantitative analyses of ecological systems using networks and network theory. An ecological network is a system that can be represented or modelled by a set of points joined by lines and these networks are both fundamental to and deeply embedded in our understanding of ecological systems: call it "network thinking." With the right characteristics and measures, network thinking allows us to develop ecological concepts and to answer ecological questions we could not study or even express without it.

Important ecological questions and the quantitative analysis of the right ecological networks are not only closely related but are inextricably interlinked. There are few simple one-to-one relationships between network characteristics and the ecological processes; and so, the book has presented a wide range of network measures which can be collectively associated with ecological characteristics. Here, we will highlight some key concepts and their properties that are particularly useful to ecology, stemming from the criteria for network categories listed in Chapter 1.

5.1.1 Structure and Function

The structure of an ecological network includes both topological properties of what node is connected to what other nodes and configuration properties of weights and labels. Network function is what the network does and its dynamics are how its elements change and how its structure changes through time. The basic understanding is that the network structure constrains its function and dynamics and that the dynamics affect the structure as the network evolves. In this relationship, connectivity and complexity are important structural properties that determine function at the more global scale, and modularity, assortativity, keystoneness, node centrality, and neighbour hierarchies (subgraphs) are important characteristics at the local scale.

The interplay of structure and function takes a special form in causal networks. These are usually described by directed acyclic graphs (DAGs) in which the nodes are variables (observed or not) and the edges show the direction of causality. The inclusion or omission of any particular edge in the DAG is equivalent to a hypothesis but is best to be informed by background knowledge of the system. There is no substitute for such detailed knowledge. For a given causal digraph, the structure is interpreted for conditional dependence or independence among the variables using the rules of directional separation (d-sep). In interpreting causal structures for ecological systems, characteristics that need to be considered carefully are loops and autocorrelation, and the inclusion of interaction terms among the variables.

The best progress in understanding will be made when correlational, structural and causal models are integrated into the statistical analysis of networks. Quantification of network structure and function should be part of our "statistical rethinking" as applied to ecological systems (McElreath 2020); some parallel "network ecology rethinking" may also be in order.

5.1.2 Ecological Interactions

It is easy to classify ecological interactions as antagonistic (predation, parasitism) or mutualistic (pollination, seed dispersal) whether these relationships are between species or between individuals. It is, however, well documented that many interactions are actually subtle trade-offs of positive and negative effects, and that these can shift even over very short periods. In addition, there is the possibility that indirect interactions are important factors in these systems, with third parties modifying or switching the primary interactions. Important properties in these interaction networks include the signs and directions of the edges, symmetry or asymmetry of the outcomes, edge weights of interspecific effects, and the transitivity of relationships such as competitive dominance in "tournaments" of pairwise competitive tests. In this case, the tournament can be seen as providing a competitive hierarchy based on win–loss records, but in some tests the competitor with the inferior record may still prevail, creating a "competitive reversal." The hierarchy still gives ranks to the competitors, but the order of dominance is not strictly observed in the results.

In our review of network analysis, we have encountered many other examples of hierarchical structure, from food webs to spatial neighbours and DAGs, and might wonder what characterises a structure as **hierarchical**. One root of this word, "hier-" means holy or sacred (think of *Hierochloe*, sacred sweetgrass). The other part of this word, "archos" means "chief" and words ending in "-archy" refer to the structure of authority ("oligarchy" is rule by the few; "anarchy" is no rule at all). Therefore, "hierarchy" would have originally referred to a sacred ordering such as the grades of priesthood or the divisions of angels. We usually think of a hierarchy in terms of vertical arrangement or ordering: higher versus lower, up versus down, with distinct levels, although each level may contain many objects. This gives a partially ordered set (shortened to "POSet"), which is a well-studied mathematical object with many applications (Brüggeman & Carlsen 2006). A hierarchy is therefore a partial ordering from high to low, with the meaning of low and high depending on context.

While hierarchical structure arises in various network topologies, including meta-networks, we have not discussed alternatives very much. The "typical" web of hubs and spokes (Figure 1.9) is basically several hierarchies aggregated together into a polycentric structure with the hubs being of a higher precedence than the spokes. A reticulate structure would provide something different, a "flat" structure, with all the nodes more-or-less equal in status (something like a square lattice or a triangulation, but not necessarily as regular). A third possibility is based on a flexible interpretation of a *heterarchy* as a structure of complex architecture with both vertical (top-down or bottom-up) and lateral (peer-to-peer) axes (Cummings 2016). This concept deserves further consideration, starting with an explanation based on the account given by Cummings (2016).

Following on from the discussion of the meaning of "hierarchy," we note that "hetero" means "different" and so heterarchies provide different orderings of the objects being studied, perhaps allowing the same objects to be ranked or left unranked in several different ways. In his discussion, Cummings (2016) used "network" for a flat ("egalitarian" or peer-to-peer) reticulated structure to be contrasted with a vertically layered hierarchy; in our vocabulary, a hierarchy is still a kind of network, and so we will use "reticulate network" as the contrast for clarity. In addition to providing the possibility of ordering the same set of objects according to more than one criterion, a heterarchy allows both the edges that create the hierarchy and the edges of a reticulate structure within and between the levels of the hierarchy (Figure 5.1). The ecological example provided by Cummings (2016) is a set of wetlands that can be arranged top-to-bottom on an elevational gradient into three height levels with unidirectional hierarchical movement of propagules downhill (directed edges). Mutualist interactions like pollination or propagule dispersal provide bidirectional edges between sites at the same or different heights. In the terminology of Chapter 4, this is a multilayer network, layers being elevations, with multitype edges, here two types: passive dispersal and mutualist intervention. The goal is to provide a flexible format to show more and different information about complex networks, and to use the multitype edges to identify nodes of functional importance even when they are not dominant in any hierarchy. Unlike vines or hypergraphs, a heterarchy maintains only the original nodes of the network, and edges join pairs of these nodes, even when there are several

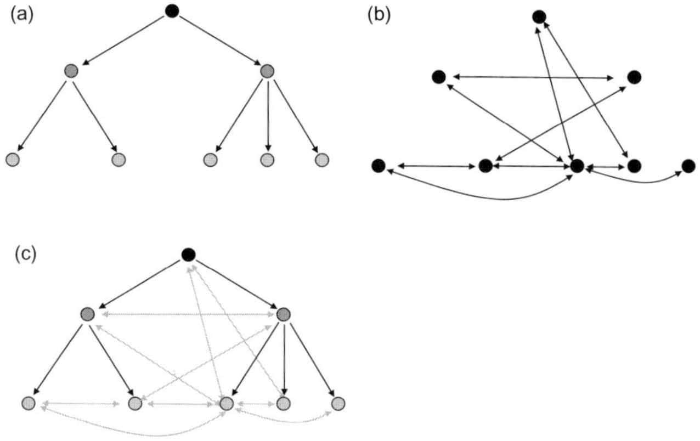

Figure 5.1 (a) A hierarchical network. (b) A reticulate network on the same nodes. (c) A heterarchy with both sets of edges. The two sets of edges can be assigned to separate layers of a bilayer multiplex network.

different types of edges, allowing the creation and analysis of a complex multirelationship structure. This is another area of network analysis for further conceptual and methodological innovation.

5.1.3 Space and Time

Although all ecological events actually take place at particular locations and at particular times (e.g. a predator's kill or a bee's flower visit) the events are often aggregated to allow summaries of rates (e.g. predation) or net outcomes (e.g. seed set efficiency). In some cases, the locational information is sufficiently important that it is maintained and the nodes of the network have positions in spatial or temporal dimensions. In these networks, the important properties will be those related to flow, dispersal, and spread, as in epidemiology. These would include connectivity and modularity, path costs, and partial orders in directional networks. In epidemiology applications, the characteristics of individual nodes (susceptible, infected, recovered, . . .) are critical to these dynamics, as well as the movement by the individuals, if mobile, or disease vectors. For dispersal in landscapes and the resulting metapopulations, metacommunities and metaecosystems, the dynamic network properties of robustness, stability, and resilience are central to any network analysis.

Even when the events are aggregated and the nodes represent individuals or species that do not themselves have locations, the resulting network may be associated with a particular time or place. For example, the visits of bee, B, to flower, F, are aggregated into counts, but the count that determines the edge

B–F belongs to a particular site and a particular time, which can then be compared to the existence or weight of the same edge at other sites or other times in a network-of-networks.

5.1.4 Multilayer

Whatever the layers represent, the focus of analysis will be on similarities and differences between the various layers, and the similarities and differences between intralayer relationships and interlayer relationships (i.e. weighted edges between the layers). Therefore, properties of interest in general will be modularity, clusters and clustering within and among layers, centrality within and among (versatility), node degrees for different edge classes, and random walks in the multilayer structure. Characteristics related to dynamics will be of interest, too, as appropriate: resilience, robustness, stability, or fragility. In special cases, such as phenological networks, measures related to flow and event spread will also be essential: orders and partial orders imposed by directed edges, "contagious" neighbour effects, synchrony, and time lags.

Multilayer networks should provide clarity by simplifying the contents of each layer, while providing order for the entire system by the way it is divided into separate layers. With very complex data, prioritising within-layer clarity may lead to large numbers of layers, endangering the clarity in the whole structure. There is an obvious trade-off between the information and informativeness within and among layers. As with Structural Equation Models (SEMs), where the claim is that each directed edge in the model represents a hypothesis to be evaluated, the organisation of a multilayer network and the choices made for what the layers are, and how they are related, also represent a form of hypothesis generation.

5.2 Networks: Construction and Reconstruction

Models of network construction (growth and evolution) range from completely random edge attachment at nodes to preferential attachment or random attachment constrained by the degree distribution (detailed in Glossary under "Network models"). The rules for construction are often reflected in the degree distribution of the nodes, and it is therefore a valuable characteristic to study, but the signatures cannot be relied upon to be unique to single models.

Reconstructing a network from data also has its challenges. In spatial or temporal networks, edge effects need to be accounted for. What are the effects of objects or events just outside the boundaries of the study, and how strong are those effects? Creating a buffer zone with data used only as secondary information is one technique to deal with the problem but may be wasteful of data. Looping the data in any one dimension to join up the two extremes is an option (it makes a 1-D time series into a circle, and a 2-D plane into a torus), but requires the assumption of stationarity to provide assurance that "the end is like the beginning." For aspatial and atemporal networks, edge effects are the equivalent of missing data, and that is a familiar problem: unobserved species, unrecognised taxonomic distinctness, interaction events too rare to be recorded, etc. There are solutions, or at least estimators for the magnitudes of the effects, for some of these kinds of problems that affect network construction or reconstruction. Chao's method (1987) for estimates with mark-recapture data was mentioned earlier in the context of species missed in the study by Timóteo et al. (2018). Another approach is to sample the incomplete data and calculate the measures of interest for each subsample to determine the robustness of the measures to missing information (Naujokaitis-Lewis et al. 2013). This is different from starting with a complete network and sampling the network to use properties of that sample to estimate properties of the whole network (Kolaczyk 2009). It is somewhat like the technique known as bootstrapping (Manley 2006), but not intended to determine statistical significance, just to address the possible effects of missed observations.

While sophisticated methods are helpful in understanding the effects of missing data, we must acknowledge that missing data will always affect our ability to interpret the network properties we observe and thus the ecological processes we study.

Network reconstruction may also aim at understanding the characteristics of ecological networks from the near or distant past. To what extent did the pollination interaction networks of English meadows in Victorian times resemble the current version (many of the "players" are the same) or the pollination networks in the paleolithic (when the players were not all the same)? How do we determine the structure of ancient networks? In what ways are they the same or different from the current ones? Obviously, the details of reconstructions are essentially untestable and with many uncertainties, but the search for general features or common constraints in comparable networks is worthwhile in the search for ecological order.

5.3 Network Complexity and Inference

The ecological networks we study are complex in many different ways, and also just plain difficult! Consider a small patch of grassland. The first problem is that we cannot see all the organisms; we probably could not give firm taxonomic labels to many of them; we cannot see many of the interactions between different taxa; and we may not even know what to look for or how to detect it, even those we can see may have missing occasions, and so forth. There is taxonomic complexity, complexity of interactions (including those that are multiactor or indirect), complexity of interaction structure . . . and then the complexities of the dynamics as they change through a day, through a season, or from year to year as ecological succession takes place or as environmental conditions change. It can be overwhelming. Background understanding of the organisms or the system is always helpful in bringing order to this complexity, and sometimes essential.

Part of the answer in interpreting the results of network analysis is the optimism that perhaps all the levels of detail are not so important to the overall picture. While it may seem that there is an almost infinite regress of level upon level ("It's turtles all the way down."), it may be that only the first level or two we see are sufficient for what we need to know and do.

5.4 Dynamics *on* and *of* Ecological Networks

From the first chapter, we have emphasised the difference between the dynamics that occur on networks (structure not changing, just the weights, labels, etc.) and the dynamics of networks in which nodes or edges are lost or gained. Figure 5.2 provides an example in the context of a spatial network of landscape patches (the network nodes for patches as shown; the edges of proximity that connect nearby patches). In the left column (Figure 5.2a), there are no changes through time. In the middle column (Figure 5.2b), the locations, sizes, and shapes of the patches do not change (and thus the proximity edges stay the same), but the quality of the patches changes, perhaps increasing cover, or density, or changing species composition. These are dynamics ***on*** the spatial network. In the right column (Figure 5.2c), not only does the quality change, but new patches (nodes) appear in the network, which will obviously change the proximity relationships as well as the set of nodes. These are dynamics ***of*** the network.

Despite the emphasis we have placed on this distinction, we admit that in practice it may not always be so clear, particularly when it comes to species interactions or

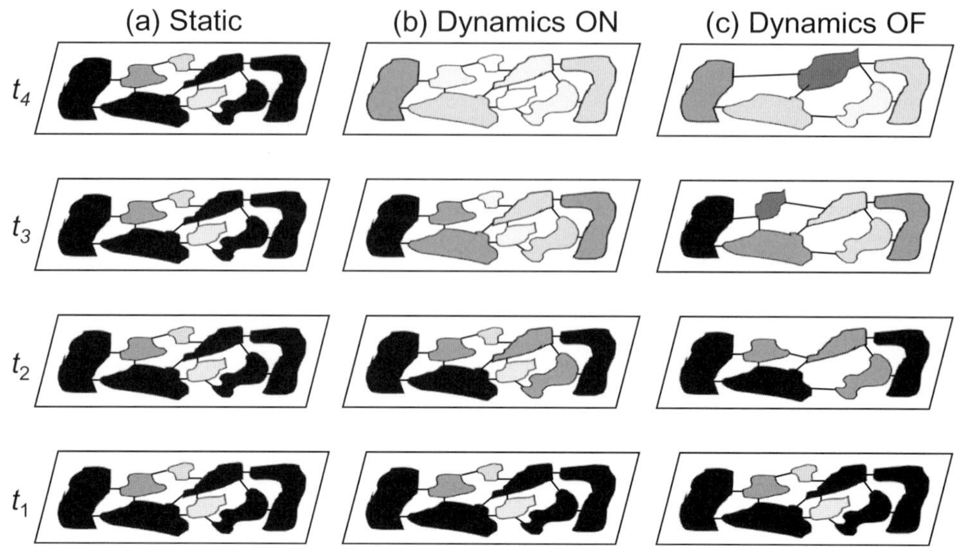

Figure 5.2 A spatial network of landscape patches through time. (a) Static. (b) Dynamics ON the network: the nodes and edges remain, but their qualities change through time. (c) Dynamics OF the network: nodes and edges are not constant but gained or lost.

species associations. There are many good examples but think of the Yucca plant and the Yucca moths that may pollinate the flowers and then lay their eggs in the flower to eat some of the developing seeds. That looks like a balanced mutualism. In a bipartite network of desert plants and their pollinators we would have a positive edge between Yuccas and Yucca moths. The wrinkle is that some of the Yucca moths are cheaters, laying their eggs without pollinating the plant (Addicott 1996). That is more like parasitism. At some point in the evolutionary history a positive edge became negative: dynamics of the network, rather than the dynamics on a network, although the basic structure of the network did not change.

5.5 A Conceptual Atlas of Network Concepts, Structures, and Methods

5.5.1 It All Ties Together

Like its topic, Figure 5.3a and b has much complexity in various forms, but it serves to draw together the important materials covered in the book, to show the relationships among concepts, structures, and methods.

The obvious place to start is on the left of the figure halfway down, with **network** and the **graph** that abstracts its structure. The basic graph has no weights on nodes or edges, but many kinds of networks require them, and so radiating out from **network** are the modified versions: weighted graph, signed graph (edges positive or negative), digraph (directed edges). Social networks are often signed (like versus dislike) and

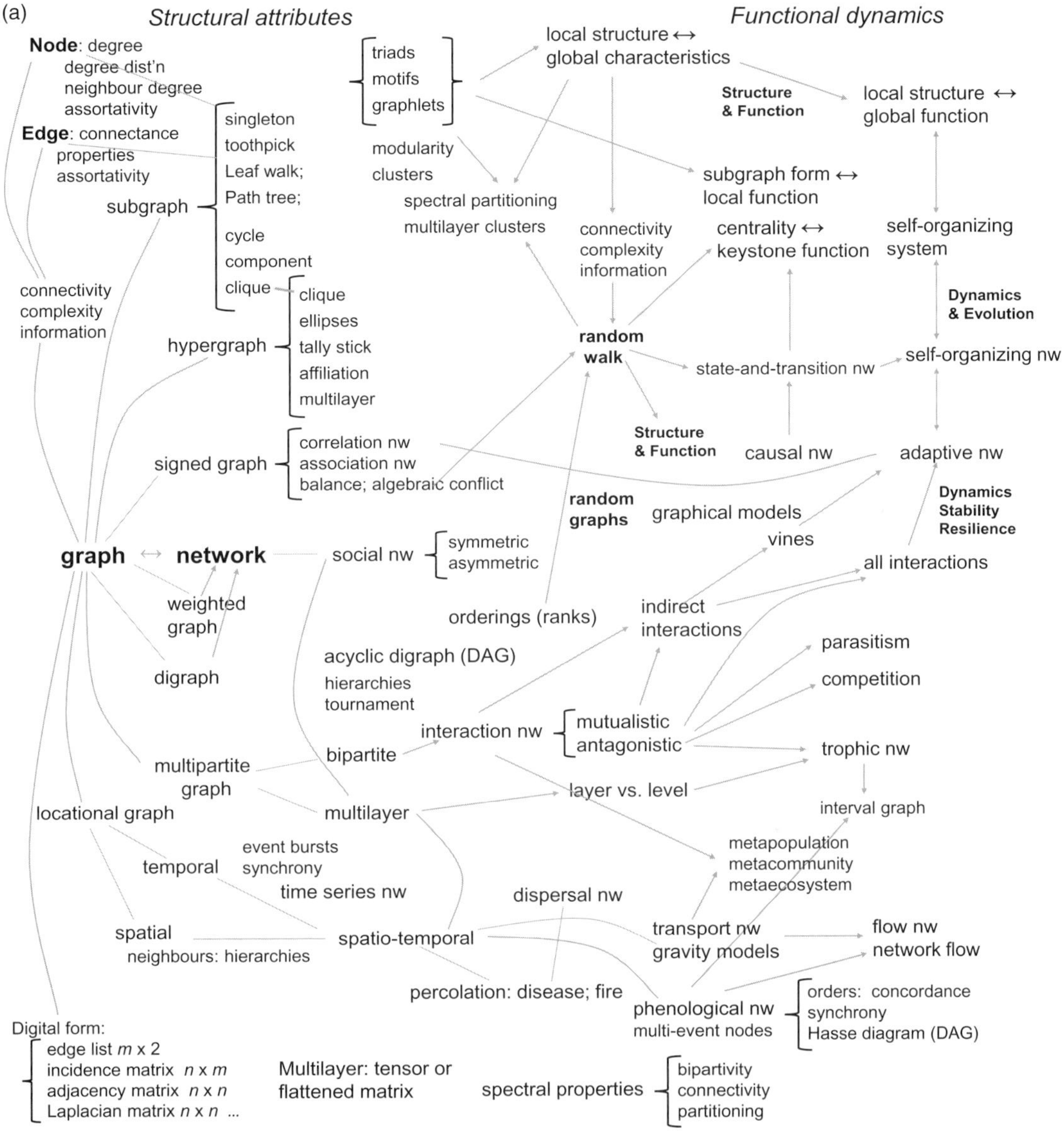

Figure 5.3 Two versions of the atlas of concepts, structures, and methods. Only the most important connections are included. (a) A version without miniature figures. (b) A version with miniature structural figures, which help the memory but make the diagram more cluttered. Choose the one that works best for you. (A black and white version of this figure will appear in some formats. For the colour version, please refer to the plate section.)

sometimes directed to allow asymmetry (A $\rightarrow$ B $\neq$ B$\rightarrow$A); other signed networks are those of association and correlation. Directed networks also occur in examples with orderings or partial orderings (rankings), dominance or other hierarchies, and causal structure models called DAGs (directed acyclic graphs).

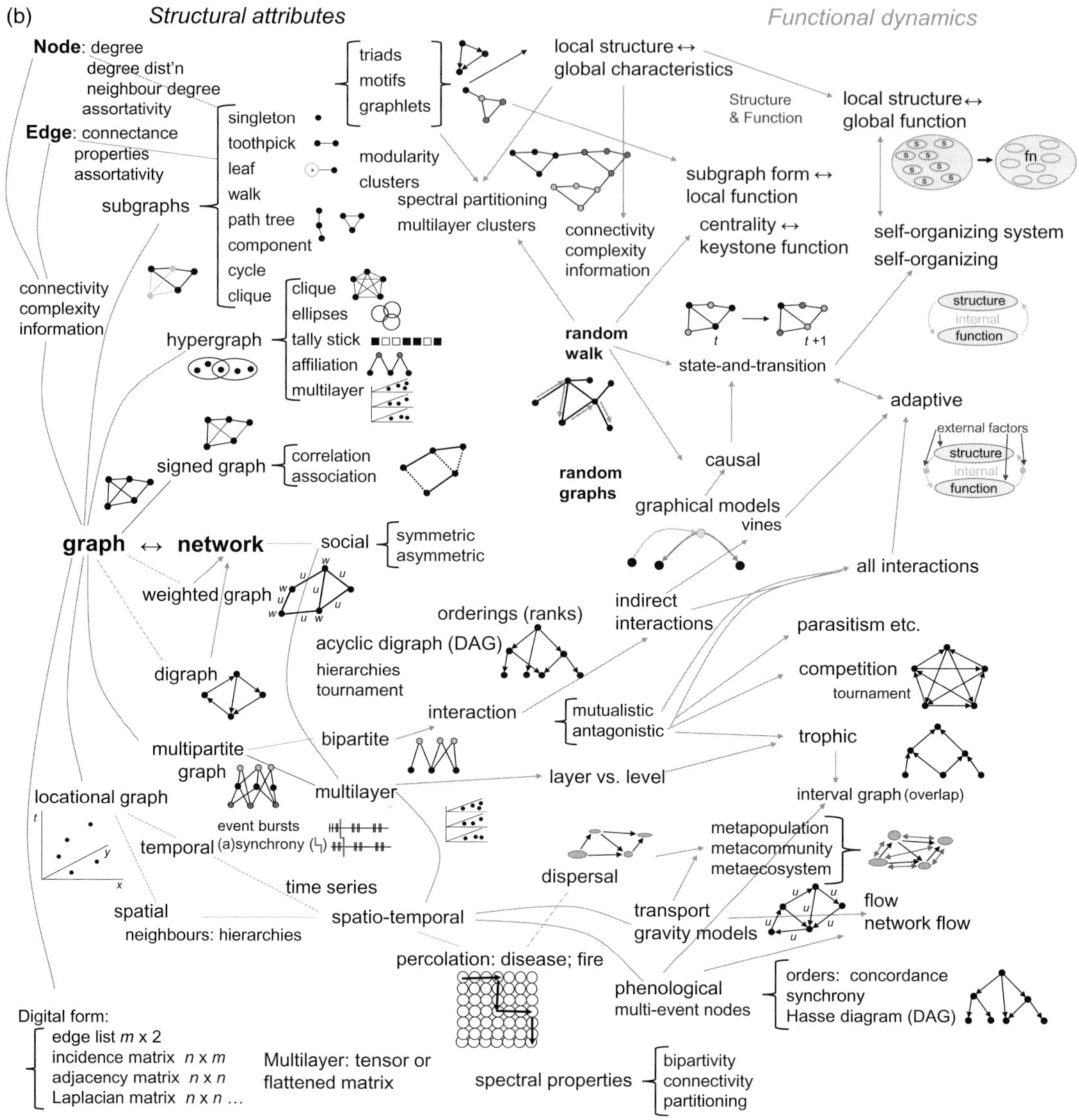

Figure 5.3 (cont.)

Further structural variations are multipartite (nodes in subsets) including bipartite (two subsets), and multilevel; with the possibility of more than two nodes per edge giving a structure called a hypergraph, which can be portrayed in several different formats.

Continuing with structural attributes (upper left), characteristics associated with nodes include degree and degree distributions, and properties associated with edges include connectance. For any graph, a subgraph is made up of subsets of its node and edges, and named subgraph structures include paths, cycles, and components. Specific designated subgraph structures can be used to analyse and summarise the local structural elements of any network: triads (triplets of nodes), motifs (often directed) as counted elements, and graphlets in which nodes positions are tallied. Larger subgraphs of interest are the clusters into which a network may be partitioned, and a property of this division into subgraphs is modularity.

In many important ecological applications, the networks are located in space or time or both, producing a locational graph: spatial, temporal, or spatiotemporal (lower left). One feature of spatial networks in ecology is the hierarchy of neighbours for each node, which can be defined in many different ways. Time-only (temporal) networks have characteristics that include the clustering of events in bursts, and synchrony of different kinds of events or of different series of the same kind. Spatiotemporal networks can be applied to studies of flow and transport, percolation and dispersal, the spread of fire or disease, and seasonal patterns like plant phenology.

The bottom left of the figure takes a detour to note the matrices that give the networks' structures in digital form, simple or multilayer. Expressing a network in this form allows that use of spectral analysis, based on the eigenvectors and eigenvalues, that provides insight into many properties of network structure and response to disturbance (algebraic connectivity and spectral gap).

The centre of Figure 5.3a and b is occupied by **random graphs** and **random walks.** Random graphs are central to our understanding of the structural characteristics of networks and how they evolve as they grow, completely random edge attachment to nodes versus preferential attachment or random attachment constrained by the degree distribution. Observed network characteristics can be evaluated by free or restricted randomisation of their structural graphs or those graphs' labels or weights, as appropriate. The idea of a random walk on a network is fundamental to a range of network concepts, including connectivity, complexity, node centrality, and many spectral graph methods and properties. Just as we use methods based on random numbers in modern computing to solve what would otherwise be very difficult (or impossible) problems requiring calculus, random walks enable the resolution of difficult problems in network analysis.

Moving to the right in Figure 5.3a and b, changes the emphasis from structural attributes to functional dynamics, such as how local structure affects global function (top right of the diagram). As one example, the bipartite structures used for ecological interaction networks lead to the interactions themselves: mutualistic, trophic, competitive, parasitic, and so forth, treated both in isolation and in combinations. These considerations lead also to indirect species interactions which are often portrayed by layered tree structures called vines. A vine is a multilevel network because the edges on the first level become the nodes on the second level, and the edges on the second level become the nodes for the third level, and so on. Other forms of graphical models are used state-and-transition dynamics, causal relationships, and self-organising and adaptive systems (upper right). The combination of the various species interactions, spatial networks and dispersal contribute to the dynamics of metapopulations and metacommunities, and thus by including abiotic factors, those of metaecosystems (lower right of the figure).

5.6 Concluding Comments

What will be the areas of focus for advancing network ecology? Multilayer networks, certainly, and networks-of-networks beyond spatial or temporal metanetworks are areas for further exploration and application.

Why multilayer networks? In part because they arise naturally in so many ways; the layers can be sites, or times, or interaction classes, and so forth and there is an understanding of the need to study ecological networks that are more than one time, one place, one interaction, in order to assess the variability of the systems we study. Including more can also make the networks more realistic by including more of the complexities, but that makes them more difficult to derive from data and more difficult to interpret from the analyses.

Why networks of networks? A similar reason: more than one place and more than one time, for a start, but spatiotemporal metanetworks are not the only form of networks-of-networks for ecological studies. The organising principle can be other than space or time: phylogeny is an obvious example. This will allow testing hypotheses based on these networks with complex structure. For the same reasons, networks with multitype nodes or multitype edges (like the heterarchies described earlier, Figure 5.1), also represent a "hot" research area to be pursued further in the near future. We can also look ahead to the integration of network thinking into conceptual ecology and how that will affect the theoretical underpinnings of our science.

Given the many forms of topological structure that ecological networks can take, from hierarchical to heterachies, a key ecological quest is to understand the factors or processes that generated the sampled ecological networks. Observed topological structure of ecological networks can be evaluated and compared using statistical approaches and probabilistic models (Poisot et al. 2016a; Lau et al. 2017; Rohr et al. 2018; Wills & Meyer 2020). We can also look forward to more progress on the interplay of "form $\leftrightarrow$ function" or "structure $\leftrightarrow$ dynamics" in the ecological context. In describing graphlet analysis, we cited the world trade example, but there should be ecological applications just as sophisticated and just as informative. For that to happen, multilayer ecological networks with multitype nodes or multitype edges need a common currency for the different interactions. In addition, new methods based on random walks and randomisations should be developed for network analysis, especially as a means to assess emergent properties of networks (see Poisot et al. 2016b).

There are other difficult questions that need answers:

- How best to compare networks?
- Should we compare whole networks or should we focus on one or very few key properties at a time?

We can compare networks based on their structure at the node-level or the network-level, or based on their function (or what we can see of it), or based on their observed dynamics (Wills & Meyer 2020).

Structures may be the easiest to compare, but similar structure does not guarantee similar function or dynamics, and so if the ultimate goal is predicting function or forecasting dynamics, comparing structures may not be sufficient. How to make inferences from such comparisons, when the processes that generated the networks are unknown? To do so, one must determine the processes that generate observed networks. To get to the key processes that generated ecological networks, mechanistic niche models (Rohr et al. 2018) need to be used such as niche models (Williams & Purves 2011) and latent-traits (Rohr et al. 2010), among others (Valdovinos 2019).

In this book, by detailing the quantitative analysis of ecological networks, we have promoted and perhaps even celebrated network thinking (or rethinking) in ecology and by association what can be called "network ecology." Will network ecology continue as a high-profile subdiscipline of ecological science? Perhaps not for very long. While network thinking is now an embedded feature of ecological research, "network ecology" will probably cease to be recognised as distinct, just as mathematical ecology, statistical ecology, and spatial ecology are no longer singled out as they once were, because network thinking will have become just an everyday component of good ecological thinking and good ecological studies. We hope that the concepts and methods presented in this book will provide a substantial contribution to that good ecology both now and into the future.

Glossary

Network Analysis and Graph Theory Glossary

Introduction

This glossary does not include every technical term found in the book, but it provides the basic vocabulary for network analysis and the graph theory on which it is based. It is somewhat informal in style, explaining clearly but keeping mathematical notation and equations to a minimum.

A ***graph*** is a mathematical object made up of two sets:

1. {nodes}. These are points or dots in a diagram representing the graph; they are also called vertices (often denoted by v); and
2. {edges} = {joining pairs of nodes}. These are the lines in a graph diagram, also called arcs or links. In a diagram, they can be straight or curved without meaning, because they represent relationships. Edges may cross with no node at the intersection; the crossings are not structural, just for convenience (Figure A1).

A ***network*** is a graph that includes quantitative information: the nodes usually represent *things* (e.g., people, species, islands. . .) with weights indicating size, abundance, potential, and so forth. The edges typically represent *connections* (e.g., interactions such as cooperation, relationships such as similarity, physical links such as contiguity or migration routes. . .) with weights their indicating characteristics such as intensity, rates, capacity, etc. (Figure A2).

Graph = G = {nodes} and {edges joining pairs of nodes}
= (V_G, E_G)
= sets $\{v_i \mid i = 1 \text{ to } n\}$ and $\{e_{ij}\}$.
Network = N = {nodes} and {edges joining pairs of nodes} and {weights for nodes or edges}
= (V_N, E_N, W_N)
= sets $\{v_i \mid i = 1 \text{ to } n\}$ and $\{e_{ij}\}$, and possibly $\{u_i \mid i = 1 \text{ to } n\}$ and $\{w_{ij}\}$

In any graph or network, the number of edges is m. In a basic graph, the pairs are unordered, with e_{ij} equivalent to e_{ji}, and the maximum number of edges is

(*cont.*)

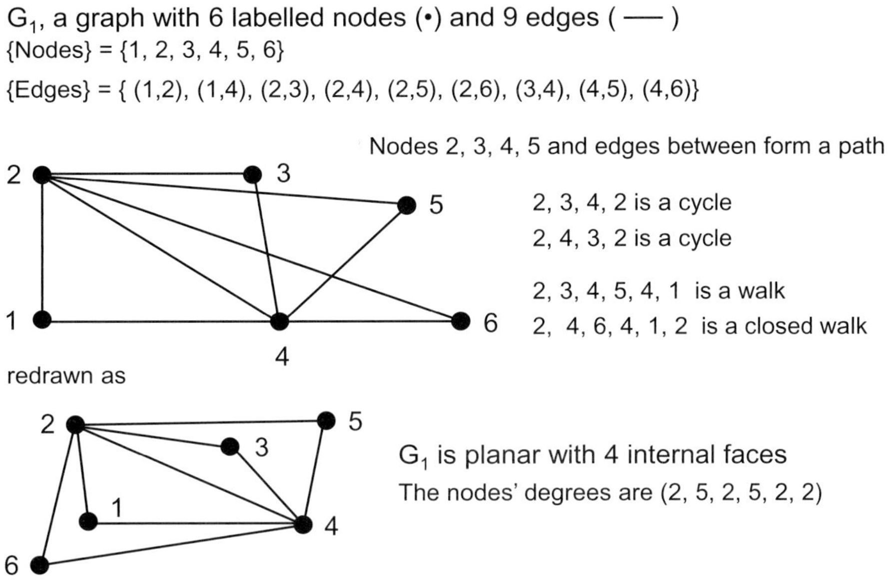

Figure A1 G_1, a graph with six labelled nodes (•) and nine edges (—).
{Nodes} = {1, 2, 3, 4, 5, 6}
{Edges} = {(1, 2), (1, 4), (2, 3), (2, 4), (2, 5), (2, 6), (3, 4), (4, 5), (4, 6)}. Nodes 2, 3, 4, 5 and edges between form a path. 2, 3, 4, 2 is a cycle. 2, 4, 3, 2 is a cycle. 2, 3, 4, 5, 4, 1 is a walk. 2, 4, 6, 4, 1, 2 is a closed walk. G_1 is planar with four internal faces. The nodes' degrees are (2, 5, 2, 5, 2, 2).

$n(n-1)/2$, which we may write as $n \wr 2$, which is read as "n select 2." In a directed graph, or ***digraph***, the edges have directions; e_{ij} is not equivalent to e_{ji} and the maximum number of edges is $n(n-1)$, and it is possible to have both e_{ij} and e_{ji} (Figure A3). In a network graph, the nodes can have labels or weights, the u_i, and the edges can have weights, the w_{ij}, (including signs, $+$ or $-$) as well as directions, distinguishing e_{ij} from e_{ji}.

One common kind of network is a *flow network*, representing the movement of some resource from one part of a system to another. In a flow network, all the edges have directions and positive weights representing, for example, their capacity to carry a flow of energy, information, or material from one node to another (see Figure A3, lower panel). Each edge may also have a cost associated with it. A flow network usually has a node that is the source (positive outdegree and zero indegree) and a node that is the sink (zero outdegree and positive indegree); these are labelled **S** and **T** in Figure A3. Usually, all other nodes have inflow equal to outflow (the conservation criterion). The analysis of such networks often deals with maximising the flow from source to sink or minimising the cost of doing so.

(cont.)

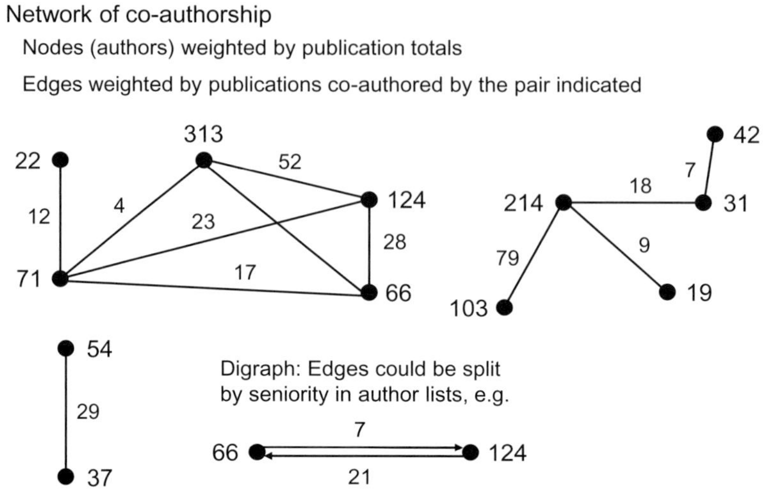

Figure A2 Network of co-authorship. Nodes (authors) weighted by publication totals. Edges weighted by publications co-authored by the pair indicated. In digraph form, the edges could be split into directed pairs weighted by seniority in author lists.

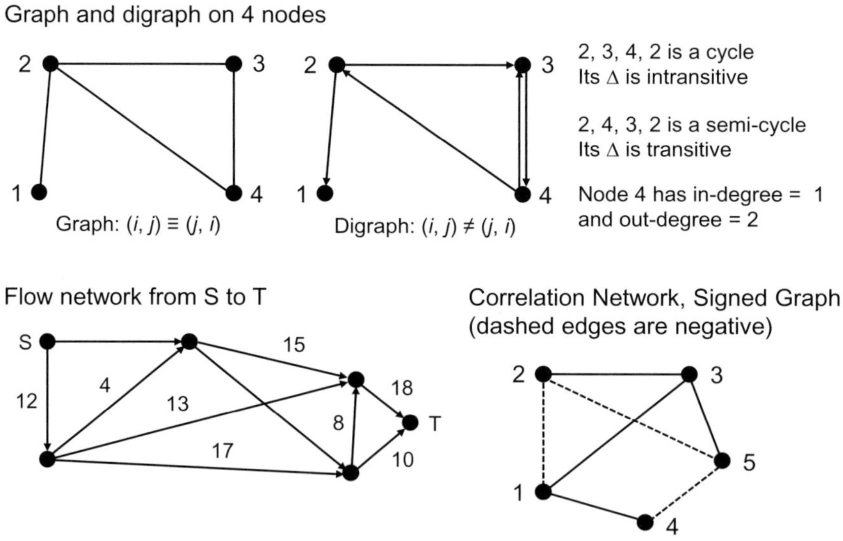

Figure A3 Upper panel: Graph and digraph on 4 nodes. In the digraph, 2, 3, 4, 2 is a cycle; its Δ is intransitive. 2, 4, 3, 2 is a semi-cycle; its Δ is transitive. Node 4 has indegree = 1 and outdegree = 2. Lower panel: Flow network from S to T and a correlation network, which is a signed graph. The dashed edges are negative.

(*cont.*)

Another common kind of network is a *correlation network* based on the measured correlation (or similarity) between pairs of nodes representing, for example, biological entities like species or proteins. The edges in a correlation network do not have direction (similarity is symmetric) but they usually have signs to differentiate between significantly positive correlation and significantly negative correlation (see Figure A3, lower right). In some examples of correlation networks, only the strong positive correlations are included in the subsequent analysis.

Related to the correlation network is a *social network* in which signed edges correspond to interactions or emotions in human and other social structures indicating positive versus negative association, liking versus disliking, and so forth.

Currently, there is much active interest in the analysis of *causal networks*, in which the nodes are variables joined by edges of hypothesised causal effects, thus producing (usually) directed acyclic networks. These causal networks are investigated using a set of techniques that can be categorised generally as *causal analysis*.

Despite these specific examples, "network" and "graph" are now often treated as almost synonymous, but *graph theory* still seems to be more about structure and *network theory* seems to be more about flow or transmission. In addition, "network" may be the structure depicted and "graph" its depiction, as in "the graph of the transportation network."

Terms in Alphabetical Order

Acyclic A graph with no cycles; most commonly seen as *directed acyclic graph (DAG)*, a digraph with no directed cycles, a common form for causal networks.

Adaptive network In an adaptive network, there is dynamic interplay between the structure or topology of the network and its functions and the states of its parts (nodes and edges). The structure can modify the functioning of the network and its functions can modify the structure. The analogy of a road or railway system is often cited (Gross & Blasius 2008; Dale 2017). When the reciprocal evolution of structure and function is driven by factors internal to the network or system, it is called a self-organising network or self-organising system (Dale 2017, figure 11.2). The alternative is that the reciprocal evolution is driven, at least in part, by external factors (Dale 2017, figure 11.3).

***Adjacency matrix*, A** A square matrix, $n \times n$, representing an undirected graph, with elements $a_{ij} = 1$ when nodes i and j are joined by an edge, and 0 otherwise.

Affiliation network A bipartite network of nodes representing (a) individual entities and (b) groupings of these entities; an edge joins an individual to a grouping to which it belongs. They may belong to more than one. For example, an affiliation network could join ecologists to the journals in which they publish their research. Because the edges always join nodes in different subsets, they have implicit direction.

Algebraic connectivity The second eigenvalue of the *Laplacian matrix*; it has predictive properties for graph connectivity and robustness to (edge) deletions.

Alocational graph The placing of the elements, in particular the nodes, are not interpreted as spatiotemporal locations, even if the structure originally had a spatial and temporal context or source; space and time are ignored in the graph itself.

Ancestor (a node) A relative term: the ancestor of a given node in a directed graph (e.g. of a causal network) is any other node from which a given node can be reached by a directed path (see also *Descendent*, *Parent*, *Child*).

Arborescence A directed rooted tree, and a digraph with no semi-cycles. There is exactly one directed path from the root to any other node, and at most one between any two nodes (Figure A4).

Aspatial graph The positions of the elements, in particular the nodes, are not interpreted as spatial locations, even if the structure originally had a spatial context or source; space is ignored in the graph itself. The network in Figure A2 is aspatial; the positions of the nodes are not related to the locations of the authors, although the authors may be closely associated with the locations of their home institutions.

Association network The nodes of an association network are entities or classes, such as species, and the edges are positive, indicating pairs that tend to be found together more often than expected, or negative, indicating pairs that tend to be found together

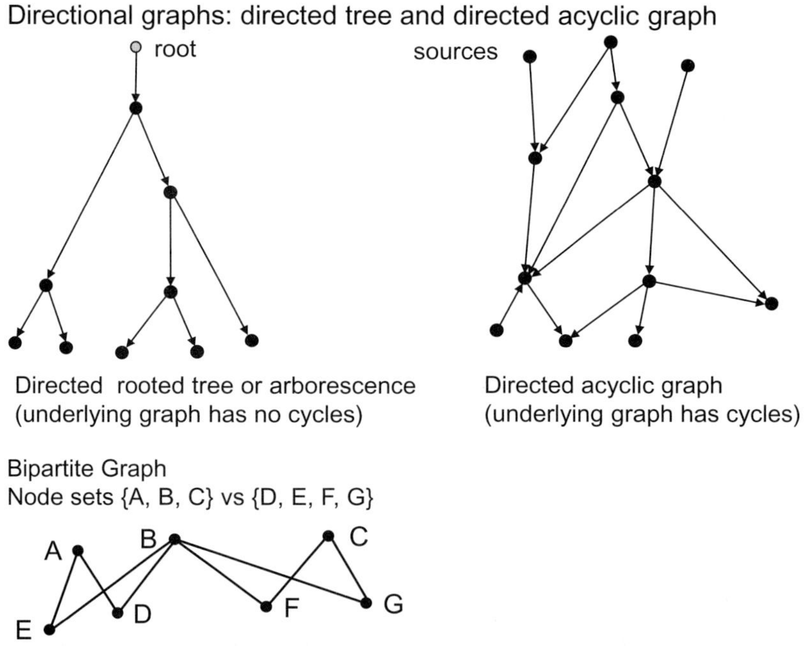

Figure A4 Upper panel: Directional graphs: directed rooted tree (arborescence; underlying graph has no cycles) and directed acyclic graph (underlying graph has cycles). Lower panel: Bipartite graph, with node sets {A, B, C} vs. {D, E, F, G}.

less often than expected. The edges are therefore signed but not directional. If the relationship of association can be refined to distinguish (causal) effects of A on B versus B on A, the result is a signed digraph.

Assortative* versus *disassortative General terms describing higher versus lower frequencies of like-to-like edges for any characteristic associated with nodes, including degree. In social networks, the term *homophily* refers to stronger and more frequent associations between nodes of greater similarity; its opposite is *heterophily*.

Assortativity coefficient A correlation measure, r, with values between -1 and $+1$, based on the bivariate frequencies of the degrees of neighbour nodes; it distinguishes between assortative networks with $r > 0$ indicating positive autocorrelation and disassortative networks with $r < 0$ indicating negative autocorrelation of neighbouring node degrees. This concept is related to *modularity*, which may measure the frequency of "like-to-like" joins.

Atemporal graph The placing of the elements, in particular the nodes, is not interpreted as a location in time even if the structure originally had a temporal context or source.

Automorphism For a graph or graphlet, an automorphism is a mapping of the object onto itself. For a node to be thus mapped onto another node in the same structure, the two nodes must occupy positions that are structurally indistinguishable.

Backdoor The backdoor criterion (shared cause) helps to identify variables to control or to condition upon if confounding is to be avoided. It refers to the lack of independence between two variables in a fork triplet that share a common cause (e.g., $X \leftarrow Z \rightarrow Y$) (see Figure 3.28).

Balance (in signed graphs or signed digraphs) The proportion of cycles that are positive, having an even number of negative edges (see Figure A3, lower right; the signed graph is not balanced because many of the cycles have one or three negative edges). It is considered to be an important characteristic in social networks, but less so in signed networks based on correlation or similarity.

Betweenness Measures the frequency with which a node occurs in the shortest paths between other pairs of nodes: one version of *centrality*.

Bipartite graph The nodes occur in two subsets with edges only between the subsets, not within (see Figure A4, lower panel).

Bipartivity A measure of how close a graph is to being fully bipartite. It is the proportion of all closed walks that are even. Spectral graph theory provides a direct measure based on the spectrum of the adjacency matrix, because the spectrum of a completely bipartite graph is symmetric about 0.

Burstiness The clustering of events in time, giving apparent bursts of activity. In temporal networks, this is the equivalent of clusters in one dimension.

Causal network A network in which the relationships between the nodes that are included as edges (usually directed) are those which have been determined to be causal interactions.

Centrality A measure of the importance of a node's position in the shortest paths between other pairs of nodes. There are many such measures; see Table 2.1.

Chain (triplet) In a path network model of variables and their interactions, a chain triplet of variable nodes is one with both edges pointing in the same direction (e.g. $X \rightarrow Z \rightarrow Y$), also called a "pipe."

Characteristic path length The median of the mean of all path distances from any node i to any other node j.

Child (a node) A relative term: the child of a given node in a directed graph (e.g. of a causal network) is any other node which can be reached by a single directed edge (see also *Ancestor*, *Descendent*, *Parent*).

Circuit theory A body of knowledge that permits the calculation of currents and voltage in an electrical or electronic network based on its components and structure.

Clique A complete graph, a subgraph with all possible edges. One method of portraying a hyperedge joining more than two nodes is to draw a clique of two-node edges joining all nodes included in the hyperedge.

Closed walk A walk that begins and ends at the same node.

Cluster analysis Methods that delineate groupings of entities, such as species or samples, based on their quantitative characteristics, based on greater similarity between entities within the same cluster than between those in different clusters. This approach to classification is often used in ecological studies; see Legendre & Legendre (2012) for a survey of clustering methods.

Clustering coefficient A measure of the probability that two neighbours of a given node (say j and k as neighbours of node i; with edges e_{ji} and e_{ik}) are themselves neighbours (joined by e_{jk}). The matrix version is this: given adjacency matrix entries $a_{ij} = 1$ and $a_{ik} = 1$, the clustering coefficient is the probability that $a_{jk} = 1$. This is also called *triad completion* or *triadic closure*.

Collider In a path network model of variables, a collider is the middle node of a triplet of variable nodes with both edges directed towards it (e.g. $X \rightarrow Z \leftarrow Y$). It is sometimes referred to as a "common effect" variable.

Complete graph There is an edge between every pair of nodes (also called a *clique*).

Complexity The complexity of a network is determined by the diversity of its composition and the lack of simplicity in its structure. It can also be related to the amount of information contained in the network or the improbability of its evolution. Many measures of network complexity have been proposed. A conceptually appealing measure is the number of different walks between pairs of nodes in the network graph: the "total walk count." A local measure of structural complexity at a node can be based on the diversity of graphlets in which the node participates.

Component A graph that is not connected consists of a number of maximally connected subgraphs called components. Many network graphs are not connected and consist of one very large component, the *giant component*, and many smaller ones, often as small as single unconnected nodes (*singletons*; see Figure 1.9).

Composite variable In a path network model of variables, a composite variable is one that is made up by combining other variables, whether these are observed or measured (manifest variables) or those that are not observed or not measured (latent variables).

Computational complexity Measures how the length of time (or amount of space) required for a particular algorithm increases with the number of nodes (or edges) in the network being analysed. For example, it can be linear, polynomial, or exponential.

Concordance A measure of the agreement of two or more orderings of the same set of objects, based on their ranks. The most familiar is Spearman's ρ, a non-parametric correlation coefficient for two samples. Because these are based on ranks, the treatment of ties is a consideration. Concordance is an important property in networks related to orderings, such as those with tournaments, DAGs and Hasse diagrams (see Section 4.4.1).

Configuration graph (state transitions) The nodes represent the possible states of a system and the edges are single-move transitions between these states that are allowed or observed (Figure A5). Originally from the study of games and puzzles (e.g. tic-tac-toe, Rubik's cube), they are used in graphical and simulation models of changing ecological networks (e.g. Daniel et al. 2018). A configuration graph shows the relationship between the graph's structure and the system's dynamics by codifying which transitions can occur.

Configuration model (random graph) A random graph model of n nodes and m edges, which is constrained by a given degree sequence for the nodes, $\{k_i | i = 1 \text{ to } n\}$. A probabilistic variant does not fix the realised degree sequence created but uses the set degree sequence to determine the probability of a particular edge: the probability of edge e_{ij} is $k_i k_j / 2m$.

Confounding In statistical analysis this is the situation where the effects of different variables cannot be distinguished because they are somehow mixed together (e.g. age and body size in juvenile animals).

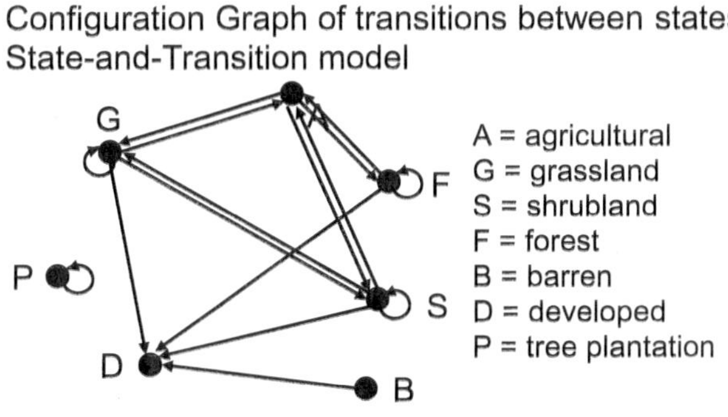

Figure A5 Configuration graph of transitions between states: a state-and-transition model. A = agricultural; G = grassland; S = shrubland; F = forest; B = barren; D = developed; P = tree plantation.

Connectance While ***connectedness*** and ***connectivity*** are closely related, connectance is different; it is the proportion of the possible positions for edges that actually have an edge. For a simple graph with n nodes and m edges, it is $2m/[n(n-1)]$. Connectance tells little about graph topology because two graphs with the same connectance can be structurally very different.

Connected graph There is a path between any two nodes. If two nodes in the graph do not have a path between them, the graph is disconnected (Figure A6).

Connectedness: Edge connectedness is the smallest number of edges that can be removed that will disconnect the network graph. ***Node connectedness*** is the smallest number of nodes that can be removed that will disconnect it.

Connectivity A graph has high ***connectivity*** if either the edge connectedness or node connectedness is large. This means that many nodes or edges must be removed in order to disconnect the graph.

Copula functions Copulas provide scale-free measures of dependence among variables; they are bi- or multivariate distribution functions with one-dimensional marginal distributions that are uniform on [0, 1]. As measures of variables' dependence, they do not assume a linear response and do not rely on the Normal distribution.

Correlation network The nodes usually represent biological entities like species or proteins, and the edges are based on the measured correlation (or similarity) between pairs of nodes. The edges have signs to differentiate between positive

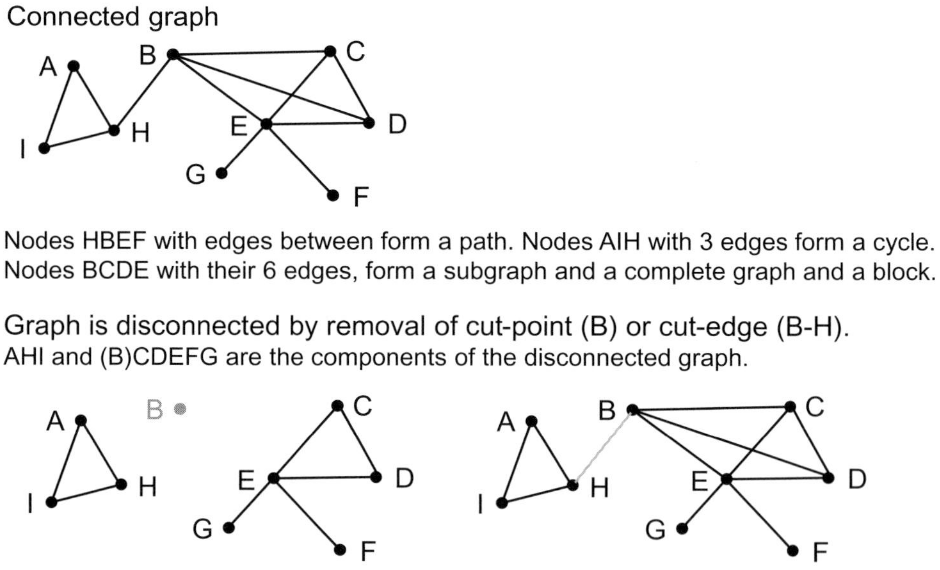

Figure A6 Connected graph and Disconnection elements. Nodes HBEF with edges between form a path. Nodes AIH with three edges form a cycle. Nodes BCDE with their six edges, form a subgraph and a complete graph and a block. Graph is disconnected by removal of cut-point (B) or cut-edge (B–H). AHI and (B) CDEFG are the components of the disconnected graph.

and negative correlation; they do not have directions. Edges for inclusion in an analysis may be determined by a threshold for the absolute value of the correlation between the two nodes, by the approximate significance of the value, or by a combination of the two.

Coupling edge (multilayer) An edge between layers that joins nodes that represent the same object in the two layers, equivalent to *edge of identity.*

Cut-edge (or ***bridge***) An edge in a connected graph the removal of which disconnects the network (see Figure A6).

Cut-point A node in a connected graph the removal of which disconnects the network (see Figure A6).

Cut-set A set of edges the removal of which disconnects the graph; the ***cut size*** is the number of edges between the two clusters of a partitioned graph whose removal will disconnect them.

Cycle A path that begins and ends at the same node, with no node or edge used more than once (sometimes called a *loop*). A cycle is a path that is a closed walk. In a digraph, a cycle must follow the directions of the edges; a polygon of edges in a digraph that do not form a cycle is a *semi-cycle* (see Figure A1).

DAG Directed acyclic graph; see entry below.

Degree The number of edges attached to a node, often designated by k_i. In a digraph, each node has both an indegree, the number of edges of which it is the sink, and an outdegree, the number of edges of which it is the source (see Figures A1 and A3). In a signed graph, each node has a positive degree and a negative degree. For comparing networks of different sizes, the *normalised degree* is (usually) k_i/n, or possibly k_i/m.

Degree distribution Summarises the degrees of all the network's nodes as the frequency or proportion of nodes with any given degree.

***Degree matrix,* D** : A square $n \times n$ matrix that has 0's everywhere except on the main diagonal, which lists the degrees of the nodes, k_i. For a digraph this can be either the indegree or the outdegree, and modifications will also be required for a signed graph like a correlation network.

Dense (edge density) A network's edge fill is considered *dense* if most of the entries in the adjacency matrix are non-zero, or if m increases more quickly than n, so that $o(m) > o(n)$, and the edge density, ρ, does not decrease with increasing n.

Descendent (a node) A relative term: the descendent of a given node in a directed graph (e.g. of a causal network) is any other node which can be reached from the given node by a directed path (see also *Ancestor*, *Parent*, *Child*).

Diameter A network's diameter is the maximum distance between any two nodes, where distance is the length of the shortest path between the two nodes (Figure A7).

Digraph: A ***directed graph*** consists of a set of nodes with a set of edges that have direction, so that the edge from A to B is different from the edge from B to A. In a ***digraph***, $e(i, j)$ is not equivalent to $e(j, i)$, meaning that $A \rightarrow B \neq B \rightarrow A$ (see Figure A3).

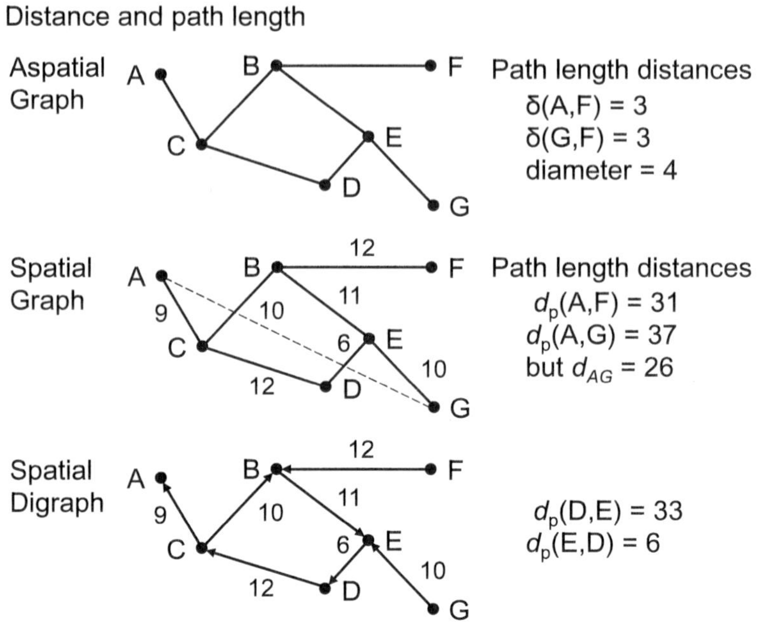

Figure A7 Distance and path length in spatial and aspatial graphs, as indicated. Top: δ(A, F) = 3; δ(G, F) = 4; diameter = 4. Middle: $d_p(A, F) = 31$; $d_p(A, G) = 37$; but $d_{AG} = 26$. Bottom: $d_p(D, E) = 33$; $d_p(E, D) = 6$.

It is possible that a particular pair of nodes is joined by two edges, one in each direction, that is, both A → B and B → A (see Figure A3). The edges may also have weights, as in a flow network, or signs as in a species interaction network.

Dimensional reduction Collapsing multidimensional information from more dimensions to fewer, while preserving as much of the original important information as possible. This includes techniques such as Principal Components Analysis and Multidimensional Scaling from multivariate analysis, but also the spectral embedding of network graphs.

Directed acyclic graph (DAG) A graph with directed edges that contains no directed cycles but may have semi-cycles; that is, its underlying graph may have non-directed cycles. Figure A4 shows a directed tree and a directional acyclic graph. In the directed tree, there is at most one directed path between any two nodes; in the acyclic digraph, there can be several. Every aborescence is a directed acyclic graph but not every acyclic graph is an aborescence.

Directional separation ("d-sep") An algorithm that determines the conditional dependence or conditional independence between any pair of variables in a graphical or causal model, based on conditioning on any other variables.

Distance: (between nodes) Can be measured in several ways, number of edges in the shortest path, physical distance in a spatial graph, or costs in a transportation network (see Figure A7).

Do operator Compare with $P(Y = y \mid X = x)$, which means the probability that variable Y takes a particular value y, given that variable X takes a particular value x. The "*do* operator" indicates that variable X is forced to take the particular value x by some intervention: $P(Y = y \mid do(X = x))$ is the probability that variable Y takes a particular value y, given that variable X has been set at the value x.

Dynamics A network's dynamics are its changes through time and the forces or constraints that determine them. We can distinguish between the dynamics *on* the network, through which the network structure remains unchanged, and the dynamics *of* the network, through which the structure does change (see Figure 1.10).

Edge matching coefficient (multilayer) The number or proportion of edges in common between any pair of layers, also called *edge overlap*.

Embedding (spectral) Dimensional reduction of data (a network graph) from a high dimensional space (e.g. n dimensions) into fewer dimensions **within** the original space, preserving as much as possible the information (e.g. distances between nearby nodes) from the higher dimensional structure.

Endogenous variable In a path network model of variables, an endogenous variable is one that has a parent node in the model (see *exogenous variable*).

Entropy In the Shannon formulation, the entropy of frequency distribution of n objects in Z categories of a classification is based on the proportions in each category, p_i, as the **negative** sum of $p_i \log p_i$, (usually base 2). This form of entropy can be used as a measure of the information content of a graph (*category* information), but it is not clear that this is a good measure of graph complexity. In Physics, entropy is a measure of the disorder of a system.

Exogenous variable In a path network model of variables, an exogenous variable is one that has no parent node in the model (see *Endogenous variable*).

Exponential Random Graph Models (ERGM) [called p^* models in the social network literature]. These are models each for a single dichotomous network relationship, with parameters reflecting a broad range of structural characteristics. The probability that there is an edge between nodes i and j can be determined by the characteristics of individual nodes (e.g. species, gender, age), the "dyadic" characteristics of the pair (e.g. distance, relatedness, shared history), and local or global structural characteristics of the network (e.g. triad completion). ERGMs predict edges using a process that resembles multivariate logistic regression for any pair of nodes, with several factors influencing the probability of a one or a zero: edge or no edge.

Factor analysis In studying the relationships among variables in a causal network, factor analysis looks at the effects of variables not measured (latent variables) on those that have been measured.

Flatland Originally a speculative or satirical novel about a two-dimensional world (Abbott 1884), the term is now used to refer to the topology and restraints of phenomena in two dimensions, often as a guide by extension to thinking about more than three dimensions.

Flow network A structure related to the movement of some resource among nodes; all the edges have directions and positive weights representing capacity to carry information or material. One node is the *source* (with positive outdegree and zero indegree) and one node is the *sink* (with zero outdegree and positive indegree). Usually, all other nodes have inflow equal to outflow (a conservation criterion).

Forest Acyclic graph that is not connected; it consists of components that are trees.

Fork In a path network model of variables, a fork in a triplet of variable nodes is the central node with both directed out from it (e.g. $X \leftarrow Z \rightarrow Y$). This is sometimes referred to as a "common cause" variable.

Frontdoor The frontdoor criterion helps to identify variables to control or to condition upon if confounding is to be avoided. It refers to the lack of independence between two variables in a chain triplet with a mediator and no unblocked backdoor path between them (e.g. $X \rightarrow Z \rightarrow Y$) (see Figure 3.28).

Geodesic path The shortest or minimum weight path between two nodes.

Graph A mathematical object made up of two sets: 1. {nodes}, the points in a graph diagram, and 2. {edges} = {pairs of nodes}, represented by the lines joining nodes in pairs (see Figure A1).

Graphlet An induced subgraph of up to five nodes. In graphlet analysis, what is counted is not the whole graphlets themselves, but the frequency of nodes at each distinct location in all graphlets. In general, analysis using *motifs* counts subgraph motifs even it is not an *induced subgraph* of the graph being analysed; in graphlet analysis, only induced subgraphs are considered (Figure A8). Graphlets are a sophisticated way of understanding local network structure and relating it to network function. Graphlets have been enumerated for simple basic graphs, and also for directed graphs and for signed graphs.

Graph-of-graphs A structure in which each node of a graph at one level consists of, or has associated with it, a graph of its own. When these are network graphs at both levels, it is also a *network-of-networks.*

Graphical model A graph that represents the relationships among variables (the nodes) such as joint probabilities, correlations, causal effects, and mathematical functions (the edges).

Gravity model A model of the interaction between two nodes in a network that has the intensity of the interaction proportional to the *product of their weights*, whatever that represents, *divided by the square of the distance* between them, however measured. The name derives from analogy with Newton's formulation for gravitational attraction. A simple example would be a model of the flow of seeds or pollen between two small landscape patches as proportional to the product of their areas divided by the square of the geographic distance between them.

Heterarchy A structure in which the objects may be ordered or ranked according to different criteria, creating a complex network ordering of the objects.

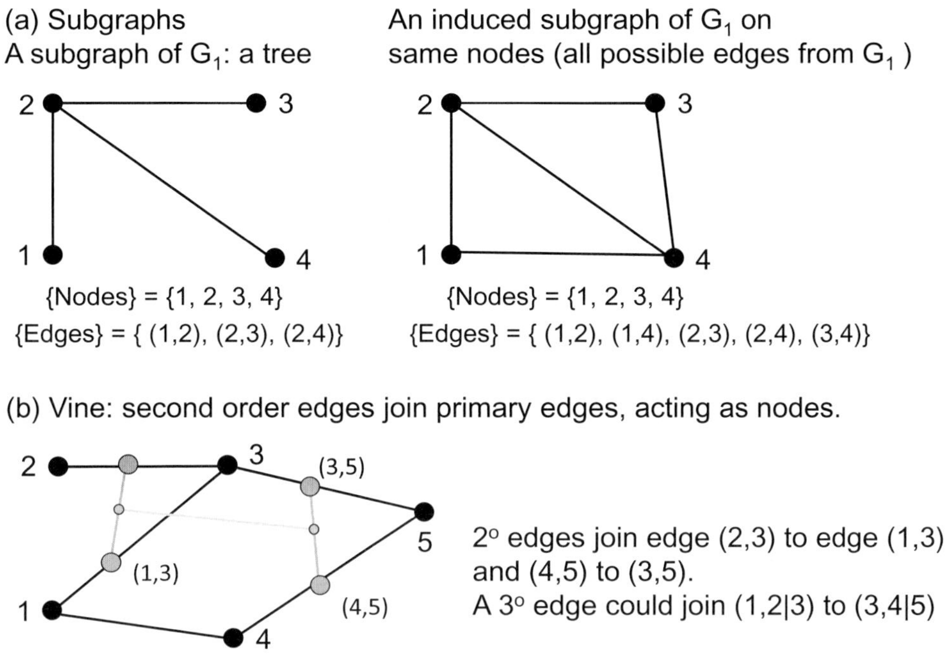

Figure A8 Upper panel: Subgraphs. The subgraph of G_1 is a tree. The right figure is an induced subgraph of G_1 on same nodes (all possible edges from G_1). Lower panel: Vine. second order edges join primary edges acting as nodes. 2° edges join edge (2, 3) to edge (1, 3) and (4, 5) to (3, 5). A 3° edge could join (1, 2|3) to (3, 4|5).

Heterophily and homophily In social networks, *heterophily* refers to stronger and more frequent associations between pairs of nodes that are dissimilar with respect to the characteristic of focus, for example, node degree; the opposite is *homophily*.

Hyperedge An edge (in a hypergraph) that may join more than two nodes; there are several ways of portraying these including ellipses and tally sticks.

Hypergraph A *hypergraph* consists of a set of nodes, V, and a set of edges, E, but the edges can join the nodes into unordered sets of two or more (these are *hyperedges* or *multinode edges*). There are several different ways of illustrating a hypergraph, including ellipses for edges and as a bipartite affinity network with one group node per hyperedge (see Figure 4.12).

Hyperpredation An *increase* in predation intensity experienced by a prey species, caused by the presence of a new alternative prey for the predator.

Hypopredation A *decrease* in predation intensity experienced by a prey species, caused by the presence of a new alternative prey for the predator.

Incidence matrix , **B** A rectangular matrix, $n \times m$, in which the rows are the nodes and the columns are the edges; for an undirected graph, the elements $b_{ik} = 1$ when edge k is attached to node i, and 0 otherwise. For a digraph, $b_{ik} = 1$ when node i is the source of edge k, $b_{jk} = -1$ when node j is the sink, and 0 otherwise. For a signed graph like a correlation network, several conventions are possible, such as positive edges indicated by $b_{ik} = b_{jk} = +1$, and negative edges by $b_{ik} = b_{jk} = -1$.

Indirect interaction The sign (+ or −) or strength of the interaction of two species, A and B, may be affected by the presence or intensity of a third species, C. The effect of species C on either A or B is mediated by its effect on their interaction rather than directly on either species; thus, its interaction is *indirect.* Similarly, the interaction of species C and D can affect the interaction of species A and B, also producing an *indirect interaction.* Such interactions can be portrayed and analysed using *vines*, in which primary edges between the original nodes act as nodes for a second order of edges (see Figure A8; cf. Figure 4.14). Higher order edges are possible.

Indirect interaction Mediation. Indirect interactions may be mediated by the species *densities* (DMII) or by other characteristics, *traits*, such as behaviour, giving TMII.

Information In this context, *information* refers to the reduction of uncertainty. For example, designating a single element from a finite set can be seen as providing information proportional to the log of the number of elements in the set. The information in a graph can come from the distribution of units into categories and those proportions, *compositional* or *category information*, and from the structural characteristics of the graph, *topological information.* These are not independent and may be confounded, for example when the categories are the nodes' automorphism orbits. Measures of information for graphs are often based on Shannon's entropy.

Interaction The relationship between two species, A and B, in which each has an effect on the other. The effect may be mutually positive, e.g. pollination mutualism, mutually negative, for example, competition, or positive for one and negative for the other, for example, parasitism.

Interaction network The network of mutually affective relationships among species, usually of one kind (e.g. pollination mutualism), but possibly including several.

Interaction term If factors X and Y are being tested for their effects on outcome Z, perhaps in the form of $Z = \alpha + \beta_x X + \beta_y Y$, the interaction of X and Y can be included by the interaction term: $Z = \alpha + \beta_x X + \beta_y Y + \beta_{xy}(XY)$. This term allows the possibility that Z's dependence on X may change with the value of Y, and vice versa.

Interdependence (multilayer) The proportion of all shortest paths between nodes in a multilayer network that include nodes in more than one layer or, equivalently, that include at least one interlayer edge.

Interval graph An interval graph is derived from a set of line segments arranged in parallel in one dimension. Each segment is represented by a node and nodes have an edge between them if and only if their corresponding line segments overlap (Dale 2017, figures 4.3 and 4.4). For example, the line segments may represent the range of food item sizes taken by each species in a community of grazers or predators.

Join count A method of analysis that begins by counting the categories of adjacent objects (network nodes), often as simple as counting "like" joins where the adjacent objects belong to the same category and "unlike" when they do not. The counts are then compared to expected counts based on random intermingling of the categories or on some restricted randomisation.

Joint degree distribution Tabulates the bivariate distribution of the degrees of node pairs joined by an edge, $f(k_i, k_j)$ where $e_{ij} \neq 0$, to determine whether the degrees of neighbouring nodes are positively or negatively autocorrelated. The frequency can be converted into the assortativity coefficient, which distinguishes between positive autocorrelation and negative autocorrelation of neighbouring node degrees. These measures can be calculated from the adjacency matrix, A.

Kernel function (1) A *kernel function* compares two structures by mapping the objects of interest from their original data space into a "feature space." Comparing two objects, x and x', does not require converting each one explicitly, but can be accomplished by a kernel function, k, acting on the cross product. This direct step is the "kernel trick" and its function essentially measures the similarity of x and x', because it takes higher values the more features the objects share. (2) The same term is used for a smoothing function, a window function, or as a method of modelling or estimating a variable's probability density function. Confusing!

Keystoneness A term that has migrated from ecology into the conceptual landscape of network analysis. It refers to the *functional* centrality of a particular node in a system, which is determined to be a keystone node in the network, parallel to the ecological *keystone species* in a trophic network (sea urchin, wolves,. . .) or the *ecosystem engineer* that modifies the habitat for many other organisms (beavers, elephants, . . .). Loss or removal of the node, like the loss of a critical species from an ecosystem, causes large and wide-spread changes in the network's dynamics. *Structural* centrality is often used as a surrogate for this property.

***Laplacian matrix,* L** A square matrix derived by subtracting the *adjacency matrix* from the *degree matrix*: $\mathbf{L} = \mathbf{D} - \mathbf{A}$. The *signed Laplacian* is $\mathbf{L}_{\pm} = \mathbf{D} - \mathbf{A}_{\pm}$ where the subscript indicates the *signed adjacency matrix* which counts all edges of either sign. [Even more confusing: there is also something called the *signless Laplacian*, which is $\mathbf{Q} = \mathbf{D} + \mathbf{A}$.]

Latency In a temporal graph, *latency* is the total time elapsed between source and sink.

Latent variable In a path network model of variables, a latent variable is one that is not observed or not measured (see *manifest variable*).

Leaf (or leaf node) A node of a tree of degree one; a node of degree one and its edge that joins it to a larger component of a network (see Figure 1.9).

Length A path's length is the sum of the weights of the edges in the path; in simple cases, the number of edges. The graph theory *distance* between any two nodes in a graph is just the shortest path between them (see Figure A7). In some applications, the shortest or minimum weight path is called a *geodesic* path. In a spatial graph or spatial digraph, paths also have physical distance lengths or lengths determined by transport costs (see Figure A7).

Line graph : G′ derived from graph G; nodes in G′ are the edges of G, edges in G′ join edges from G that share a node in G.

Locational network or graph The nodes have locations in time or space: temporal, spatial or spatiotemporal. The edges may or may not have locations other than those of their end-point nodes.

Manifest variable In a path network model of variables, a manifest variable is one that is observed or measured (see *Latent variable*).

Mediator In a path network model of variables, a mediator in a triplet of variable nodes is the central node with the causal effect flowing through it and around it (e.g. $X \rightarrow Z \rightarrow Y$ as well as $X \rightarrow Y$). With a mediator, Z, there can be both a direct effect, $X \rightarrow Y$, and an indirect effect, $X \rightarrow Z \rightarrow Y$, transmitted by Z.

Meta- A prefix denoting a spatially dispersed ecological system in patches maintained by dispersal between the patches. Hence:

metapopulation, metacommunity, metaecosystem, metanetwork, metaweb. . .

Modularity The non-uniformity of the distribution of edges in a network graph. One form of modularity is the tendency of network neighbours to have similar characteristics, indexed by the frequency of "like-to-like" joins in the network graph (related to a join-count statistic). As described earlier, modularity can also refer to how easily the graph can be *partitioned* into distinct modules (clusters of nodes) with high internal but low external *connectivity,* with a higher density of edges within clusters than between.

Module A subgraph characterised by high relative frequency of joins between nodes within and few outside; it can also refer to a high frequency of edges between nodes with similar characteristics (e.g. labels). *Modules* of nodes are also called "clusters" or "communities." In a graph or digraph with signed edges, well-defined modules have only positive edges within them and negative edges external to them (Figure A9).

Motif A small subgraph of defined structure that recurs in the networks being studied; these are with or without edge directions. Motifs usually have three or four nodes and the frequency of each is used to make inferences about structure.

Multi-feature effects = multiway interactions In addition to interactions between pairs of factors or features or entities like species, interactions among three, four or more need to be considered and our methods of network analysis need to be adjusted to account for these as possibilities.

Multigraph A *multigraph* is a structure with nodes and edges like an ordinary graph, but any pair of nodes can be joined by more than one edge.

Multilayer network A *multilayer* network consists two or more subnetworks as layers each with a graph of relationships within the layer, possibly with sets of edges between layers. For example, individuals in a population are mapped at several times: each time slice is a layer. A ***multiplex network*** is a multilayer network with the same

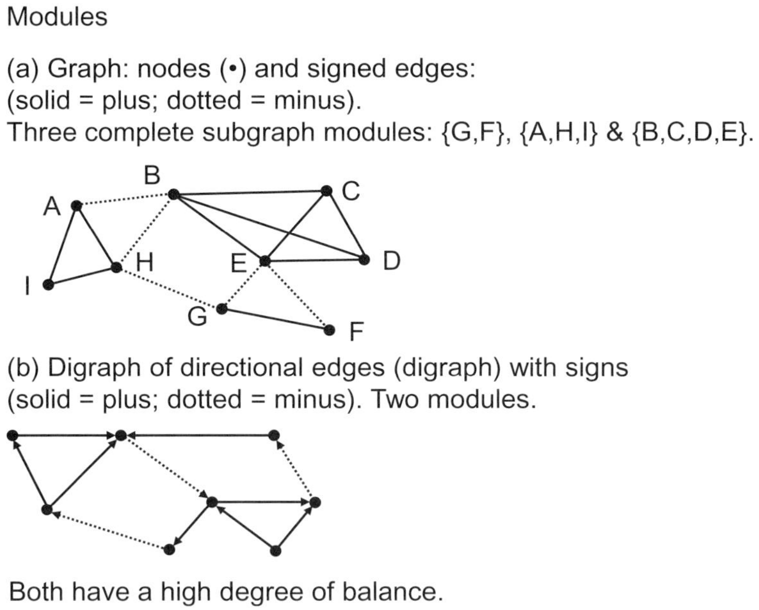

Figure A9 Modules. (a) Graph: nodes (•) and signed edges (solid = plus; dotted = minus). There are three complete subgraph modules: {G, F}, {A, H, I} and {B, C, D, E}. (b) Digraph of directional edges (digraph) with signs (solid = plus; dotted = minus). Two modules, both with a high degree of balance.

set of nodes in each layer but usually different interactions in separate layers; interlayer edges only link instances of the same node (identity edges).

Multilevel network A *multilevel network* also consists of several layers, but the nodes in the different layers belong to different levels of a natural hierarchy, for example, molecule, cell, organism, population, community, ecosystem. The relationships among what would be the layers are inclusion not serial dependence.

Multipartite graph The nodes belong to two or more subsets and (most) edges are between nodes in different subsets, with (almost) none within subsets.

Multiplex network A multilayer network with the same set of nodes in each layer, usually with different interactions in separate layers; interlayer edges only link instances of the same node (edges of identity).

Multiplicity The multiplicity of an eigenvalue is the number of times that a particular value occurs for the same matrix. For example, in the seven eigenvalues listed as $\{[-2]^3, 0, [2]^3\}$, -2 and $+2$ both occur three times and thus both have a multiplicity of 3.

Network (1) A graph that includes quantitative information, usually representing aspects of a real system (see Figure A2): often, the nodes represent *things* (e.g. species, islands. . .) with weights indicating size, abundance, potential etc.; and the edges represent *connections* (e.g. interactions like predation, relationships such as similarity, physical links like contiguity or migration routes . . .) with weights indicating their characteristics such as intensity, rates, capacity, and so forth (2) In many biological applications, "network" and term "graph" are treated as synonyms, although it makes sense to refer to the graph of a particular network. Hence, (3) the system of entities and their interactions being studied, from which a graph of the relationships can be abstracted.

Network models

1. *Erdös–Rényi* (random): N nodes and the sequential addition of K edges in random positions. The degree distribution is binomial to Poisson (i.e. bell shaped); path lengths are short at high edge densities; low clustering coefficients; no clusters; no autocorrelation of node characteristics.
2. *Barabási–Albert* (scale-free): Sequential node addition, each with c edges and preferential edge insertion. Degree distribution has low mode and long tail, governed by a power law. Path lengths increase logarithmically with N, low clustering coefficient, no clusters but assortative for node degree.
3. *Watts–Strogatz* (small world): Random rewiring or edge addition to a homogeneous regular graph. Degree distribution is compact, asymmetric, and unimodal, with no long tail. Path lengths are short, high clustering coefficient; clusters of nodes, but not assortative for node degree.
4. *Configuration* (Bender–Canfield): Random network evolution but with the number of nodes and their degree sequence fixed (a restricted randomisation).
5. *Probabilistic Configuration*: Random network evolution with the number of nodes fixed, but their degree distribution is probabilistic although based on the original degree sequence.
6. *Stochastic Block*: The nodes occur in several blocks; each block has a set probability of edge occurrence for each pair of nodes within it.
7. *Exponential Random Graph Model*: The probability that there is an edge at any particular pair of nodes is determined from the characteristics of the node, of the pair of nodes, of the neighbourhood in the network, or of the network as a whole.

Network-of-networks A structure in which each node of a network has a network at another structural or organisational level associated with it.

Neural network Based on the concept of finding mathematical representations of information processing in natural neural systems, this approach provides efficient non-linear models for data analysis in a spatial context, for images or for machine learning. Usually, it is a multilayer structure with layers of nodes (input and output layers with one or more hidden layers between) and weighted directed edges between the layers representing the flow of information.

Normalised degree For comparing networks of different sizes, the *normalised degree* of node i is derived from its degree, k_i, usually k_i/n but possibly k_i/m.

Orbit (automorphism orbit) An automorphism is a mapping of a network graph (or subgraph) onto itself. Any nodes that can be thus mapped onto each other belong to the same automorphism orbit. In essence, nodes in the same orbit occupy positions in the graph or graphlet that are structurally indistinguishable. Alternatively, an orbit is the set of nodes that can be mapped onto each other by an automorphism.

PageRank A method (algorithm) for ranking the importance of nodes based on an iterative weighting of their incoming and outgoing traffic. As well as following paths between adjacent neighbours, the method allows occasional random jumps between possibly distant pairs with no edge joining them ("teleporting").

Parent (a node) This is relative term: the parent of a given node in a directed graph (e.g. of a causal network) is any node that has a directed edge to the given node (see also *Ancestor*, *Descendent*, *Child*).

Partial order (of objects) A relationship " →" creates a partial order when A→B and B→C implies A→C, and A→B and B→A means A=B. Not all pairs of objects in the set are required to be ordered by " →"; that is, neither A→B or B→A may be true. Also called a partially ordered set or "poset."

Partitioning The process or problem of dividing a graph into subgraphs based on specific characteristics or criteria. For example, to divide a graph into modules or node clusters maximising edges within clusters and minimising edges between them.

Path An alternating sequence of nodes and edges in which each node is attached to the next node in the sequence by an edge; and each edge in the path is joined to the subsequent edge at a shared node (see Figure A1). No node or edge may be used more than once.

Path analysis One approach to quantifying the causal relationships among observed variables based on correlation and regression.

Path graph An unbranched tree; for n nodes, the two end-nodes have degree 1, and all the other $n - 2$ nodes have degree 2. Adding an edge between the two end-nodes converts a path graph to a ring graph, and removing any edge from a ring graph creates a path graph.

Percolation This is a process of movement or spread of a property through a network. *Site* (*node*) *percolation* occurs when a property of nodes spreads from node to adjacent node if the node is open to the change (think of a fire spreading from forest stand to some of the adjacent stands). In *edge percolation*, it is the edges' property (open or closed) that permits or prevents the spread of change.

Phenology (or phenological) network Records the dates of phenological marker events (such as first flower opening) for shared species over a range of geographic sites and over a series of years. The events can be treated as locational nodes in both space and time, producing partial orders of events within and among sites and across years.

Planar graph Can be drawn in a two-dimensional plane with no edges crossing. The edges do not have to be drawn as straight lines. G_1 in Figure A1 is planar as is the full graph in Figure A3. A *minimum planar graph* is a spatial planar graph (no edges crossing) that uses the smallest total length of edges.

Projection (multilayer) The projection graph of a multilayer network contains all nodes from all layers, and all edges are aggregated in some way: by the union of edges

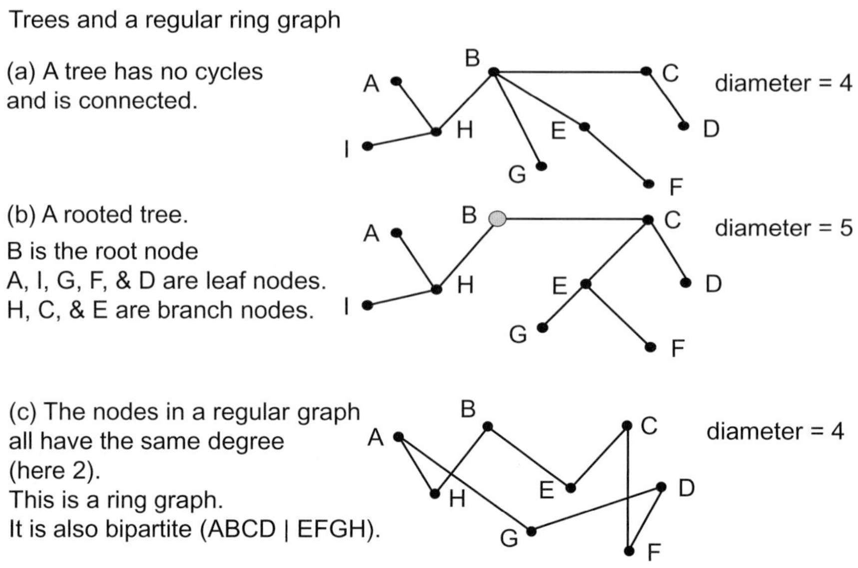

Figure A10 Trees and a regular ring graph. (a) A tree is a connected graph with no cycles. (b) A rooted tree has one node designated as the root. B is the root node. A, I, G, F, and D are leaf nodes. H, C, and E are branch nodes. (c) The nodes in a regular graph all have the same degree (here 2). This is a ring graph. It is also bipartite (ABCD I EFGH).

(only one edge per node pair) or by the inclusion of all edges in a multigraph (possibly many edges for a node pair).

Random walk A walk from node to node along edges in which, at each iteration, the next edge to be included is chosen at random according to some rule. A simple rule is to use equal probability for all edges attached to the current node. This is a key conceptual approach to studying network structure and characteristics.

Random walk: times *Access time* is the expected number of iterations of a random walk to move between any two nodes in the network. *Cover time* is the expected number of iterations for the walk to reach all nodes in the network.

Reciprocity (of a digraph) The proportion of all edges that are members of reciprocal pairs of edges between the same two nodes. In the digraph in Figure A2 two edges of five form a reciprocal pair; one of six node pairs have both reciprocal edges.

Regular graph All nodes have the same degree. A *complete graph*, which has all possible edges (each node has degree $n - 1$) is a regular graph; so is a *ring graph* where the nodes all have degree 2 (e.g. Figure A10c).

Ring graph Each node has two neighbours (degree 2) and there is only one cycle: the whole graph (Figure A10c).

Self-organising network; self-organising system In an adaptive network, when the reciprocal evolution of structure and function, or of topology and states, is driven by factors internal to the system, it is said to be self-organising (see Dale 2017, figure 11.2). The alternative is that the reciprocal evolution of structure and function is driven, at least in part, by external factors (Dale 2017, figure 11.3).

Semi-cycle In a digraph, walks, paths and cycles must follow the directions of the edges they include; similar-looking subgraphs that do not follow the edge directions are called semi-walks, semi-paths, and semi-cycles. Therefore, a *semi-cycle* is a closed polygon of nodes and edges without repetition that does not follow all the directions of the edges it includes (see Figure A3).

Signed digraph The edges have both signs and directions. In the adjacency matrix, each element, a_{ij} or a_{ji}, takes one of three values, +1, 0, or -1; for any pair of nodes there are nine edge combinations possible. Edge signs are often indicated by solid (or black) for positive and broken (or grey) for negative (see Figure A3).

Signed graph The edges have signs, $+$ or $-$, indicating the nature of the relationship portrayed, such as reinforcing versus antagonistic interactions or positive versus negative correlation. Edge signs are usually indicated by solid (or black) for positive and dashed (or grey) for negative (see Figure A3).

Singleton An informal name for a network component that is an unconnected node.

Spanning tree A connected subgraph of G that is a tree (i.e. has no cycles) and includes all nodes of G; a *minimum spanning tree* does so with the smallest total edge length. In Figure A8, the subgraph with three edges is a spanning tree of the induced subgraph with five edges.

Sparse (*edge density*) A network is *sparse* if most of the entries in the adjacency matrix are zero, or if m does not increase with than n so that $o(m) = o(n) \ll o(n^2)$. Often, in a sparse network, the average node degree is constant and so the edge density, ρ, decreases with increasing n.

Spatial graph and digraph The nodes have spatial locations; this means that edges, and thus paths, have weights that are physical distances, or weights related to transport or communication (see Figure A7).

Spatiotemporal graph and digraph The nodes of these graphs have locations in both time and space, which means the edges, and thus paths, have physical and temporal distances.

Spectral graph theory The study of the structural (and other) properties of a network's graph based on the set of eigenvalues (the spectrum) of the adjacency, Laplacian, or other matrixes.

Spectral properties A graph's ***spectrum*** is the set of eigenvalues of the adjacency, Laplacian, or other matrixes. For example, the ***spectral gap*** is the difference between the first and second eigenvalues of the adjacency matrix. The ***algebraic connectivity*** is the second eigenvalue of the Laplacian matrix. Both are descriptive of current connectivity and robustness to potential deletions.

Spurious correlation The result of a common cause in a causal model. X and Y may seem to be correlated if they are both dependent on a third variable, Z: $X \leftarrow Z \rightarrow Y$. Because the variables are correlated, it would be better to refer to this feature as induced correlation.

State-and-transition model For a single site assumes that the site is in one of a finite number of states that describe its characteristics or condition, for example, "undisturbed shortgrass prairie," with a set of probabilities that any site in state A will convert to state B, C, or D, in a particular time period. A configuration graph can illustrate those transitions (see Figure 1.11). In a spatial network of sites those transitions between states at any node can be affected by the states and transitions for neighbouring nodes.

Structural Equation Model (SEM) An equation describing the relationship between variables, for example, $Z = \alpha + \beta_x X + \varepsilon$ is *structural* if there is strong evidence that the relationship is causal. A structural equation model consists of two or more structural equations to account for multivariate relationships (Grace 2006).

Subgraph J is a subgraph of graph G, if it includes only nodes that are a subset of the nodes of G, and edges that are a subset of the graph's edges that join pairs of nodes in J. More formally, V_J is a subset of V_G and E_J is a subset of $E_G \cap V_J \times V_J$. A subgraph is *induced* if it contains all the edges of G it possibly can, so that $E_J = E_G \cap V_J \times V_J$. See Figure A8a.

Subgraph dilation and erosion The dilation of a subgraph is the subgraph itself together with all the edges and nodes that are first order neighbours of its original nodes. (The subgraph annexes the nodes on its periphery.) Its erosion is the opposite: the removing the nodes that are on the "outside" border of the subgraph (those with neighbours not in the subgraph), retaining only nodes that have no neighbours outside the original subgraph.

Subnetwork J is a subnetwork of network N, if it includes only nodes that are a subset of the nodes of N, and edges that are a subset of the N's edges that join pairs of nodes in J. The quantitative characteristics of a subnetwork can be very different from those of the whole network.

Supra-adjacency matrix The supra-adjacency matrix represents a multilayer network of n nodes in M layers. This is a $(n \times M) \times (n \times M)$ matrix, consisting of an $M \times M$ array of $n \times n$ adjacency matrices with the intralayer edges coded on the main diagonal of the array. The other matrices give the interlayer edges. The matrix entry $a(i, \alpha, j, \beta)$ is 1 if there is an edge between node i in layer α and node j in layer β, and 0 if there is none, with weights, and so forth, accommodated in the usual way. An intralayer edge is given by $a(i, \alpha, j, \alpha) = 1$; and interlayer edge is $a(i, \alpha, j, \beta) = 1$ with $\alpha \neq \beta$.

Supra-Laplacian matrix The equivalent of the standard Laplacian matrix for multilayer networks, based on the supra-adjacency matrix subtracted from a diagonal supra-degree matrix of the nodes' intralayer degrees.

Supra-walk A walk in a multiplex network that allows one, but not more than one, interlayer edge between any two intralayer edges in the walk.

Tally stick A diagrammatic summary of the presence and absence in k classes like species, using a string of full and empty boxes or circles, thus keeping tally of composition. For example, ■■■□□■□, indicates the presence of species 1, 2, 3, and 6, with species 4, 5, and 7 absent.

Teleport In a random walk with steps usually between adjacent nodes, a *teleport* step is a jump to any randomly chosen node. This allows the random walk to break out of the traps of long cycles and is a feature of PageRank methods.

Temporal* or *"time-only" graph The nodes have locations in time, which means that there is directionality inherent in the whole structure and "distance" is measured by time.

Tensor A *tensor* is a multidimensional array that extends the concept of a vector. In network applications, the basic entries are 1 for an edge and 0 for no edge, but weights and signs and directions can be accommodated just as in the familiar adjacency matrix. For a multilayer network of n nodes in M layers, the connections are coded by a $n \times M \times n \times M$ tensor with entries $A_{j\beta}{}^{i\alpha}$ for edges between node i in layer α and node j in layer β. An alternative representation is the "supra-adjacency" matrix: a square $(n \times M) \times (n \times M)$ matrix, which is an $M \times M$ array of $n \times n$ adjacency matrices with the intralayer edges coded on the main diagonal of the array.

Time series network A locational network of sites located in space, each with an associated time series, often of environmental data such as stream flow or air temperature. These are analysed for spatiotemporal characteristics such as trends, cycles, and synchrony. Such time series data may be critical in understanding phenological network changes across sites and years.

Toothpick An informal name for the network component consisting of two nodes and their connecting edge, unconnected to the rest of the network (see Figure 1.9).

Tournament A digraph with one directed edge for each pair of n nodes is a tournament: it resembles the complete results of one game per pair of n players, with outcome win or loss. The underlying undirected graph is a complete simple graph.

Transience* and *transient behaviour When a network is perturbed by a change to a node or edge weight or by changes in topology, it may move from one steady state to another. That transition may be quick and smooth, or the system may exhibit *transient behaviour* of mild or large fluctuations for a length of time before stabilising; that is, it bounces around before it settles.

Transitive reduction In a partial order or a digraph, the transitive reduction is created by removing edges according to the rule that the edge $x \rightarrow y$ is included only if $x < y$ and there is no node z such that $x < z < y$; otherwise the edge is removed. Figure 4.11b gives an example. Transitive reductions of directed acyclic graphs are *Hasse diagrams.*

Transitivity (of a triangle in a digraph) This is based on the directional interactions among three nodes: if A → B and B → C, then A → C, the triangle of edges is transitive (see Figure A3). If A → B and B → C, but C → A, then it is intransitive.

Tree A connected graph that contains no cycles; see Figure A10a. A *rooted tree* has a single node designated as the root; all other nodes can be characterised by their distance (path length) from the root, creating an intrinsic ordering for the tree; see Figure A10b.

Triangulation A planar graph in which the edges create a plane tiling of contiguous triangles, not necessarily identical. Spatial graphs based on proximity are often represented as Delaunay or other triangulations.

Versatility (multilayer) The multilayer equivalent of centrality (any of several versions) calculated over the entire multilayer structure, not an average over single-layer centralities.

Vine A *vine* is a tree-like graphical model in which primary edges between the original nodes then act as nodes for a second order of edges. A *regular vine* for n variables is a set of nested trees in which the edges of tree j are used as the nodes of tree $j + 1$, and two edges in tree j are joined by an edge in tree $j + 1$ only if they share a node in common. An *irregular vine* is a set of nested trees in which the edges of tree j are used as the nodes of tree $j + 1$, but there are no strict rules as to how the structure evolves with iteration. A vine is a multilevel network: the edges at the first level become the nodes at the second level, and so on to the next and higher levels. Vines are ideal forms for describing and analyzing indirect interactions between species (see Figures A8, 4.14, and 4.15).

Walk Any alternating sequence of nodes and edges in which each node is attached to the next node in the sequence by an edge; unlike a path, a walk can use the same elements more than once (see Figure A1).

Weight degree A node property that is the sum of neighbours‘ edge weights or of neighbours' node weights. In a weighted digraph, this may be split between weight indegree and weight outdegree in the obvious way.

Weighted graph Nodes can have weights, such as counts, and edges can have numerical weights, such as relationship intensity or measures of similarity or “distance“ (see Figure A2 and spatial graphs in Figure A7). A *flow network* is a weighted digraph with the edge weights being the capacities of the edges to carry information or material.

References

Abbot, E. A. (1884). *Flatland: A Romance of Many Dimensions*. London: Seeley & Co.

Abrams, P. A. (1995). Implications of dynamically variable traits for identifying, classifying, and measuring direct and indirect effects in ecological communities. *American Naturalist*, **146**, 112–134.

Ackerly, D. D., & Stuart, S. A. (2009). Physiological ecology: Plants. In S. A. Levin (ed.), *The Princeton Guide to Ecology*, pp. 20–26. Princeton, NJ: Princeton University Press.

Addicott, J. F. (1996). Cheaters in yucca/moth mutualism. *Nature*, **380**, 114–115.

Ahn, Y.-Y., Bagrow, J. P., & Lehman, S. (2010). Link communities reveal multiscale complexity in networks. *Nature*, **466** (7307), 761–764.

Ahuja, R. K., Magnanti, T. L., & Orlin, J. B. (1993). *Network Flows: Theory, Algorithms, and Applications*. Upper Saddle River, NJ: Prentice-Hall.

Airoldi, E., Blei, D. M., Fienberg, S. E., Goldenberg, A., Xing, E. P., & Zheng, A. X. (eds.) (2007). *Statistical Network Analysis: Models, Issues, and New Directions*. Berlin: Springer.

Aldous, J. M., & Wilson, R. J. (2000). *Graphs and Applications*. London: Springer.

Allesina, S., & Pascual, M. (2008). Network structure, predator-prey models, and stability in large food webs. *Theoretical Ecology*, **1**, 55–64.

Altaf-Ul-Amin, M., Saito, A., Koma, T., et al. (2003). Prediction of protein function based on *k*-cores of protein-protein interaction networks and amino acid sequences. *Genome Informatics*, **14**, 498–499.

Alvarez-Vega, M. (2011). *Graph Kernels and Applications in Bioinformatics*. PhD dissertation, Utah State University, https://digitalcommons.usu.edu/etd/1185.

Anderson, T. K., & Sukhdeo, M. V. K. (2011). Host centrality in food web networks determines parasite diversity. *PLoS ONE*, **6**(10), e26798.

Aparício, D., Ribiero, P., & Silva, F. (2015). Network comparison using directed graphlets. arXiv:1511.01964v1[cs.SI].

Aref, S., & Wilson, M. C. (2018). Measuring partial balance in signed networks. *Journal of Complex Networks*, **6**, 566–595.

Baiser, B., Elesha, R., & Kahveci, T. (2015). Motifs in the assembly of food web networks. *Oikos*, **125**, 480–491.

Bang-Jensen, J., & Gutin, G. (2009). *Digraphs: Theory, Algorithms and Applications*, 2nd ed. London: Springer.

Barrat, A., Barthélemy, M., & Vespignani, A. (2008). *Dynamical Processes on Complex Networks*. Cambridge: Cambridge University Press.

Bastolla, U., Fortuna, M. A., Pascual-Garcia, A., Ferrera, A., Luque, B., & Bascompte, J. (2009). The architecture of mutualistic networks minimizes competition and increases biodiversity. *Nature*, **458**, 1018–1021.

Bate, A. M., & Hilker, F. M. (2012). Rabbits protecting birds: Hypopredation and limitations of hyperpredation. *Journal of Theoretical Biology*, **297**, 103–115.

Batista-Foguet, J. M., Coenders, G., Saris, W. E., & Bisbe, J. (2004). Simultaneous estimation of indirect and interaction effects using structural equation models. *Metodoloski Zvezki*, **1**, 163–184.

Battiston, F., Nicosia, V., Chavez, M., & Latora, V. (2016). Multilayer motif analysis of brain networks. *Chaos*, **27**, 047404.

Batushansky, A., Toubiana, D., & Fait, A. (2016). Correlation-based network generation, visualization, and analysis as a powerful tool in biological studies: A case study in cancer cell metabolism. *BioMed Research International*, **2016**, 8313272.

Baur, M., & Benkert, M. (2005). Network comparison. In U. Brandes & T. Erlebach (eds.), *Network Analysis: Methodological Foundations*, pp. 318–340. Berlin: Springer-Verlag.

Bedord, T, & Cooke, R. M. (2004). Vines: A new graphical model for dependent random variables. *The Annals of Statistics*, **30**, 1031–1068.

Bell, T. (2009). Ecology of microbial populations. In S. A. Levin (ed.), *The Princeton Guide to Ecology*, pp. 239–246. Princeton, NJ: Princeton University Press.

Bender, E., & Canfield, E. (1978). The asymptotic number of labeled graphs with given degree sequence. *Journal of Combinatorial Theory, A*, **24**, 296–307.

Benedek, Z., Jordán, F., & Báldi, A. (2007). Topological keystone species complexes in ecological interaction networks. *Community Ecology*, **8**, 1–7.

Bianconi, G. (2018). *Multilayer Networks: Structure & Function*. Oxford: Oxford University Press.

Blanchet, F. G., Legendre, P., & Borcard, D. (2008). Modelling directional spatial processes in ecological data. *Ecological Modelling*, **215**, 325–336.

Blanchet, F. G., Legendre, P., Maranger, R., Monti, D., & Pepin, P. (2011). Modelling the effect of directional spatial ecological processes at different scales. *Oecologia*, **166**, 357–368.

Bliss, C. A., Danforth, C. M., & Dodds, P. S. (2014). Estimation of global network statistics from incomplete data. *PLoS ONE*, **9**, e108471.

Blüthgen, N. (2010). Why network analysis is often disconnected from community ecology: A critique and an ecologist's guide. *Basic & Applied Ecology*, **11**, 185–195.

Boccaletti, S., Bianconi, G., Criado, R., et al. (2014). The structure and dynamics of multilayer networks. *Physics Reports*, **544**, 1–122.

Bode, M., Burrage, K., & Possingham, H. (2008). Using complex network metrics to predict the persistence of metapopulations with asymmetric connectivity patterns. *Ecological Modelling*, **214**, 201–209.

Bonald, T., Hollocou, A., & Lelarge, M. (2018). Weighted spectral embedding of graphs. arXiv:1809.11115v2.

Bonchev, D., & Buck, G. A. (2007). Quantitative measures of network complexity. In D. Bonchev & D. H. Rouvray (eds.), *Complexity in Chemistry, Biology, and Ecology*, pp. 191–235. New York: Springer Science+Business Media.

Bonchev, D., & Rouvray, D. H. (eds.) (2007). *Complexity in Chemistry, Biology, and Ecology*. New York: Springer Science+Business Media.

Borcard, D., Legendre, P., Avois-Jacquet, C., & Tuomisto, H. (2004). Dissecting the spatial structure of ecological data at multiple scales. *Ecology*, **85**, 1826–1832.

Borgwardt, K. M., Petri, T., Vishwanathan, S. V. N., & Kriegel, H.-P. (2007). An efficient sampling scheme for comparison of large graphs. In *Proceedings of Conference on Mining and Learning with Graphs*, Florence, Italy, 1–3 August 2007.

Borrelli, J. J. (2015). Selection against instability: Stable subgraphs are most frequent in empirical food webs. *Oikos*, **124**, 1583–1588.

Bossenbroek, J. M., Kraft, C. E., & Nekola, J. C. (2001). Prediction of long-distance dispersal using gravity models: Zebra mussel invasion of inland lakes. *Ecological Applications*, **11**, 1778–1788.

Bray, J. R., & Curtis, J. T. (1957). An ordination of upland forest communities of southern Wisconsin. *Ecological Monographs*, **27**, 325–349.

Bressan, M., Chierichetti, F., Kumar, R., Leucci, S., & Panconesi, A. (2017). Counting graphlets: Space vs time. In *Conference on Web Search & Data Mining*, 6–10 February 2017. Association for Computing Machinery, 557–566.

Brouwer, A. E., & Haemers, W. H. (2012). *Spectra of Graphs*. New York: Springer Science +Business Media.

Brualdi, R. A. (2010). Spectra of digraphs. *Linear Algebra and its Applications*. **432**, 2181–2213.

Brüggemann, R., & Carlsen, L. (2006a). Introduction to partial order theory exemplified by the evaluation of sampling sites. In Brüggemann, R. & Carlsen, L. (eds), *Partial Order in Environmental Sciences and Chemistry*, pp. 61–110. Berlin: Springer-Verlag.

(2006b). *Partial Order in Environmental Sciences and Chemistry*. Berlin: Springer-Verlag.

Buck, J. C., Rohr, J. R., & Blausten, A. R. (2016). Effects of nutrient supplementation on host-pathogen dynamics of the amphibian chytrid fungus: A community approach. *Freshwater Biology*, **61**, 110–120.

Burkle, L. A., Marlin, J. C., & Knight, T. M. (2013). Plant-pollinator interactions over 120 years: Loss of species, co-occurrence, and function. *Science*, **339**, 1611–1615.

Burnham, K. P., & Anderson, D. R. (2004). Multimodel inference: Understanding AIC and BIC in model selection. *Sociological Methods & Research*, **33**, 261–304.

Byrnes, J. E., Reed, D. C., Cardinale, B. J., Cavanaugh, K. C., Holbrooks, S. J., & Schmitts, R. J. (2011). Climate-driven increases in storm frequency simplify kelp forest food webs. *Global Change Biology*, **17**, 2513–2524.

Chao, A. (1987). Estimating the population size for capture-recapture data with unequal catchability. *Biometrics*, **43**, 783–791.

Chave, J. (2009). Competition, neutrality and community organization. In S. A. Levin (ed.), *The Princeton Guide to Ecology*, pp. 264–273. Princeton, NJ: Princeton University Press.

Chen, H.-W., Liu, W-c., Davis, A. J., Jordán, F., Hwang, M-J., & Shao, K-T. (2008). Network position of hosts in food webs and their parasite diversity. *Oikos*, **117**, 1847–1855.

Chisholm, C., Lindo, Z., and Gonzalez, A. (2010). Metacommunity diversity depends on connectivity and patch arrangement in heterogeneous habitat networks. *Ecography*, **34**, 415–424.

Chuine, I., & Régnière, J. (2017). Process-based models of phenology for plants and animals. *Annual Review of Ecology Evolution & Systematics*, **48**, 159–182.

Chung, F. R. K. (1997). *Spectral Graph Theory*. Providence, RI: American Mathematical Society.

Clauset, A., Moore, C., & Newman, M. E. J. (2007). Structural inference of hierarchies in networks. In E. Airoldi, D. M. Blei, S. E. Fienberg, A. Goldenberg, E. P. Xing, & A. X. Zheng (eds.), *Statistical Network Analysis: Models, Issues, and New Directions*, pp. 1–13. Berlin: Springer.

Connor, N., Barberán, A. & Clauset, A. (2017). Using null models to infer microbial co-occurrence networks. *PLoS ONE*, **12**, e0176751.

Conover, W. J. (1980). *Practical Nonparametric Statistics*, 2nd ed. New York: Wiley.

Constantini, G., & Perugini, M. (2014). Generalization of clustering coefficients to signed correlation networks. *PLoS ONE*, **9**, e88669.

Costa, L. S. da F., Rodrigues, F. A., Travieso, G., & Villas Boas, P.R. (2007). Characterization of complex networks: A survey of measurements. *Advances in Physics*, **56**, 167–242. arXiv:cond-mat/0505185.

Coux, C., Rader, R., Bartomeus, I., & Tylianakis, J. M. (2016). Linking species functional roles to their network roles. *Ecological Letters*, **19**, 762–770.

Cozzo, E., Kivelä, M., De Domenico, M., et al. (2015). Structure of triadic relations in multiplex networks. *New Journal of Physics*, **17**, 073029.

Craft, M. E., & Caillaud, D. (2010). Network models: An underutilized tool in wildlife epidemiology. *Interdisciplinary Perspectives on Infectious Diseases*. **2011**, Article ID 676949. DOI: 10.1155/2011/676949.

Craft, M. E., Volz, E., Packer, C., & Meyers, L.A. (2010). Disease transmission in territorial populations: The small-world network of Serengeti lions. *Journal of the Royal Society Interface*, **8**, 776–786.

Crane, H. (2018). *Probabilistic Foundations of Statistical Network Analysis.* Boca Raton, FL: CRC Press.

Cummings, G. (2016). Heterarchies: Reconciling networks and hierarchies. *Trends in Ecology & Evolution*, **31**, 622–632.

Dale, M. R. T. (1985). Graph theoretical methods for comparing phytosociological structures. *Vegetatio*, **63**, 79–88.

(1999). *Spatial Pattern Analysis in Plant Ecology*. Cambridge: Cambridge University Press.

(2017). *Applying Graph Theory in Ecological Research*. Cambridge: Cambridge University Press.

Dale, M. R. T., & Fortin, M.-J. (2009). Spatial autocorrelation and statistical tests: Some solutions. *Journal of Agricultural, Biological, and Environmental Statistics,* **14**, 188–206.

(2010). From graphs to spatial graphs. *Annual Review of Ecology, Evolution, and Systematics,* **41**, 21–38.

(2014). *Spatial analysis. A guide for ecologists*, 2nd ed. Cambridge: Cambridge University Press.

Daniel, C. J., Frid, L., Sleeter, B. M., & Fortin, M.-J. (2016). State-and-transition simulation models: A framework for forecasting landscape change. *Methods in Ecology & Evolution*, **7**, 1413–1423.

Daniel, C. J., Sleeter, B. M., Frid, L., & Fortin, M.-J. (2018). Integrating continuous stocks and flows into state-and-transition simulation models of landscape change. *Methods in Ecology and Evolution*, **9**, 1133–1143.

Danon, L., Ford, A.P., House, T., etb al. (2011). Networks and the epidemiology of infectious disease. *Interdisciplinary Perspectives on Infectious Diseases*, **2011**, Article ID 284909. DOI: 10.1155/2011/284909.

de Aguiar, M. A. M., Newman, E. A., Pires, M. M., et al. (2019). Revealing biases in the sampling of ecological interaction networks. *PeerJ.* e7566. https://doi.org/10.7717/peerj.7566

De Domenico, M., Solé-Ribalta, A., Cozzo, E., et al. (2013). Mathematical formulation of multilayer networks. *Physical Review X*, **3**, 041022.

De Domenico, M., Solé-Ribalta, A., Ormodei, E., Gomez, S., & Arenas, A. (2015). Ranking in interconnected multilayer networks reveals versatile nodes. *Nature Communications*, **6**: 6868.

DeFord, D. R., & Pauls, S. D. (2017). Spectral clustering methods for multiplex networks. arXiv:1703.05355v1.
Dehmer, M., & Emmert-Strieb, F. (2015). *Quantitative Graph Theory*. Boca Raton, FL: CRC Press.
Delmas, E., Besson M., Brice, M.-H., et al. (2019). Analyzing ecological networks of species interactions. *Biological Reviews*, **94**, 16–36.
Dempsey, K., & Ali, H. (2011). Evaluation of essential genes in correlation networks using measures of centrality. In *Proceedings of IEEE International Conference on Bioinformatics and Biomedicine Workshops.*
Deng, Y., Jiang, Y. H., Yang, Y., He, Z., Luo, F., & Zhou, J. (2012). Molecular ecological network analysis. *BMC Bioinformatics*, **13**, 113.
Donetti, L., Neri, F., & Munoz, M. A. (2006). Optimal network topologies: Expanders, Cages, Ramanujan graphs, entangled networks and all that. *Journal of Statistical Mechanics*, **8**, P08007. arXiv:cond-mat/0605565v2.
Dong, X., Frossard, P., Vandergheynst, P., & Nefedov, N. (2011). Clustering with multi-layer graphs: A spectral perspective. *IEEE Transactions on Signal Processing*, **60**, 5820–5831.
Doreian, P., & Mrvar, A. (2009). Partitioning signed social networks. *Social Networks*, **31**, 1–11.
Dormann, C. F., Fründ, J., & Schaefer, H. M. (2017). Identifying causes of patterns in ecological networks: Opportunities and limitations. *Annual Review of Ecology, Evolution, and Systematics*, **48**, 559–584.
Doyle, P. G., & Snell, J. L. (1984). *Random Walks and Electric Networks.* Washington, DC: Mathematical Association of America.
Dray, S., Legendre, P., & Peres-Neto, P. R. (2006). Spatial modelling: A comprehensive framework for principal coordinate analysis of neighbour matrices (PCNM). *Ecological Modelling*, **196**, 483–493.
Dray S., Pélissier,R., Couteron, P., et al.(2012). Community ecology in the age of multivariate multiscale spatial analysis. *Ecological Monographs*, **82**, 257–275.
Duboscq, J., Romano, V., Sueur, C., & MacIntosh, A. J. J. (2015). Network centrality and seasonality interact to predict lice load in a social primate. *Scientific Reports*, **6**, 22095.
Dunne, J. A., Williams, R. J., & Martinez, N. D. (2002). Network structure and biodiversity loss in food-webs: Robustness increases with connectance. *Ecology Letters*, **5**, 558–567.
Duran-Pinedo, A. E., Paster, B., Teles, R., & Frias-Lopez, J. (2011). Correlation network analysis applied to complex biofilm communities. *PLoS ONE*, **6**, e28438.
Durrett, R. (2007). *Random Graph Dynamics*. Cambridge: Cambridge University Press.
Dutilleul, P. (1993). Modifying the *t* test for assessing the correlation between two spatial processes. *Biometrics*, **49**, 305–314.
El Maftouhi, A., Harutyunyan, A., & Manoussakis, Y. (2015). Weak balance in random signed graphs. *Internet Mathematics*, **11**, 143–154.
Ellington, E. H., Bastille, A., Rousseau, G., et al. (2015). Using multiple imputation to estimate missing data in meta-regression. *Methods in Ecology and Evolution*, **6**, 153–163.
Emer, C., Galetti, M., Pizo, M. A., Guimaraes Jr, P. R., Moraes, S., Piratelli, A., & Jordano, P. (2018). Seed-dispersal interactions in fragmented landscapes–a metanetwork approach. *Ecology Letters*, **21**, 484–493.
Erickson, R. O. (1943). Population size and geographical distribution of *Clematis fremontii* var. *riehlii*. *Annals of the Missouri Botanical Garden*, **30**, 63–68.

Estrada, E. (2007). Characterization of topological keystone species: Local, global and 'meso-scale' centralities in food webs. *Ecological Complexity*, **4**, 46–57.

(2012). *The Structure of Complex Networks: Theory and Applications*. Oxford: Oxford University Press.

Estrada, E., & Bodin, Ö. (2008). Using network centrality measures to manage landscape connectivity. *Ecological Applications*, **18**, 1810–1825.

Estrada, E., & Rodriguez-Velazquez, J. A. (2006). Subgraph centrality and clustering in complex hyper-networks. *Physica A*, **364**, 581–594.

Facchetti, G., Iacono, G., & Altafini, C. (2011). Computing global structural balance in large-scale signed social networks. *Proceedings of the National Academy of Science*, **108**, 20953–20958.

Fall, A., Fortin, M.-J., Kneeshaw, D. D., et al.(2004). Consequence of various landscape-scale ecosystem management strategies and fire cycles on age-class structure and harvest in boreal forests. *Canadian Journal of Forest Research*, **34**, 310–322.

Fall, A., Fortin, M.-J., Manseau, M., & O'Brien, D. (2007). Spatial graphs: Principles and applications for habitat connectivity. *Ecosystems*, **10**, 448–461.

Fan, Y., Chen, J., Shirkey, G., John, R., Wu, S. R., Park, H., & Shao, C. (2016). Applications of structural equation modeling (SEM) in ecological studies: An updated review. *Ecological Processes*, **5**, 19.

Fausett, L. V. (1994). *Fundamentals of Neural Networks: Architectures, Algorithms & Applications.* Upper Saddle River, NJ: Prentice-Hall.

Faust, K., & Raes, J. (2012). Microbial interactions: From networks to models. *Nature Reviews*, **10**, 538–550.

Fernandez, M., Riveros, J. D., Campos, M., Mathe, K., & Narasimhan, G. (2015). Microbial "social networks." *BioMedCentral Genomics*, **16**, Suppl. 11, S6.

Finn, K. R., Silk, M. J., Porter, M. A., & Pinter-Wollman, N. (2017). Novel insights into animal sociality from multilayer networks. arXiv:1712.01790v1.

Fischer, M. M. (2009). Neural networks for spatial data analysis. In A. S. Fotheringham & P. A. Rogerson (eds.), *The SAGE Handbook of Spatial Analysis*, pp. 375–396. Los Angeles: SAGE.

Fisher, N. I. (1997). Copulas. In *Encyclopedia of Statistical Sciences*, Update Vol. 1, pp. 159–163. New York: John Wiley & Sons.

Fletcher, R. J., Revell, A., Reichert, B. E., Kitchens, W. M., Dixon, J. D., & Austin, J. D. (2013). Network modularity reveals critical scales for connectivity in ecology and evolution. *Nature Communications*, **4**, 2572.

Fontaine, C., Guimaraes, P. R., Kéfi, S., et al. (2011). The ecological and evolutionary implications of merging different types of networks. *Ecology Letters*, **14**, 1170–1181.

Friedman, J., & Alm, E. J. (2012). Inferring correlation networks from genomic survey data. *PLoS Computational Biology*, **9**, e1002687.

Gabow, H. N., & Myers, E. W. (1978). Finding all spanning trees of directed and undirected graphs. *SIAM Journal of Computing*, **7**, 280–287.

Gago, S., Hurajova, J. C., & Madaras, T. (2015). Betweenness centrality in graphs. In M. Dehmer & F. Emmert-Streib (ed.), *Quantitative Graph Theory*, pp. 233–257. Boca Raton, FL: CRC Press.

Gallier, J. (2013). Notes on elementary spectral graph theory. Applications to graph clustering using normalized cuts. ResearchGate 258374324.

(2017). Spectral theory of unsigned and signed graphs. Applications to graph clustering: a survey. arXiv:1601.04692v1.

Gao, J., Barzel, B., & Barabási, A.-L. (2016). Universal resilience patterns in complex networks. *Nature*, **530**, 307–312.

Gao, J., Buldyrev, S. V., Stanley, H. E., & Havlin, S. (2012). Networks formed from interdependent networks. *Nature Physics*, **8**, 40–48.

Gao, J., Li, D., & Havlin, S. (2014). From a single network to a network of networks. *National Science Review*, **1**, 346–356.

García-Callejas, D., Malowny-Horas, R., & Araújo, M. B. (2018). Multiple interactions networks: Toward more realistic descriptions of the web of life. *Oikos*, **127**, 5–22.

Gärtner, T. (2003). A survey of kernels for structured data. *ACM SIGKDD Explorations Newsletter*, **5**, 49–58.

(2008). *Kernels for Structured Data*. Singapore: World Scientific.

Gaüzère, B., Brun, L., & Willemin, D. (2015). Graph kernels in chemoinformatics. In M. Dehmer & F. Emmert-Streib, (eds), *Quantitative Graph Theory*, pp. 425–469. Boca Raton, FL: CRC Press.

Geng, H., Tran-Gyamfi, M. B., Lane, T. W., Sale, K. L., & Yu1, E.T. (2016). Changes in the structure of the microbial community associated with *Nannochloropsis salina* following treatments with antibiotics and bioactive compounds. *Frontiers in Microbiology*, **7**, Article 1155. https://doi.org/10.3389/fmicb.2016.01155.

Gera, R., Alonso, L., Crawford, B., House, J., Mendez-Bermudez, J. A., Knuth, T., & Miller, R. (2018). Identifying network structure similarity using spectral graph theory. *Applied Network Science*, **3**, 2.

Gerstmann, H., Doktor, D., Glässer, C., & Möler, M. (2016). PHASE: A geostatistical model for the Kriging-based spatial prediction of crop phenology using public phenological and climatological observations. *Computers and Electronics in Agriculture*, **127**, 726–738.

Ghosh, S., Das, N., Gonçalves, T., Quaresma, P., & Kundu, M. (2017). The journey of graph kernels through two decades. *Computer Science Review*, **27**, 88–111.

Gilarranz, L. J., & Bascompte, J. (2012). Spatial network structure and metapopulation persistence. *Journal of Theoretical Biology*, **297**, 11–16.

Gilarranz, L. J., Hastings, A., & Bascompte, J. (2014). Inferring topology from dynamics in spatial networks. *Theoretical Ecology*, **8**, 15–21.

Gilbert, J. A., Steele, J. A., Caporaso, J. G., et al. (2012). Defining seasonal marine microbial community dynamics. *ISME Journal*, **6**, 298–308.

Girvan, M., & Newman, M. E. J. (2002). Community structure in social and biological networks. *Proceedings of the National Academy of Sciences of the USA*, **99**, 7821–7826.

Glaz, J., Naus, J., & Wallenstein, S. (2001). *Scan Statistics.* New York: Springer Science +Business Media.

Godfrey, S. S. (2013). Networks and the ecology of parasite transmission: A framework for wildlife parasitology. *International Journal for Parasitology: Parasites and Wildlife*, **2**, 235–245.

Godfrey, S. S., Moore, J. A., Nelson, N. J., & Bull, C. M. (2010). Social network structure and parasite infection patterns in a territorial reptile, the Tuatara (*Sphenodon punctatus*). *International Journal of Parasitology*, **40**, 1575–1585.

Godsil, C., & Royle, G. (2001). *Algebraic Graph Theory.* New York: Springer Science +Business Media.

Goldreich, O. (2010). *P, NP, and NP-Completeness.* Cambridge: Cambridge University Press.

Golubski, A. J., & Abrams, P. A. (2011). Modifying the modifiers: What happens when interspecific interactions interact? *Journal of Animal Ecology*, **80**, 1097–1108.

Golubski, A. J., Westlund, E. E., Vandermeer, J., & Pascual, M. (2016). Ecological networks over the edge: Hypergraph trait-mediated indirect interaction (TMII) structure. *Trends in Ecology and Evolution*, **31**, 344–354.

Gomez, S., Diaz-Guilera, A., Gomez-Gardeñes, J., Perez-Vicente, C. J., Moreno, Y., & Arenas, A. (2013). Diffusion dynamics on multiplex networks. *Physical Review Letters*, **110**, 028701.

Grace, J. B. (2006). *Structural Equation Modeling and Natural Systems*. Cambridge: Cambridge University Press.

Grace, J. B., & Irvine, K. M. (2019). Scientist's guide to developing explanatory statistical models using causal analysis principles. Extended pre-publication draft of Grace & Irvine (2020).

(2020). Scientist's guide to developing explanatory statistical models using causal analysis principles. *Ecology*, **101**, e02962.

Gravel, D., Massol, F., & Leibold, M. A. (2016). Stability and complexity in model meta-ecosystems. *Nature Communications*, **7**, 12457.

Gray, M., Stansberry, M. J., Lynn, J. S., Williams, C. F., White, T. E., & Whitney, K. D. (2018). Consistent shifts in pollinator-relevant floral coloration along Rocky Mountain elevation gradients. *Journal of Ecology*, **106**, 1910–1924.

Gross, T., & Blasius, B. (2008). Adaptive coevolutionary networks: A review. *Journal of the Royal Society Interface*, **5**, 259–271.

Gross, T., & Sayama, H., eds. (2009). *Adaptive Networks*. Cambridge, MA: NECSI and Springer.

Gu, S., Johnson, J., Faisal, F. E., & Milenković, T. (2018). From homogeneous networks to heterogeneous network alignment via colored graphlets. *Nature Scientific Reports*, **8**, 12524.

Gunderson, L. H. (2000). Ecological resilience- on theory and application. *Annual Review of Ecology & Systematics*, **31**, 425–439.

Haegeman, B., Arnoldi, J.-F., Wang, S., de Mazancourt, C., Montoya, J. M., & Loreau, M. (2016). Resilience, invariability, and ecological stability across levels of organization. bioRxiv, DOI: http://dx.doi.org/10.1101/085852.

Hamede, R. K., Bashford, J., McCallum, H., & Jones, M. (2009). Contact networks in a wild Tasmanian devil (*Sarcophilus harrisii*) population: Using social network analysis to reveal seasonal variability in social behaviour and its implications for transmission of devil facial tumour disease. *Ecology Letters*, **12**, 1–11.

Harary, F. (1953). On the notion of balance of a signed graph. *Michigan Mathematical Journal*, **2**, 143–146.

(1959). On the measurement of structural balance. *Behavioral Science*, **4**, 316–323.

(1969). *Graph Theory*. Reading, MA: Addison-Wesley.

Hastie, T., Tibshirani, R., & Friedman, J. (2009). *The Elements of Statistical Learning: Data Mining, Inference, and Prediction*, 2nd ed. New York: Springer Science+Business Media.

Hastings, A. (2001). Transient dynamics and the persistence of ecological systems. *Ecological Letters*, **4**, 215–220.

(2004). Transients: The key to long-term ecological understanding. *Trends in Ecology and Evolution*, **19**, 39–45.

Hastings, A., Abbott, K. C., Cuddington, K., et al. (2018). Transient phenomena in ecology. *Science*, **361**, aat6412.

Heckathorn, D. D. (1997). Respondent-driven sampling: A new approach to the study of hidden populations. *Social Problems*, **44**, 174–199.

Helm, D. (2006). Evaluation of biomonitoring data. In R. Brüggemann & L. Carlsen, (eds.), *Partial Order in Environmental Sciences and Chemistry*, pp. 285–307. Berlin: Springer-Verlag.

Holling, C. (1973). Resilience and stability of ecological systems. *Annual Review of Ecology & Systematics*, **4**, 1–23.

Horn, H. S. (1975). Markovian properties of forest succession. In M. L. Cody & J. M. Diamond (eds.), *Ecology and Evolution of Communities*, pp. 196–211. Cambridge, MA: Belknap.

(1976). Succession. In R. M. May (ed.), *Theoretical Ecology: Principles and Applications*, pp. 187–204. Oxford: Blackwell.

Horvath, S. (2011). *Weighted Network Analysis: Application in Genomics and Systems Biology*. New York: Springer.

Hsieh, H.-Y., Liere. H., Soto, E. J., & Perfecto, J. (2012). Cascading trait-mediated interactions induced by ant pheromones. *Ecology & Evolution*, **2**, 2181–2191.

Hulovatyy, Y., Chen, H., & Milenković, T. (2015). Exploring the structure and function of temporal networks with dynamic graphlets. *Bioinformatics*, **31**, 1171–1180.

Ilany, A., Barocas, A., Koren, L., Kam, M., & Geffen, E. (2013). Structural balance in the social networks of a wild mammal. *Animal Behaviour*, **85**, 1397–1405.

Ings, T. C., Montoya, J. M., Bascompte, J., et al. (2009). Ecological networks – beyond food webs. *Journal of Animal Ecology*, **78**, 253–269.

Ives, A. R., & Carpenter, S. R. (2007). Stability and diversity in ecosystems. *Science*, **317**, 58–62.

Jiang, L. Q., & Zhang, W. J. (2015). Determination of keystone species in CSM food web: A topological analysis of network structure. *Network Biology*, **5**, 13–33.

Joe, H. (1997). *Multivariate Models and Dependence Concepts*. Boca Raton, FL, CRC Press.

Johansson, J., Kristensen, N. P., Nilsson, J.-A., & Jonz, N. (2015). The eco-evolutionary consequences of interspecific phenological asynchrony—a theoretical perspective. *Oikos*, **124**, 102–112.

Jordán, F. (2009). Keystone species and food webs. *Philosophical Transactions of the Royal Society, B*, **364**, 1733–1741.

Jordán, F., Takaács-Sánta, A., & Molnár, I. (1999). A reliability theoretical quest for keystones. *Oikos*, **86**, 453–462.

Jordano, P. (2016). Sampling networks of ecological interactions. *Functional Ecology*, **30**, 1883–1893.

Junker, B. H., & Schreiber, F. (2008). *Analysis of Biological Networks*. Hoboken, NJ: John Wiley & Sons.

Kamisinski, A., Cholda, P., & Jajszczyk, A. (2015). Assessing the structural complexity of computer and communication networks. *ACM Computing Surveys*, **47**, Article 66. DOI: 10.1145/2755621.

Kao, T.-C., & Porter, M. A. (2017). Layer communities in multiplex networks. arXiv:706.04147v1.

Kashtan, N., Itzkovitz, S., Milo, R., & Alon, U. (2004). Topological generalizations of network motifs. *Physical Review E*, **70**, 031909.

Kéfi, S., Berlow, E. L., Wieters, E. A., et al. (2015). Network structure beyond food webs: Mapping non-trophic and trophic interactions on Chilean rocky shores. *Ecology*, **96**, 291–303.

Kéfi, S. Thébault, E., Eklöf, A., et al. (2018). Toward multiplex ecological networks: Accounting for multiple interaction types to understand community structure and dynamics. In J. C. Moore, P. C. de Ruiter, K. S. McCann, & V. Wolters (eds). *Adaptive Food Webs: Stability and Transitions of Real and Model Ecosystems*, pp. 74 –87. Cambridge: Cambridge University Press.

Kelly, W. P., Thorne, T., & Stumpf, M. P. H. (2010). Statistical null models for biological network analysis. In M. P. H. Stumpf, & C. Wiuf (eds.), *Statistical and Evolutionary Analysis of Biological Networks*, pp. 145–165. London: Imperial College Press.

Kemp, J. E., Evans, D. M., Augustyn, W. J., & Ellis, A. G. (2017). Invariant antagonistic network structure despite high spatial and temporal turnover of interactions. *Ecography*, **40**, 1315–1324.

Kepaptsoglou, K., Karlaftis, M. G., & Tsamboulas, D. (2010). The gravity model specification for modeling international trade flows and free trade agreement effects: A 10–year review of empirical studies. *The Open Economics Journal*, **3**, 1–13.

Kim, H. J., & Kim, J. M. (2005). Cyclic topology in complex networks. *Physical Review E*, **72**, 036109.

Kim, J., & Lee, J.-G. (2015). Community detection in multi-layer graphs: A survey. *SIGMOD Record*, **44**, 37–48.

Kivelä, M., Arenas, A., Barthelemy, M., Gleeson, J. P., Moreno, Y., & Porter, M. A. (2014). Multilayer networks. *Journal of Complex Networks*, **2**, 203–271.

Klau, G. W., & Weiskercher, R. (2005). Robustness and resilience. In U. Brandes, & T. Erlebach (eds.) *Network Analysis: Methodological Foundations*, pp. 417–437. Berlin: Springer-Verlag.

Kolaczyk, E. D. (2009). *Statistical Analysis of Network Data*. New York: Springer.

Kolaczyk, E. D., & Csárdi, G. (2014). *Statistical Analysis of Network Data with R*. New York: Springer Science+Business Media.

Kondor, R. I., & Lafferty, J. (2002). Diffusion kernels on graphs and other discrete input spaces. In *Proceedings of the International Conference on Machine Learning* (ICML), Sydney, 2002.

Koschützki, D. (2008). Network centralities. In B. H. Junker, & F. Schreiber, (eds.), *Analysis of Biological Networks*, pp. 65–84. Hoboken, NJ: John Wiley & Sons.

Krivitsky, P. N., & Kolaczyk, E.D. (2015). On the question of the effective sample size in network modelling: An asymptotic inquiry. *Statistical Science*, **30**, 184–198.

Kunegis, J. (2014). Applications of structural balance in signed social networks. arXiv:1402.6865v1.

Kunegis, J., Schmidt, S., Lommatzsch, A., Lerner, J, De Luca, E. W., & Albyrak, S. (2010). Spectral analysis of signed graphs for clustering, prediction and visualization. In *Proceedings of the SIAM International Conference on Data Mining*, Columbus, 2010, pp. 559–570.

Kurowicka, D., & Cooke, R. M. (2002). The vine copula method for representing high dimensional dependent distributions: Application to continuous belief nets. In E. Yücesan, C. H. Chen, J. L. Snowdon, & J. M. Charnes (eds.), *Proceedings of the 2002 Winter Simulation Conference*, San Diego, pp. 270–278.

Lasserre, J., Chung, H. R., & Vingron, M. (2013). Finding associations among histone modifications using sparse partial correlation networks. *PLoS Computational Biology*, **9**, e1003168.

Lau, M. K., Borrett, S. R., Baiser, B., Gotelli, N. J., & Ellison, A. M. (2017). Ecological network metrics: Opportunities for synthesis. *Ecosphere*, **8**, p.e01900.

Layeghifard, M., Hwang, D. M., & Guttman, D. S. (2017). Disentangling interactions in the microbiome: A network perspective. *Trends in Microbiology*, **25**, 217–228.

Lazega, E., Jourda, M.-T., Mounier, L., & Stofer, R. (2008). Catching up with big fish in the big pond? Multi-level network analysis through linked design. *Social Networks*, **30**, 159–176.

Lee, A. B., Luca, D., & Roeder, K. (2010). A spectral graph approach to discovering genetic ancestry. *The Annals of Applied Statistics*, **4**, 179–202.

Lee, K.-M., Miu, B., & Goh, K.-I. (2015). Towards real-world complexity: An introduction to multiplex networks. *European Physics Journal*, arXiv:1502.03909v1.

Lee, S. H., Kim, P.-J., & Jeong, H. (2006). Statistical properties of sampled networks. *Physical Review*, *E*. 016102. arXiv:cond-mat/0500232v4.

Lefcheck, J. S. (2016). Piecewise structural equation modelling in R for ecology, evolution and systematics. *Methods in Ecology and Evolution*, **7**, 573–579.

Legendre, P., & Legendre, L. (2012). *Numerical Ecology*, 3rd English ed. Amsterdam: Elsevier.

Levin, S. A. (1998). Ecosystems and the biosphere as complex adaptive systems. *Ecosystems*, **1**, 431–436.

Li, B., Springer, J., Bebis, G., & Gunes, M. H. (2013). A survey of network flow applications. *Journal of Network and Computer Applications*, **36**, 567–581.

Li, Y., Wu, X., & Lu, A. (2016). On spectral analysis of directed signed graphs. *International Journal of Data Science and Analytics*, **6**, 147–162.

Libralato, S., Christensen, V., & Pauly, D. (2006). A method for identifying keystone species in food web models. *Ecological Modelling*, **195**, 153–171.

Little, R. J. A., & Rubin, D. B. (1987). *Statistical Analysis with Missing Data*. New York: John Wiley & Sons.

Lugo-Martinez, J., & Radivojac, P. (2014). Generalized graphlet kernels for probabilistic inference in sparse graphs. *Network Science*, **2**, 254–276.

Lupatini, M., Suleiman, A. K. A., Jacques, R. J. S., et al. (2014). Network topology reveals high connectance levels and few key microbial genera within soils. *Frontiers in Environmental Science*, **2**, 10.

Lurgi, M., Galiana, N., Lopez, B. C., Joppa, L. N., & Montoya, J. M. (2014). Network complexity and species traits mediate the effects of biological invasions on dynamic food webs. *Frontiers in Ecology and Evolution*, **2**, 36.

Malmros, J., Masuda, N., & Britton, T. (2013). Random walks on directed networks: Inference and respondent-driven sampling. *Journal of Official Statistics*, **32**, 433–459.

Malod-Dognin, N., & Pržulj, N. (2019). Network alignment. In N. Pržulj (ed.) *Analyzing Network Data in Biology and Medicine*, pp. 369–413. Cambridge: Cambridge University Press.

Manley, B. J. F. (2006). *Randomization, Bootstrap and Monte Carlo Methods in Biology*, 3rd ed. Boca Raton, FL: CRC Press.

Marcus, D. A. (2008). *Graph Theory: A Problem Oriented Approach.* Washington, DC: Mathematical Association of America.

Marsh, M. K., McLeod, S. R., Hutchings, M. R., & White, P. C. L. (2011). Use of proximity loggers and network analysis to quantify social interactions in free-ranging wild rabbit populations. *Wildlife Research*, **38**, 1–12.

Martensen, A. C., Saura, S., & Fortin, M.-J. (2017). Spatio-temporal connectivity: Assessing the amount of reachable habitat in dynamic landscapes. *Methods in Ecology and Evolution*, **8**, 1253–1264.

Martinez, N. D. (1992). Constant connectance in community food webs. *American Naturalist*, **139**, 1208–1281.

Matsui, T. (1993). A flexible algorithm for generating all the spanning trees in undirected graphs. *Algorithmica*, **18**, 530–543.

May, R. M. (1973). *Stability and Complexity in Model Ecosystems*. Princeton, NJ: Princeton University Press.

Maynard, D. S., Crowther, T. W., & Bradford, M. A. (2017). Competitive network determines the direction of the diversity-function relationship. *PNAS*, **114**, 11464–11469.

McCann, K. S. (2000). The diversity-stability debate. *Nature*, **405**, 228–233.

McElreath, R. (2020). *Statistical Rethinking*, 2nd ed. Boca Raton, FL: CRC Press.

McRae, B. H., Dickson, B. G., Keitt, T. H., & Shah, V. B. (2008). Using circuit theory to model connectivity in ecology, evolution, and conservation. *Ecology*, **89**, 2712–2724.

Menge, B. A. (1995). Indirect effects in a marine rocky intertidal interaction webs: Patterns and importance. *Ecological Monographs*, **65**, 21–74.

Miki, T., & Jacquet, S. (2010). Indirect interactions in the microbial world: Specificities and similarities to plant-insect systems. *Population Ecology*, **52**, 475–483.

Milenkovic, T., & Pržulj, N. (2008). Uncovering biological network function via graphlet degree signatures. *Cancer Informatics*, **6**, 257–273.

Milo, R., Shen-Orr, S., Itzkovitz, S., Kashtan, N., Chklovskii, D., & Alon, U. (2002). Network motifs: Simple building blocks of complex networks. *Science*, **298**, 824–827.

Minor, E. S., & Lookingbill, T. R. (2010). A multiscale network analysis of protected-area connectivity for mammals in the United States. *Conservation Biology*, **24**, 1549–1558.

Minor, E. S., & Urban, D. L. (2007). Graph theory as a proxy for spatially explicit population models in conservation planning. *Ecological Applications*, **17**, 1771–1782.

Mondragon, R. J., Iacovacci, J., & Bianconi, G. (2018). Multilink communities of multiplex networks. *PLoS ONE*, **13**, e0193821.

Moore, J. C., de Ruiter, P. C., McCann, K. S., & Wolters, V., eds. (2018). *Adaptive Food Webs: Stability and Transitions of Real and Model Ecosystems*. Cambridge: Cambridge University Press.

Moore, C., & Mertens, S. (2011). *The Nature of Computation*. Oxford: Oxford University Press.

Mora, B. B., Gravel, D., Gilarranz, L. J., Poisot, T., & Stouffer, D. B. (2018). Identifying a common backbone of interactions underlying food webs from different ecosystems. *Nature Communications*, **9**, 2603.

Mowshowitz, A., & Dehmer, M. (2012). Entropy and the complexity of graphs revisited. *Entropy*, **14**, 559–570.

Mowshowitz, A., & Mitsou, V. (2009). Entropy, orbits and spectra of graphs. In M. Dehmer & F. Emmert-Streib (eds.), *Analysis of Complex Networks: From Biology to Linguistics*, pp. 1–22. Weinheim: Wiley-VCH Verlag.

Murphy, M. A., Dezzani, R., Pilliod, D. S., & Storfers, A. (2010). Landscape genetics of high mountain frog metapopulations. *Molecular Ecology*, **19**, 3634–3649.

Naujokaitis-Lewis, I., Rico, Y., Lovell, J., Fortin, M.-J., & Murphy, M. (2013). Implications of incomplete networks on estimation of landscape genetic connectivity. *Conservation Genetics*, **14**, 287–298.

Nelson, G. A. (2014). Cluster sampling: A pervasive, yet little recognized survey design in fisheries research. *Transactions of the American Fisheries Society*, **143**, 926–938.

Nelsen, R. B. (1992). On measures of association as measures of positive dependence. *Statistics & Probability Letters*, **14**, 269–274.

(1999). *An Introduction to Copulas*. New York: Springer.

Neuhaus, M., & Bunke, H. (2006). A random walk kernel derived from graph edit distance. *Structural, Syntactic and Statistical Pattern Recognition*, Hong Kong, August 2006. Proceedings, pp. 191–199. DOI: 10.1007/11815921_20.

Newaz, K., & Milenković, T. (2019). Graphlets in network science and computational biology. In N. Pržulj (ed.), *Analyzing Network Data in Biology and Medicine*, pp. 193–240. Cambridge: Cambridge University Press.

Newman, M. E., Watts, D. J., & Strogatz, S. H. (2002). Random graph models of social networks. *PNAS*, **99**, 2566–2572.

Newman, M. E. J. (2010). *Networks: An Introduction*. Oxford: Oxford University Press.

Newman, M. E. J., & Park, J. (2003). Why social networks are different from other types of networks. *Physics Review E*, **68**, 036122.

Nicosia, V., Tang, J., Mascolo, C., Musolesi, M., & Latora, V. (2013). Graph metrics for temporal networks. In P. Holme & J. Saramäki (eds.), *Temporal Networks*, pp. 15–40. Berlin: Springer-Verlag.

Okabe, A., & Satoh, T. (2009). Spatial analysis on a network. In A. S. Fotheringham & P. A. Rogerson (eds.), *The SAGE Handbook of Spatial Analysis*, pp. 443–464. Los Angeles: SAGE.

Okuyama, T., & Bolker, B. M. (2007). On quantitative measures of indirect interactions. *Ecological Letters*, **10**, 264–271.

Opgen-Rhein, R., & Strimmer, K. (2007). From correlation to causation networks: A simple approximate learning algorithm and its application to high-dimensional plant gene expression data. *BioMed Central Systems Biology*, **1**, 37.

Paerl, H. W., & Pinckney, J. L. (1996). A mini-review of microbial consortia: Their roles in aquatic production and geochemical cycling. *Microbial Ecology*, **31**, 225–247.

Park, J., & Barabási, A.-L. (2007). Distribution of node characteristics in complex networks. *Proceedings of the National Academy of Sciences of the USA*, **104**, 17916–17920.

Paulau, P. V., Feenders, C., & Blasius, B. (2015). Motif analysis in directed ordered networks and applications to food webs. *Scientific Reports*, **5**, 11926.

Pearl, J. (2009). *Causality: Models, Reasoning and Inference*, 2nd ed. Cambridge: Cambridge University Press.

Pearl, J., Glymour, M., & Jewell, N. P. (2016). *Causal Inference in Statistics*. Chichester: John Wiley & Sons.

Pearl, J., & MacKenzie, D. (2018). *The Book of Why*. New York: Basic Books.

Pedersen, A. B., & Fenton, A. (2006). Emphasizing the ecology in parasite community ecology. *Trends in Ecology and Evolution*, **22**, 133–138.

Pellissier, L., Albouy, C., Bascompte, J., et al. (2018). Comparing species interaction networks along environmental gradients. *Biological Reviews*, **93**, 785–800.

Perfecto, I., Vandermeer, J., & Philpott, S. M. (2014). Complex ecological interactions in the coffee agroecosystem. *Annual Review of Ecology, Evolution, and Systematics*, **45**, 137–158.

Peterson, E. E., Ver Hoef, J. M., Isaak, D. J., (2013). Modeling dendritic ecological networks in space: An integrated network perspective. *Ecology Letters*, **16**, 707–719.

Peyrard, N., Pellegrin, F., Chadoeuf, J., & Nandris, D. (2006). Statistical analysis of the spatio-temporal dynamics of rubber tree (*Hevea brasiliensis*) trunk phloem necrosis: No evidence of pathogen transmission. *Forest Pathology*, **36**, 360–371.

Phillips, J. D. (2011). The structure of ecological state transitions: Amplification, synchronization, and constraints in response to environmental change. *Ecological Complexity*, **8**, 336–346.

Phillips, S. T., Williams, P., Midgeley, G., & Archer, A. (2008). Optimizing dispersal corridors for the Cape Proteaceae using network flow. *Ecological Applications*, **18**, 1200–1211.

Pilosof, S., Porter, M. A., Pascual, M., & Kéfi, S. (2017). The multilayer nature of ecological networks. *Nature Ecology & Evolution*, **1**, 0101.

Pimm, S. L. (1984). The complexity and stability of ecosystems. *Nature*, **307**, 321–326.

Piraveenan, M., Propokenko, M., & Zomaya, A. Y. (2008). Local assortativeness in scale-free networks. *Europhysics Letters*, **84**, 28002.

(2010). Classifying complex networks using unbiased local assortativity. *Proceedings of the Alife XII Conference,* Odense, Denmark.

Pocock, M. J. O., Evans, D. M., & Memmott, J. (2012). The robustness and restoration of a network of ecological networks. *Science*, **335**, 973–977.

Poisot, T., Canard, E., Mouillot, D., Mouquet, N., & Gravel, D. (2012). The dissimilarity of species interaction networks. *Ecology Letters*, **15**, 1353–1361.

Poisot, T., Cirtwill, A. R., Cazelles, K., Gravel, D., Fortin, M. J., & Stouffer, D. B. (2016a). The structure of probabilistic networks. *Methods in Ecology and Evolution*, **7**, 303–312.

Poisot, T., Stouffer, D. B., & Kéfi, S. (2016b). Describe, understand and predict: Why do we need networks in ecology? *Functional Ecology*, **30**, 1878–1882.

Porter, M. A., & Gleeson, J. P. (2015). Dynamical systems on networks: A tutorial. arXiv:1403.7663v2 [nlin.AO].

Potapov, A., Muirhead, J. R., Lele, S. R., & Lewis, M. A. (2011). Stochastic gravity models for modeling lake invasions. *Ecological Modelling*, **222**, 964–972.

Pržulj, K., Wigle, D. A., & Jurisica, I. (2004). Functional topology in a network of protein interactions. *Bioinformatics*, **20**, 340–348.

Pržulj, N. (2007). Biological network comparison using graphlet degree distribution. *Bioinformatics*, **23**, 177–183.

Pržulj, N., & Malod-Dognin, N. (2016). Network analytics in the age of big data. *Science*, **353**, 123–124.

Radicchi, F., & Bianconi, G. (2017). Redundant interdependencies boost the robustness of multiplex networks. *Physical Review X*, **7**, 011013.

Rayfield, B., Fortin, M.-J., & Fall, A. (2011). Connectivity for conservation: A framework to classify network measures. *Ecology*, **92**, 847–858.

Reshef, D. N., Reshef, Y. A., Finucane, H. K., et al. (2011). Detecting novel associations in large data sets. *Science*, **334**, 1518–1524.

Roesch, F. A. (1993). Adaptive cluster sampling for forest inventories. *Forest Science*, **39**, 655–669.

Rohr, R. P., Naisbit, R. E., Mazza, C., & Bersier, L. F. (2018). Statistical approaches for inferring and predicting food-web architecture. In J. C. Moore, P. C. Ruiter, K. S. McCann, & V. Wolters (eds.), *Adaptive Food Webs: Stability and Transitions of Real and Model Ecosystems*, pp. 178–192. Cambridge: Cambridge University Press.

Rohr, R. P., Scherer, H., Kehrli, P., Mazza, C., & Bersier, L. F. (2010). Modeling food webs: Exploring unexplained structure using latent traits. *American Naturalist*, **173**, 170–177.

Ruan, Q., Dutta, D., Schwalbach, M. S., Steele, J. A., Fuhrman, J. A., & Sun, F. (2006). Local similarity analysis reveals unique associations among marine bacterioplankton species and environmental factors. *Bioinformatics*, **22**, 2532–2538.

Ruiz, E. J. L., Mozo, H. G., Vilches, E. D., & Galán, C. (2012). The use of geostatistics in the study of floral phenology of Vulpia geniculata (L.) Link. *The Scientific World Journal*, 624247.

Sarajlić, A., Malod-Dognin, N., Yaveroglu, O. N., & Pržulj, N. (2016). Graphlet-based characterization of directed networks. *Scientific Reports*, **6**, 35098.

Sayama, H., Pestov, I., Schmidt, J., Bush, B. J., Wong, C., Yamanoi, J., & Gross, T. (2013). Modeling complex systems with adaptive networks. *Computers and Mathematics with Applications*, **65**, 1645–1664.

Scheffer, M. (2009). Alternative stable states and regime shifts in ecosystems. In S. A. Levin. (ed.), *The Princeton Guide to Ecology*, pp. 395–406. Princeton, NJ: Princeton University Press.

Schneider, D. W., Ellis, C. D., & Cummings, K. S. (1998). A transportation model assessment of the risk to native mussel communities from zebra mussel spread. *Conservation Biology*, **12**, 788–800.

Schölkopf, B., Tsuda, K., & Vert, J.-P. (2004). *Kernel Methods on Computational Biology*. Cambridge, MA: MIT Press.

Schröder, W., Schmidt, G., & Schönrock, S. (2014). Modelling and mapping of plant phenological stages as bio-meteorological indicators for climate change. *Environmental Sciences Europe*, **26**, 5.

Schwöbbermeyer, H. (2008). Network motifs. In B. H. Junker, & F. Schreiber (eds.), *Analysis of Biological Networks*, pp. 85–111. Hoboken, NJ: John Wiley & Sons.

Shalizi, C. R. (2006). Methods and techniques of complex systems science: An overview. In T. Deisboeck & J. Y. Kresh (eds.), *Complex Systems Science in Biomedicine*, pp. 33–114. Boston, MA: Springer.

Shao, J., Zhou, X., Luo, Y., et al. (2016) Direct and indirect effects of climatic variations on the interannual variability in net ecosystem exchange across terrestrial ecosystems. *Tellus B*, DOI: http://dx.doi.org/10.3402/tellusb.v68.30575

Shervashidze, N., Schweitzer, P., van Leeuwen, E. J., Mehlhorn, K., & Borgwardt, K. M. (2011). Weisfeiler-Lehman graph kernels. *Journal of Machine Learning Research*, **12**, 2539–2561.

Shervashidze, N., Vishwanathan, S. V. N, Petri, T., Mehlhorn, K., & Borgwardt, K. (2009). Efficient graphlet kernels for large graph comparison. In *Proceedings of the 12th International Conference on Artificial Intelligence and Statistics (AISTATS) 2009*, pp. 488–495.

Shipley, B. (2016). *Cause and Correlation in Biology*, 2nd ed. Cambridge: Cambridge University Press.

Shizuka, D., & McDonald, D. B. (2012). A social network perspective on measurements of dominance hierarchies. *Animal Behaviour*, **83**, 925–934.

Shtileman, E., & Stone, L. (2015). The effects of connectivity on metapopulation persistence: Network symmetry and degree correlations. *Proceedings of the Royal Society B: Biological Sciences*, **282**, 20150203.

Silk, M. J., Croft, D. P., Delahay, R. J., et al. (2017). Using social network measures in wildlife disease ecology, epidemiology, and management. *BioScience*, **67**, 245–257.

Silvertown, J. W., & Lovett-Doust, J. (1993). *Introduction to Plant Population Biology*. Oxford: Blackwell.

Simmons, B. I., Cirtwill, A. R., Baker, N. J., Dicks, L. V., Stouffer, D. B., & Sutherland, W. I. (2018). Uncovering indirect interactions in bipartite ecological networks. bioRxiv, DOI: 10.1101/315010.

Smith, D. R., Conroy, M. J., & Brakhage, D. H. (1995). Efficiency of adaptive cluster sampling for estimating density of wintering waterfowl. *Biometrics*, **51**, 777–788.

Snijders, T. A., & Bosker, R. (2012). *Multilevel Analysis: An Introduction to Basic and Advanced Multilevel Modelling*, 2nd ed. London: SAGE.

Solé, R. V., & Montoya, J. M. (2001). Complexity and fragility in ecological networks. *Proceedings of the Royal Society London B: Biological Sciences*, **268**, 2039–2045.

Solé-Ribalta, A., De Demenico, M., Kouvaris, N. E., Diaz-Guilera, A., Gomez, S., & Arenas, A. (2013). Spectral properties of the Laplacian of multiplex networks. *Physical Review E*, **88** (3), p.032807.

Sørensen, P. B., Lerche, D. B., & Thomsen, M. (2006). Developing decision support based on field data and partial order theory. In R. Brüggemann, & L. Carlsen, (eds.), *Partial Order in Environmental Sciences and Chemistry*, pp. 259–283. Berlin: Springer-Verlag.

Sotomayor, D. A., & Lortie, C. J. (2015). Indirect interactions in terrestrial plant communities: Emerging patterns and research gaps. *Ecosphere*, **6**, Article 103.

Šourek, G., O Kuzelka,O., & Zelezný, F. (2013). Predicting top-*k* trends on twitter using graphlets and time features. *ILP*, Late Breaking Papers, 52.

Spielman, D. (2012). Spectral graph theory. In U. Naumann, & O. Schenk(eds.), *Combinatorial Scientific Computing*, pp. 495–523. Boca Raton, FL: Chapman & Hall/CRC.

Staniczenko, P. P. A., Kopp, J. C., & Allesina, S. (2012). The ghost of nestedness in ecological networks. *Nature Communications*, **4**, 1391.

Stella, M., Selakovic, S., Antonioni, A., & Andrazzi, C.S. (2018). Ecological multiplex interactions determine the role of species for parasite spread amplification. *eLife*, e32814.

Stumpf, M. P. H., & Wiuf, C. (2010). *Statistical and Evolutionary Analysis of Biological Networks*, London: Imperial College Press.

Stumpf, M. P. H., Wiuf, C., & May, R. M. (2005). Subnets of scale-free networks are not scale-free: Sampling properties of networks. *Proceedings of the National Academy of Sciences of the USA*, **102**, 4221–4224.

Tang, W., Lu, Z., & Dhillon, S. (2009). Clustering with multiple graphs. In *Proceedings of the 9th International Conference on Data Mining*, 1016–1–21, Miami.

Thébault, R. (2013). Identifying compartments in presence-absence matrices and bipartite networks: Insights into modularity measures. *Journal of Biogeography*, **40**, 759–768.

Thébault, E., & Fontaine, C. (2010). Stability of ecological communities and the architecture of mutualistic and trophic networks. *Science*, **329**, 853–856.

Thébault, E., Sauve, A. M. C., & Fontaine, C. (2018). Merging antagonistic and mutualistic bipartite webs: A first step to integrate interaction diversity into network approaches. In J. C. Moore, P. C. de Ruiter, K. S. McCann, & V. Wolters (eds.), *Adaptive Food Webs: Stability and Transitions of Real and Model Ecosystems*, pp. 62–73. Cambridge: Cambridge University Press.

Thompson, R. M., & Williams, R. (2017). Unpacking resilience in food webs: An emergent property or a sum of the parts? In J. C. Moore, P. C. de Ruiter, K. S. McCann, & V. Wolters (eds.) *Adaptive Food Webs: Stability and Transitions of Real and Model Ecosystems*, pp. 88–103. Cambridge: Cambridge University Press.

Thompson, S. K. (2006). Adaptive web sampling. *Biometrics*, **62**, 1224–1234.

(2011). Adaptive network and spatial sampling. *Survey Methodology*, **37**, 183–196. (Statistics Canada, Catalogue No. 12–001–X.)

Timóteo, S., Correia, M., Rodriguez-Echeverria, S., Freitas, H., & Heleno, R. (2018). Multilayer networks reveal the spatial structure of seed-dispersal interactions across the Great Rift landscapes. *Nature Communications*, **9**, 140.

Tomczak, M. T., Heymans, J. J., Yletyinen, J., Niiranen, S., Otto, S. A., & Blenckner, T. (2013). Ecological network indicators of ecosystem status and change in the Baltic Sea. *PLoS ONE*, **8**, e75439.

Trøjelsgaard, K., & Olesen, J. M. (2016). Ecological networks in motion: Micro- and macroscopic variability across scales. *Functional Ecology*, **30**, 1926–1935.

Trpevski, I., Dimitrova, T., Boshkovski, T., Stikov, N., & Kocarev, L. (2016). Graphlet characteristics in directed networks. *Scientific Reports*, **6**, 37057.

Trussell, G. C., Matassa, C. M., & Ewanchuk, P.J. (2017). Moving beyond linear food chains: Trait-mediated indirect interactions in a rocky intertidal food web. *Proceedings of the Royal Society B: Biological Sciences*, **284**, 20162590.

Tufte, E. R. (1990). *Envisioning Information*. Chester: Graphics Press.

Ulrich, W. (2009). Ecological interaction networks: Prospects and pitfalls. *Ecological Questions*, **11**, 17–25.

Urban, D. L., Minor, E. S., Treml, E. A., & Schick, R. S. (2009). Graph models of habitat mosaics. *Ecology Letters*, **12**, 260–273

Vacic, V., Iakoucheva, L. M., Lonardi, S., & Radivjac, P. (2010). Graphlet kernels for prediction of functional residues in protein structures. *Journal of Computational Biology*, **17**, 55–72.

Valdovinos, F. S. (2019). Mutualistic networks: Moving closer to a predictive theory. *Ecology Letters*, **22**, 1517–1534.

Vile, D., Shipley, B., & Garnier, E. (2006). A structural equation model to integrate changes in functional strategies during old-field succession. *Ecology*, **87**, 504–517.

Vishwanathan, S. V. N., Schraudolph, N. N., Kondor, R., & Borgwardt, K. M. (2010). Graph kernels. *Journal of Machine Learning Research*, **11**, 1201–1242.

Wang, P., Robins, G., Pattison, P., & Lazega, E. (2013). Exponential random graph models for multilevel networks. *Social Networks*, **35**, 96–115.

Warton, D. I., Blanchet, F. G., O'Hara, R. B., et al. (2015). So many variables: Joint modelling in community ecology. *Trends in Ecology and Evolution*, **30**, 766–779.

Wasserman, S., & Faust, K. (1994). *Social Network Analysis*. Cambridge: Cambridge University Press.

Weiss, S., et al. (16 others). (2016). Correlation detection strategies in microbial data sets vary widely in sensitivity and precision. *The ISME Journal*, **10**, 1669–1681.

Whalen, M. A., Duffy, J. E., & Grace, J. B. (2013). Temporal shifts in top-down vs. bottom-up control of epiphytic algae in a seagrass ecosystem. *Ecology*, **94**, 510–520.

White, S. R., Tannas, S., Bao, T., Bennet, J. A., Bork, E. W., & Cahill, J. F. Jr. (2013). Using structural equation modelling to test the passenger, driver and opportunist concepts in a *Poa pratensis* invasion. *Oikos*, **122**, 377–384.

Williams, M. J., & Musolesi, M. (2016). Spatio-temporal networks: Reachability, centrality and robustness. *Royal Society Open Science*, **3**(6), p.160196.

Williams, R. J., & Purves, D. W. (2011). The probabilistic niche model reveals substantial variation in the niche structure of empirical food webs. *Ecology*, **92**, 1849–1857.

Williams, R. J., Purves, D. W., Williams, R. J., Howe, A., & Hofmokel, K. S. (2014). Demonstrating microbial co-occurrence pattern analyses within and between ecosystems. *Frontiers in Microbiology*, **5**, 358.

Wills, P., & Meyer, F. G. (2020). Metrics for graph comparison: A practitioner's guide. *PLoS ONE*, **15**, p.e0228728.

Wilson, R. C., & Zhu, P. (2008). A study of graph spectra for comparing graphs and trees. *Pattern Recognition*, **41**, 2833–2841.

Windels, S. F. L., Malod-Dognin, N., & Pržulj, N. (2018). Graphlet Laplacians: graphlet-based neighbourhoods highlight topology-function and topology-disease relationships. *bioRxiv*, DOI:10.1101/460964.

Wootton, J. T. (1994). Predicting direct and indirect effects: An integrated approach using experiments and path analysis. *Ecology*, **75**, 151–165.

Yaveroğlu, Ö. N., Malod-Dognin, N., Davis, D., et al. (2014). Revealing the hidden language of complex networks. *Scientific Reports*, **4**, 4547.

Yletyinen, J., Bodin, Ö., Weigel, B., Nordström, M. C., Bonsdorff, E., & Blenker, T. (2015). Regime shifts in marine communities: A complex systems perspective on food web dynamics. *Proceedings of the Royal Society B: Biological Sciences*, **283**, 20152569.

Zaborain-Mason, J., Russ, G. R., Abesamis, R. A., Bucol, A. A., & Connolly, S. R. (2017). Network theory and metapopulation persistence: Incorporating node self-connections. *Ecology Letters*, DOI :10.1111/ele12784.

Zhang, B., & Horvath, S. (2005). A general framework for weighted gene co-expression network analysis. *Statistical Applications in Genetics and Molecular Biology*, **4**, a7.

Zhang, Q., & Shi, X. (2017). A mixture copula Bayesian network model for multimodal genomic data. *Cancer Informatics*, **16**, 1-11.

Index